Chromatography: Principles and Instrumentations

Chromatography: Principles and Instrumentations

Editor: Rufus Stafford

NYRESEARCH PRESS

New York

Published by NY Research Press
118-35 Queens Blvd., Suite 400,
Forest Hills, NY 11375, USA
www.nyresearchpress.com

Chromatography: Principles and Instrumentations
Edited by Rufus Stafford

International Standard Book Number: 978-1-63238-768-4 (Hardback)

Cataloging-in-Publication Data

Chromatography : principles and instrumentations / edited by Rufus Stafford.
     p. cm.
Includes bibliographical references and index.
ISBN 978-1-63238-768-4
1. Chromatographic analysis. 2. Chemistry, Analytic. 3. Chromatograms. I. Stafford, Rufus.
QD79.C4 C47 2020
543.8--dc23

Contents

Preface

The main aim of this book is to educate learners and enhance their research focus by presenting diverse topics covering this vast field. This is an advanced book which compiles significant studies by distinguished experts in the area of analysis. This book addresses successive solutions to the challenges arising in the area of application, along with it; the book provides scope for future developments.

The laboratory technique which deals with the separation of a mixture is known as chromatography. This technique involves the dissolution of the mixture in a fluid which is known as the mobile phase. It is then introduced to another material which is known as the stationary phase. Separation occurs when various constituents of the mixture travel at different speeds. Chromatography is mainly classified into two types, preparative and analytical. Preparative chromatography is aimed at separating the constituents of the mixture for later use. It can also be considered a form of purification. Analytical chromatography generally makes use of lesser amounts of material and is usually used to measure the relative quantity of analytes. This book is compiled in such a manner, that it will provide in-depth knowledge about the theory and practice of chromatography. It will serve as a valuable source of reference for graduate and post graduate students. This book is an essential guide for both academicians and those who wish to pursue this discipline further.

It was a great honour to edit this book, though there were challenges, as it involved a lot of communication and networking between me and the editorial team. However, the end result was this all-inclusive book covering diverse themes in the field.

Finally, it is important to acknowledge the efforts of the contributors for their excellent chapters, through which a wide variety of issues have been addressed. I would also like to thank my colleagues for their valuable feedback during the making of this book.

Editor

A New Approach for the Characterization of Organic Residues from Stone Tools Using GC×GC-TOFMS

Katelynn A. Perrault [1,*], **Pierre-Hugues Stefanuto** [1], **Lena Dubois** [1], **Dries Cnuts** [2], **Veerle Rots** [2] and **Jean-François Focant** [1]

[1]   Organic and Biological Analytical Chemistry Group, Chemistry Department, University of Liège, Allée du 6 Août 11 (Bât B6c), Quartier Agora, Sart-Tilman, Liège 4000, Belgium; phstefanuto@ulg.ac.be (P.-H.S.); lena.dubois@doct.ulg.ac.be (L.D.); jf.focant@ulg.ac.be (J.-F.F.)

[2]   Traceolab/Prehistory, University of Liège, Quai Roosevelt 1B (Bât. A4), Liège 4000, Belgium; dries.cnuts@ulg.ac.be (D.C.); veerle.rots@ulg.ac.be (V.R.)

*   Correspondence: katelynn.perrault@ulg.ac.be

Academic Editor: Shari Forbes

Abstract: Headspace solid-phase microextraction (HS-SPME) and gas chromatography-mass spectrometry (GC-MS) have traditionally been used, in combination with other analyses, for the chemical characterization of organic residues recovered from archaeological specimens. Recently in many life science fields, comprehensive two-dimensional gas chromatography-time-of-flight mass spectrometry (GC×GC-TOFMS) has provided numerous benefits over GC-MS. This study represents the first use of HS-SPME-GC×GC-TOFMS to characterize specimens from an experimental modern reference collection. Solvent extractions and direct analyses were performed on materials such as ivory, bone, antlers, animal tissue, human tissue, sediment, and resin. Thicker film column sets were preferred due to reduced column overloading. The samples analyzed by HS-SPME directly on a specimen appeared to give unique signatures and generally produced a higher response than for the solvent-extracted residues. A non-destructive screening approach of specimens may, therefore, be possible. Resin and beeswax mixtures prepared by heating for different lengths of time appeared to provide distinctly different volatile signatures, suggesting that GC×GC-TOFMS may be capable of differentiating alterations to resin in future studies. Further development of GC×GC-TOFMS methods for archaeological applications will provide a valuable tool to uncover significant information on prehistoric technological changes and cultural behavior.

Keywords: HS-SPME; GC×GC-TOFMS; volatile signature; archaeology; organic residues

1. Introduction

Gas chromatography coupled to mass spectrometry (GC-MS), in combination with non-contact sampling methods such as headspace solid-phase microextraction (HS-SPME), is a popular technique used in archaeological research for organic residue analyses [1–5]. This is a particularly challenging area of research for several reasons. First and probably most pertinent, the amount of organic residue available for analysis is most often extremely minimal. In addition, the age of archaeological specimens typically results in some or extensive degradation. To complicate the situation further, the detailed environmental history of the sample and potential exposure to contamination is unknown. Finally, an ideal method of analysis should be non-destructive in order to preserve the sample if other testing is required. Despite being a particularly challenging area of research, GC-MS has been employed successfully in numerous archaeological investigations including, but not limited to, the analysis of smoking pipe residue from ancient burials [6], ceramics [7], plasters [7], resins [2,5,8–10], and resin additives, such as beeswax [9,11].

Many of the matrices suggested above are also challenging to analyze because they are comprised of a complex range of non-volatile, semi-volatile, and volatile components [2,5]. The history of the sample only adds to this complexity, leading to the degradation of certain components and the potential for contaminant compounds to become present over time. A number of studies using one-dimensional gas chromatography (1D GC) have demonstrated the complexity of a vast range of modern reference and archaeological samples [1,3,9,12–14]. In these studies, total ion current (TIC) chromatograms typically exhibit a large number of compounds with a broad dynamic range, and therefore may suffer from insufficient baseline separation of compounds. This could potentially lead to peaks that are obscured by co-elutions and, therefore, are not detected, or potentially peak misidentification using mass spectrometry. Derivatization is often used to reduce matrix effects [2,5,9]; however, many TIC chromatograms in published literature still exhibit poor baseline separation likely due to co-elution with an unresolved complex mixture (UCM) coming from the matrix. Co-elution of peaks at the detector can result in poor mass spectral library matches, which may result in the lack of an identity assigned to a peak, or in a more unfortunate scenario, may cause peak misidentification. The large number of trace components and UCM present in these chromatograms can also obscure baseline separation of components which may result in non-optimal quantification, especially in the absence of robust deconvolution algorithms. This can potentially impact age estimations when investigating degradation markers.

It is proposed that characterization of archaeological samples would benefit substantially from the use of comprehensive two-dimensional gas chromatography-time-of-flight mass spectrometry (GC×GC-TOFMS). This comprehensive, multidimensional technique offers an improved peak capacity, leading to considerable enhancement in peak separation and subsequent quantification. GC×GC is similar to 1D GC in most principles; however, it exploits the use of both a first dimension (^{1}D) and second dimension (^{2}D) column to produce two independent mechanisms of separation. The 1D separation is comparable to a conventional 1D GC run. Short plugs of eluent from the ^{1}D column are trapped, focused, and released at the modulator. These small plugs of eluent are then injected onto the shorter ^{2}D column that exhibits a different stationary phase, where they undergo a secondary separation. This generates a chromatogram based on both a first dimension retention time (^1t_R) and second dimension retention time (^2t_R). The resulting two-dimensional plane creates a drastic increase in the number of compounds that can be separated, hence improving the overall peak capacity of the system. GC×GC additionally offers the advantage of global method sensitivity enhancement following cryogenic zone compression (CZC) [15]. The output based on three axes (^1t_R, ^2t_R, and intensity) improves the visualization of the characteristics of complex samples, therefore presenting an attractive option for end users in various life science application areas [16]. This is further enhanced by the availability of a fourth mass spectral axis that provides information about the identity of the chromatographic signals through the use of deconvoluted TOFMS fragmentation spectra that can be compared to reference spectral libraries.

Overall, this technique is highly sensitive and selective and has been shown to be extremely successful in other fields of VOC measurements such as forensic taphonomy [16–20], arson investigation [21], breath analysis [22], food and fragrance analysis [23,24], and many others. However, in most of these applications, the quantity of the sample is higher and the age of the sample is much more recent when comparing with archaeological investigations. Therefore, the aim of this proof-of-concept study was to screen several types of modern reference specimens with a generic volatile method using HS-SPME-GC×GC-TOFMS to determine whether valuable results could be obtained with the types and amount of sample typically available. This was done with the aim of presenting insights into the use of GC×GC-TOFMS in the future of archaeological research.

2. Materials and Methods

2.1. Samples

A wide range of samples available in an experimental reference collection were tested (Table 1). These samples were chosen based on availability and based on the types of samples and materials typically encountered during an archaeological recovery. Residues were from modern samples which were either collected directly (*i.e.*, from plants or animals) or were collected from experimental stone tools. The former were immediately placed in vials for analysis. The latter were derived from modern reproductions of archaeological stone tools, which were used in modern settings for a range of activities relevant for prehistoric lifeways (e.g., hunting, butchering, hideworking, *etc.*).

Table 1. Preparation of sample vials containing organic residue prior to SPME sampling.

Sample Name	Mode of Preparation	Approximate Age (If Known)
Ivory (dry)	Water extraction	3 years
Ivory powder	Manually placed in vial	
Bone (dry)	Water extraction	3 years
Bone powder	Manually placed in vial	
Sediment & bone powder	Mixture of sediment and powder manually placed in vial	
Meat & starch	Water extraction	
Meat & starch	Water extraction/ultrasonic bath	
Meat, blood & fat	Water extraction	
Meat, blood & fat	Solvent extraction	
Sample name	Mode of preparation	Approximate age (if known)
Hide (fresh)	Water extraction	4 months frozen
Hide (fresh)	Solvent extraction	4 months frozen
Hide (fresh)	Piece manually placed in tube	4 months frozen
Hand residue [1]	Water extraction	
Hand residue	Solvent extraction	
Resin [2] & beeswax [3] mixture	1:1 mixture prepared by heating until liquid and mixing together, 2 g of final mixture was manually placed in vial	
Resin & beeswax mixture (Heated 1 h)	Prepared as above but heated for an additional 1 h after mixing	
Plastiline [4]	Water extraction	
Antler	Water extraction	
Leather from binding	Water extraction	
Sediment	Water extraction/ultrasonic bath	33,000 years
Blank vial	N/A	

[1] Hand residue: A stone tool was handled by the researcher without gloves and an extraction was taken afterwards; [2] Picea abies resin, Natural, Rochefort, Belgium; [3] Beeswax, Natural, Rochefort Belgium; [4] Plastiline, J Herbin, Paris France.

In order to extract the residues from these stone tools, two types of commonly used extraction techniques [25,26] were investigated. Water extraction was performed by pipetting 4 μL of distilled water onto the surface of the item, agitating the liquid with the pipette tip, and then reaspirating the water into a headspace vial. Solvent extraction was also performed in the same manner but using a mixture of ethanol (VWR collection, Fontenay-sous-Bois, France), distilled water (Delhaize, Liège, Belgium), and acetonitrile (Merck Millipore, Darmstadt, Germany) in a 1:1:1 ratio. Pipetting was performed with a Transferpette® S (VWR International, Leuven, Belgium) with polypropylene tips. Water extraction with ultrasonication was also performed on some samples which involved placing the material in a weigh boat with 5 mL of distilled water and performing ultrasonication (Elmasonic P 120 H, Elma, Singen, Germany) for 5 min at 37 kHz, after which the liquid and material were poured into a headspace vial. A full description of extractions and treatments performed to prepare sample vials is shown in Table 1. All samples were prepared in 20 mL screw cap headspace vials with 1.3 mm PTFE septa (Gerstel®, Kortrijk, Belgium).

Although the residues were from modern references, a sediment sample on an ancient stone tool was also analyzed. This sample was derived from a stone tool found in a stratified cave site in a layer dated to approximately 33,000 years ago. This situation is important in an archaeological context in order to determine whether it is possible to differentiate between a general taphonomic signature (e.g., the sediment) and the unique signature of use-related residues on the tool itself. Although additional studies need to be performed to answer this question in its entirety, this sample was analyzed in order to incorporate a particular situation in which GC×GC-TOFMS may be useful for providing information to differentiate two volatile signatures.

2.2. Instrumental Analysis

Samples were analyzed using a Pegasus 4D GC×GC-TOFMS (LECO® Corporation, St. Joseph, MI, USA) equipped with a secondary oven and a quad-jet, dual-stage, thermal modulator. The ^{1}D column was an Rtx-5MS (Restek® Corporation, Bellefonte, PA, USA) (30 m × 0.25 mm i.d. × 0.25 μm d_f), and the ^{2}D column was an Rxi-17Sil MS (Restek® Corporation) (1.0 m × 0.15 mm i.d. × 0.15 μm d_f). The connection between the two columns was performed used a SilTite μ-union (SGE Analytical Science®, Wetherill Park, Australia). All samples were run on this column set with settings described in the following paragraphs, but selected samples were further investigated on two other column sets: Rxi-624Sil MS ^{1}D column (Restek® Corporation) (30 m × 0.25 mm i.d. × 1.40 μm d_f) × Stabilwax ^{2}D column (Restek® Corporation) (2 m × 0.25 mm i.d. × 0.50 μm d_f); Stabilwax ^{1}D column (25 m × 0.25 mm i.d. × 0.50 d_f) × Rtx-200Sil MS ^{2}D column (Restek® Corporation) (1 m × 0.25 mm id × 0.25 d_f).

The sample vials were extracted using automated HS-SPME with a multipurpose sampler (MPS) (Gerstel®). A 50/30 μm divinylbenzene/carboxen/polydimethylsiloxane (DVB/CAR/PDMS) Stableflex 24 Ga fiber was chosen (Supelco®, Bellefonte, PA, USA) based on previous literature [5] and experience characterizing similar types of expected volatiles. Each vial was first incubated at 50 °C for 10 min. The fiber was exposed for 15 min to the headspace of each sample with a penetration depth of 21.00 mm. The fiber was desorbed with a penetration depth of 54.00 mm for 180 s in a CIS4 Cooled Injection System (Gerstel®). The injector temperature was programmed at −10 °C for 0.15 min, was increased to 350 °C at 12.00 °C·s^{-1}, and then held at 350 °C for 5.00 min. The injection was performed in splitless mode with a 3 mL·min^{-1} septum purge flow and a purge time of 120 s. The inlet purge flow was 20 mL·min^{-1}.

High purity helium (Air Liquide®, Liège, Belgium) carrier gas flow was held at a constant rate of 1.00 mL·min^{-1} throughout the run. The ^{1}D oven was held initially at 50 °C for 3 min, followed by an increase to 220 °C at a rate of 5° C·min^{-1} (total 37 min). The ^{2}D oven offset was +15 °C above the ^{1}D oven temperature and the modulator offset was +20 °C from the ^{2}D oven. A 3 s modulation period (P_M) was used with a 0.40 s hot pulse time. An acquisition delay of 300 s was used. The MS transfer line was held at 220 °C. The mass acquisition range was 29–450 amu and operated with a rate of 100 Hz. The ion source temperature was 230 °C and the electron ionization energy was 70 eV. The detector voltage was 1500 V.

2.3. Data Processing

As the objective of this proof-of-concept study was to screen various sample types, tentative compound identifications were performed based on the National Institute of Standards and Technology (NIST) 2014 and Wiley 10 libraries with a match factor threshold > 700. Peaks identified below this threshold were labelled "unknown" and those above this threshold were labelled with tentative names. Sample acquisition and peak finding was performed using ChromaTOF v. 4.50.8.0 (LECO® Corporation). A baseline offset of 0.6 was used, with a 12 s ^{1}D peak width and a 0.1 s ^{2}D peak width. A signal-to-noise (S/N) ratio of 50 was used and unique mass was used for area calculations. Statistical Compare was used by importing the samples into two classes (*i.e.*, blanks and samples) in order to perform peak alignment and filtering of non-specific peaks (e.g., column bleed, background compounds, *etc.*). The analyte spectral match required a mass threshold of 10 and a minimum similarity match of 600. Peaks were searched for a second time if not identified by the initial peak find algorithm down to a S/N of 20. A Fisher Ratio (FR) was calculated for each compound and compounds with undefined FRs (*i.e.*, only present in one class) or those exceeding the calculated critical F-value (F_{crit}) (*i.e.*, present in both classes but highly variable) were kept for further processing. This approach has already been well-established in several volatile profiling studies [27–29].

3. Results and Discussions

3.1. Chromatographic Considerations and Sample Analysis

Data processing revealed anywhere from 200 to 2000 features identified within a sample. Figure 1 shows a variety of chromatograms obtained using the Rtx-5MS × Rxi-17Sil MS column set. The resin samples yielded the most complex volatile signature of all sample types analyzed. Resin is an important substance in archaeological research as it has been used during prehistory as an adhesive to haft stone tools [9,30]. Hafting is considered an invention that revolutionized stone tool use [30,31] and that necessitated abstract thought and planning [9,32]. Therefore, being able to understand the mode of preparation allows an understanding of the cognitive and technical capabilities of their makers [33]. Figure 1 (top left and top right) demonstrates the ability to differentiate the volatile signature of the two resin and beeswax mixture samples based on different preparation methods differentiated by heating length. It, therefore, follows that it may be possible, given a larger database of controlled study information, to provide more information in the future about the type of resin discovered on an archaeological tool and its potential mode of preparation.

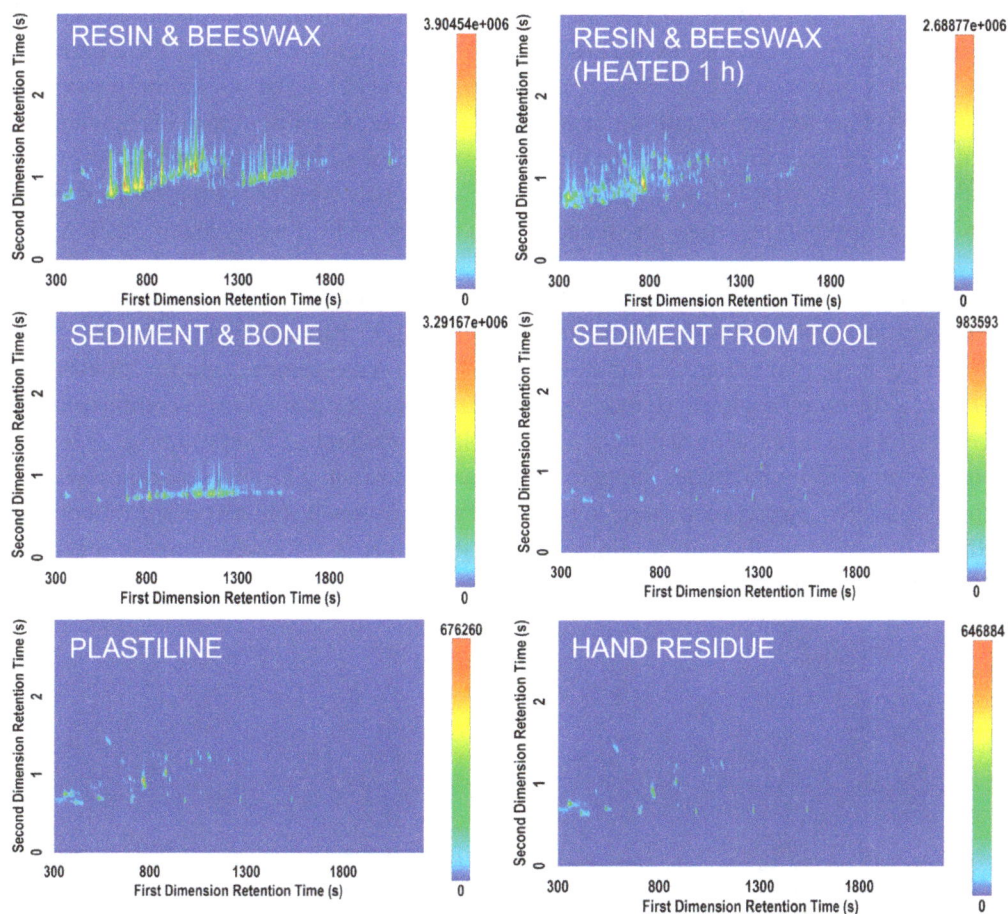

Figure 1. Selected total ion current (TIC) contour plots from six samples analyzed on the Rtx-5MS × Rxi-17Sil MS column set.

The sample of mixed sediment and bone displayed in Figure 1 also yielded a complex profile (middle left). This sample exhibited numerous hydrocarbon compounds, leading to a need for more effective optimization of the first dimension separation. There were only few trace compounds that eluted separately along the second dimension axis, suggesting that a well-optimized 1D GC-MS method may be useful in screening sediment samples if bone fragments are suspected to be contained in them (*i.e.*, differentiation of sediment from sediment mixed with bone). However, this GC×GC-TOFMS method had added value in providing a "total screening" approach, in that the same method could potentially be applied to a vast array of potential unknown residues to provide information about sample characteristics. In an archaeological context, due to the minimal amounts of samples available it may not be possible to distinguish visually (or microscopically) between a sediment sample and other possible organic residues. A total screening approach can be valuable from an investigative standpoint and when processing a large number of samples from an archaeological site. In addition, comparing results from sediment samples and extractions from stone tools can help in distinguishing between residues that can be linked to the use of a specific stone tool and residues that occur on a more general level throughout the site. The signature from a minimal amount of sediment washed from an archeological stone tool also yielded a volatile signature different than that from the mixed sediment and bone sample. In this situation, the number of components in the volatile signature was fewer but there was a higher need for the selectivity of GC×GC-TOFMS due to the fact that more compounds separated along the ^{2}D axis.

The samples of ivory, bone, meat/starch, antler, meat/starch/blood, and leather generated minimal signals that could not be visually differentiated from blank vials without further testing. This could be due to the fact that no volatile compounds existed in the samples that were available for SPME extraction. It could also be due to the fact that a longer or warmer SPME extraction was required to release characteristic volatiles from these matrices. In addition, obtaining experimental replicates would allow for the generation of statistics that could assist in determining whether a peak is discriminatory for that sample type, and therefore would lead to more robust determinations of compound importance. Although strong GC×GC-TOFMS signals were not obtained for these samples, further analytical optimization may prove valuable in the future for their characterization.

Figure 2 shows the fresh resin and beeswax mix characterized on the three different column sets. Each column set yielded a slightly different structure of chromatogram. The first column set (top) is a non-polar × polar column combination, which is a conventional column set used for initial testing. The middle chromatogram demonstrates a mid-polar × polar column combination, and the bottom chromatogram demonstrates a polar × mid-polar combination. The groupings of terpenes and terpenoids appear at different angles depending on the column set. The best separation was obtained by the Rtx-624Sil MS × Stabilwax column set in this study. However it is suspected that optimization of conditions could potentially lead to sufficient separation over the chromatographic space using any one of these phase combinations.

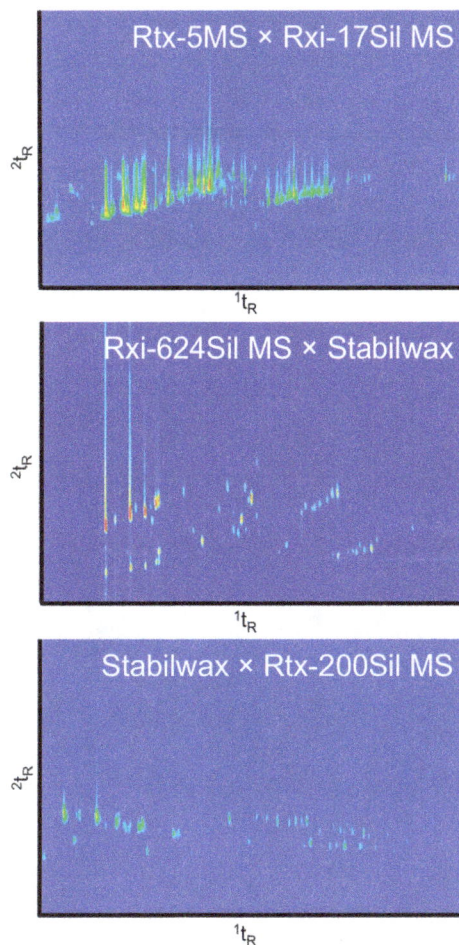

Figure 2. Column set comparison of a resin and beeswax (1:1) mixture analyzed on three different column sets. Top: Rtx-5MS (30 m × 0.25 μm i.d. × 0.25 d_f) × Rxi-17Sil MS (1 m × 0.15 μm i.d. × 0.15 d_f), P_M = 3 s; Middle: Rxi-624Sil MS (30 m × 0.25 μm i.d. × 1.4 d_f) × Stabilwax (2 m × 0.25 μm i.d. × 0.5 d_f), P_M = 4 s; Bottom: Stabilwax (25 m × 0.25 μm i.d. × 0.5 d_f) × Rtx-200Sil MS (1 m × 0.25 μm i.d. × 0.25 d_f) P_M = 2 s.

Using the first column set, there was a large amount of peak tailing in both the ^{1}D and ^{2}D. This phenomenon was attributed to the thinner films used for the columns in the Rtx-5MS × Rxi-17Sil MS set in comparison to the other column sets tested. Second dimension tailing was apparent for many of the samples, as seen in Figure 1. However, some samples also exhibited peak tailing in the ^{1}D which led to exaggerated "L"-shaped peaks. This was most apparent for samples that contained carboxylic acids, such as the hide sample displayed in Figure 3. Increasing the film thickness of the columns increased the retention time of specific compounds, yet allowed for improved resolution of the volatile compounds being analyzed. The third column combination, Stabilwax × Rtx-200Sil MS, had reduced peak tailing in the ^{1}D due to the thicker film on this column, but continued to exhibit considerable peak tailing in the ^{2}D due to the thin film on the Rtx-200Sil MS column.

On the first column set with thin film thicknesses in both dimensions, some compounds tailed across the majority of the ^{2}D, which is not ideal in a GC×GC analysis where the ^{2}D should be used as an additional dimension of selectivity. Nonetheless, due to the preliminary nature of this work, this column set was used for initial analysis of all samples as it was considered to be the conventional column set and would, therefore, provide a potential starting point for comparison to the literature. However, future analyses will likely be performed using the Rxi-624Sil MS × Stabilwax column set due to the improved peak shape and resolution for the type of compounds being analyzed.

Figure 3 also demonstrates the complexity of the volatile signature of these samples. The red insert displays the dynamic range of compounds and potential for co-elution in the first dimension. These issues are mitigated using GC×GC whereby the selectivity of the secondary phase affords improved peak capacity for separating trace compounds from high abundance compounds. Selected compounds labelled in Figure 3 demonstrates the wide range of potential compounds that can be encountered when examining volatiles from mammalian tissues. The ability to fully characterize the complexity of these types of matrices, even in trace quantities, will be extremely beneficial to improve the characterization of volatiles from archaeological residues.

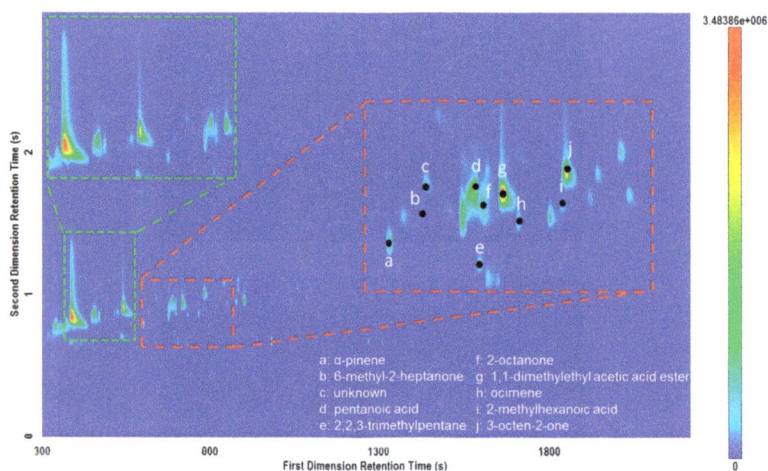

Figure 3. Total ion current (TIC) contour plot for a piece of hide sample analyzed on the Rtx-5MS × Rxi-17Sil MS column set demonstrating "L"-shaped peaks for some carboxylic acid peaks (green). Several labeled peaks demonstrate the value of the GC×GC selectivity and typical compounds obtained (red).

3.2. Sample Extraction Techniques

Figure 4 displays the comparison of different potential methods of extracting volatiles for analysis from animal hide. Based on the results of this study, improved results (*i.e.*, richer chromatograms) were generally obtained when the HS-SPME extraction was performed directly on the residue of interest rather than on an extract from an item. However, due to the common practice of performing solvent extractions from items of interest in archaeological research, both approaches were tested. Further studies with sample replication may demonstrate the potential effects of different solvents or ultrasonication to optimize extraction of specific residues and provide additional comparison to performing direct analysis of residue.

Previous studies of archaeological residues have typically relied on extraction and derivitization [5,9,10]. However, the development of a non-destructive direct HS-SPME extraction from the residue is a substantial benefit, as further analyses can then be conducted on the item in its original state after it has been screened by VOC analysis. This requires no solvents or derivatization agents and less physical handling of potentially important items and residues. It is currently unknown what amount of material is necessary in order to detect the volatile signature directly from various samples by GC×GC-TOFMS. For example, the volume of blood, mass of resin, and/or mass of sediment collected may be too low to produce a sufficiently concentrated headspace for analysis. In addition, residues may be a combination of several traces. For example, this could occur if a stone tool was used to cut both vegetal material and a piece of meat. At trace levels, it is unknown how these volatiles signatures may interact and whether this technique would be capable of differentiating such minimal traces.

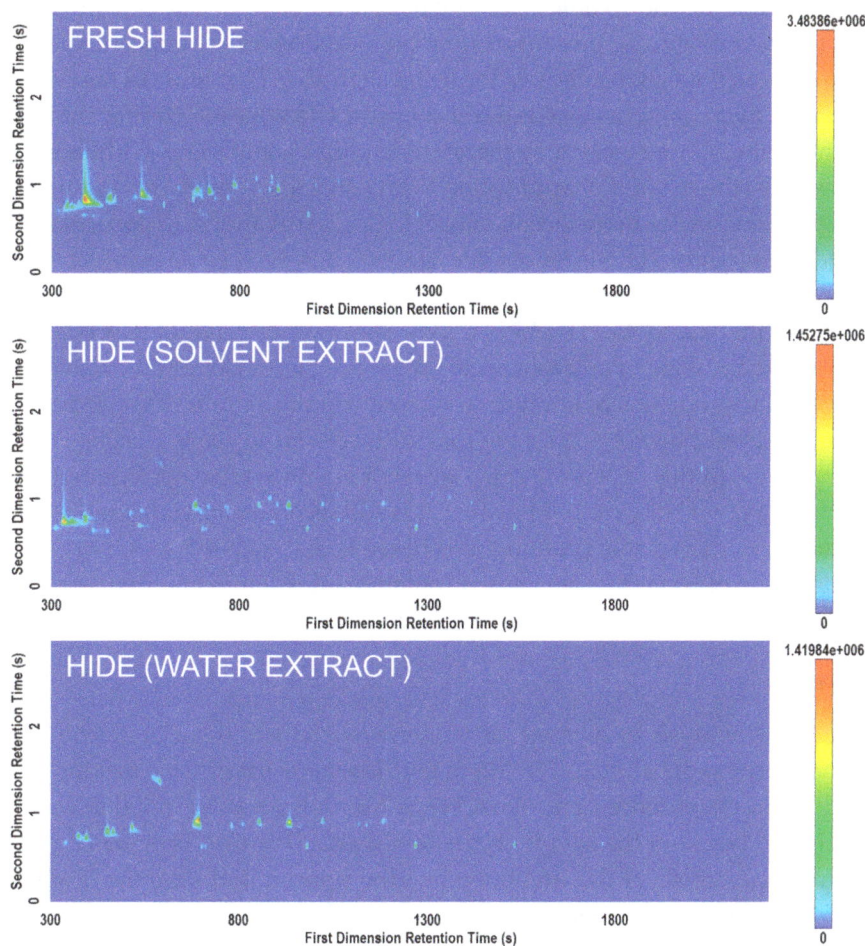

Figure 4. Comparison of total ion current (TIC) contour plots for three hide samples prepared by direct analysis (top); extraction in a mixture of solvents (middle); and extraction in water (bottom) on the Rtx-5MS × Rxi-17Sil MS column set.

3.3. Interferences

Due to the trace nature of this type of profiling, it was necessary to include samples that would allow the assessment of potential contamination or interferences with the ability to successfully obtain a volatile signature. Two samples were included in the analysis which allowed an initial assessment of such a situation.

Plastiline, an oil-based modelling clay used for mounting artefacts for microscopic analysis, was extracted and the extraction liquid was placed in a vial. Plastiline is a proprietary product whose composition is protected by patents. It generally contains oils, waxes, clay minerals, and in some cases a small portion of sulfur [34]. This is important in determining the workflow for a particular archaeological item of interest. In the situation where plastiline (or other mounting product) generates an interfering signal, the volatile signature may possibly be overwhelmed by plastiline volatiles and be unidentifiable.

The plastiline was found to contain minimal but identifiable levels of some volatiles such as hydrocarbons, aldehydes, ethers, ketones, and aromatics. No sulfur compounds were detected in the volatile profile, indicating that the ingredients of this particular brand likely did not include any sulfur additives. Based on this initial analysis, plastiline could potentially provide an interference with volatile signatures of interest, especially when sample residues are present in trace quantities. In fact, many archaeological workflows no longer include this material due to the fact that it can cause damage to the physical appearance and chemical composition of the residues on a stone tool under examination [34].

This may also be the case for the volatile signature being created by the item of interest. This type of background signal, if known to be present during the workflow of an archaeological investigation, could be subtracted from the signal of the residue being analyzed. The nature of the GC×GC selectivity also reduces the risk that compounds of interest will co-elute with interfering compounds such as those found in plastiline. Future work may therefore require a comparison of different compositions of mounting materials, the method of application to archaeological artefacts, and the interference of these. This is less of an issue for more recent archaeological excavations, in particular those that took into account the necessary precautions for residue analysis; however, for some older excavations these potential contaminations may occur.

Due to the fact that GC×GC-TOFMS is a very sensitive technique, it was also necessary to determine whether the handling of archaeological artefacts without gloves may also interfere with the volatile signature from specimens of interest. The hand residue sample was produced by handling a tool with bare hands and then washing the tool with solvent to see if a background signal from the handling appeared. In this case, only trace levels of volatiles were contributed to the profile in comparison with a blank vial (Figure 1). However, it is still recommended that gloves are worn when handling specimens prior to volatile profiling. It is possible that with the sensitivity of this technique may cause the detection of secondary transfer of contaminant residues and interfere with the volatile profile of the target residue.

3.4. Future Perspectives

This preliminary analysis of a range of modern reproductions of archaeological specimens indicates that the potential of GC×GC-TOFMS to provide value for organic residue characterization is promising. Even in the absence of a robust HS-SPME optimization, notable volatile signatures could be investigated between the sample types analyzed. This also provides a potential avenue for a non-destructive analysis of the unaltered residue upon initial discovery, without requiring solvent extraction and/or derivitization. Although marker compounds cannot be extrapolated from this sample set, the analyses presented herein allowed for an evaluation of the general use of GC×GC-TOFMS for organic residue characterization. Since the SPME-GC×GC-TOFMS analysis is non-destructive, it can provide an additional tool to be incorporated in the analytical workflow of residue characterization, in addition to the vast array of instrumentation currently used for these purposes.

There are a number of future avenues for this work to be further investigated. First and foremost, it would be valuable to begin analyzing replicate experimental samples to be able to apply robust statistical methods for the objective differentiation of residues. This must be done on lower quantities of residue in order to ensure the robustness of characterization strategies. Applying artificial aging of controlled experimental samples may also aid in improving characterization approaches in the future. In addition, the application of these techniques to additional archaeological specimens with complex histories would be interesting in order to determine if the experimental data can be extrapolated to real scenarios.

In addition, the detection of compounds in this study was performed by low-resolution TOFMS. This affords the ability to perform mass spectral library searching and produce matches above a certain quality. However, in order to improve these identification, high-resolution mass spectrometry will be employed in the future. Some of the samples analyzed in this study contained extremely complex profiles, such as the resin samples that contained hundreds of components likely originating from plant materials within the resin or other additives. Differentiating compounds of this nature by low-resolution mass spectrometry can be difficult due to similarities in their produced spectra. GC×GC coupled with high-resolution mass spectrometry has the potential to provide a higher degree of discrimination between these compounds and improve the identification confidence.

Acknowledgments: We wish to thank Restek® Corporation, Trajan® Scientific and Medical, and Supelco Sigma-Aldrich® for providing us with GC phases and various consumables. Christian Lepers is further acknowledged for his creation of the reference collection used in this study, and for providing the resin and beeswax mixtures. The research of K.A.P. is supported by Wallonie-Bruxelles International. The research of D.C. and V.R. is supported by the European Research Council under the European Union's Seventh Framework Programme (FP/2007-2013)/ERC Grant Agreement No. 312283.

Author Contributions: J.-F.F. and V.R. conceived and designed the experiments; K.A.P., P.-H.S., L.D. and D.C. performed the experiments; K.A.P., P.-H.S. and L.D. analyzed the data; J.-F.F. and V.R. contributed reagents/materials/analysis tools; K.P. wrote the paper and K.A.P., J.-F.F., V.R., P.-H.S, L.D. and D.C. contributed to preparation and revisions of the paper.

Conflicts of Interest: The authors declare no conflict of interest.

Abbreviations

The following abbreviations are used in this manuscript:

1D GC	One-dimensional gas chromatography
^{1}D	First dimension
^{2}D	Second dimension
$^{1}t_R$	First dimension retention time
$^{2}t_R$	Second dimension retention time
CZC	Cryogenic zone compression
d_f	Film thickness
FR	Fisher ratio
F_{crit}	Critical F-value
GC-MS	Gas chromatography-mass spectrometry
GC×GC-TOFMS	Comprehensive two-dimensional gas chromatography-time-of-flight mass spectrometry
i.d.	Inner diameter
HS-SPME	Headspace solid-phase microextraction
NIST	National Institute of Standards and Technology
S/N	Signal-to-noise ratio
TIC	Total ion current
UCM	Unresolved complex mixture

References

1. Hamm, S.; Bleton, J.; Connan, J.; Tchapla, A. A chemical investigation by headspace SPME and GC-MS of volatile and semi-volatile terpenes in various olibanum samples. *Phytochemistry* **2005**, *66*, 1499–1514. [CrossRef] [PubMed]

2. Hamm, S.; Lesellier, E.; Bleton, J.; Tchapla, A. Optimization of headspace solid phase microextraction for gas chromatography/mass spectrometry analysis of widely different volatility and polarity terpenoids in olibanum. *J. Chromatogr. A* **2003**, *1018*, 73–83. [CrossRef] [PubMed]

3. Jerković, I.; Marijanović, Z.; Gugić, M.; Roje, M. Chemical profile of the organic residue from ancient amphora found in the Adriatic Sea determined by direct GC and GC-MS analysis. *Molecules* **2011**, *16*, 7936–7948. [CrossRef] [PubMed]

4. Mcgovern, P.E.; Mirzoian, A.; Hall, G.R. Ancient egyption herbal wines. *Proc. Natl. Acad. Sci. USA* **2009**, *106*, 7361–7366. [CrossRef] [PubMed]

5. Regert, M.; Alexandre, V.; Thomas, N.; Lattuati-Derieux, A. Molecular characterisation of birch bark tar by headspace solid-phase microextraction gas chromatography-mass spectrometry: A new way for identifying archaeological glues. *J. Chromatogr. A* **2006**, *1101*, 245–253. [CrossRef] [PubMed]

6. Rafferty, S.M. Identification of Nicotine by Gas Chromatography/Mass Spectroscopy Analysis of Smoking Pipe Residue. *J. Archaeol. Sci.* **2002**, *29*, 897–907. [CrossRef]

7. Pecci, A.; Giorgi, G.; Salvini, L.; Ontiveros, M.Á.C. Identifying wine markers in ceramics and plasters using gas chromatography-mass spectrometry. Experimental and archaeological materials. *J. Archaeol. Sci.* **2013**, *40*, 109–115. [CrossRef]

8. Modugno, F.; Ribechini, E.; Colombini, M.P. Aromatic resin characterisation by gas chromatography-mass spectrometry. Raw and archaeological materials. *J. Chromatogr. A* **2006**, *1134*, 298–304. [CrossRef] [PubMed]

9. Regert, M. Investigating the history of prehistoric glues by gas chromatography-mass spectrometry. *J. Sep. Sci.* **2004**, *27*, 244–254. [CrossRef] [PubMed]

10. Helwig, K.; Monahan, V.; Poulin, J.; Antiquity, A. The Identification of Hafting Adhesive on a Slotted Antler Point from a Southwest Yukon Ice Patch. *Am. Anqituity* **2014**, *73*, 279–288.

11. Regert, M.; Colinart, S.; Degrand, L.; Decavallas, O. Chemical alteration and use of beeswax through time: Accelerated ageing tests and analysis of archaeological samples from various environmental contexts. *Archaeometry* **2001**, *43*, 549–569. [CrossRef]

12. Mathe, C.; Culioli, G.; Archier, P.; Vieillescazes, C. Characterization of archaeological frankincense by gas chromatography-mass spectrometry. *J. Chromatogr. A* **2004**, *1023*, 277–285. [CrossRef] [PubMed]

13. Boëda, E.; Connan, J.; Dessort, D.; Muhesen, S.; Mercier, N.; Valladas, H.; Tisnérat, N. Bitumen as a hafting material on Middle Palaeolithic artefacts. *Nature* **1996**, *380*, 336–338. [CrossRef]

14. Evershed, R.P.; Heron, C.; Goad, J. Analysis of organic residues of archaeological origin by high-temperature gas chromatography and gas chromatography mass spectrometry. *Analyst* **1990**, *115*, 1339–1342. [CrossRef]

15. Patterson, D.G.; Welch, S.M.; Turner, W.E.; Sjödin, A.; Focant, J.F. Cryogenic zone compression for the measurement of dioxins in human serum by isotope dilution at the attogram level using modulated gas chromatography coupled to high resolution magnetic sector mass spectrometry. *J. Chromatogr. A* **2011**, *1218*, 3274–3281. [CrossRef] [PubMed]

16. Perrault, K.A.; Nizio, K.D.; Forbes, S.L. A comparison of one-dimensional and comprehensive two-dimensional gas chromatography for decomposition odour profiling using inter-year replicate field trials. *Chromatographia* **2015**, *78*, 1057–1070. [CrossRef]

17. Stefanuto, P.-H.; Perrault, K.; Stadler, S.; Pesesse, R.; Brokl, M.; Forbes, S.; Focant, J.-F. Reading cadaveric decomposition chemistry with a new pair of glasses. *ChemPlusChem* **2014**, *79*, 786–789. [CrossRef]

18. Perrault, K.A.; Stefanuto, P.-H.; Stuart, B.H.; Rai, T.; Focant, J.-F.; Forbes, S.L. Reducing variation in decomposition odour profiling using comprehensive two-dimensional gas chromatography. *J. Sep. Sci.* **2015**, *38*, 73–80. [CrossRef] [PubMed]

19. Perrault, K.A.; Rai, T.; Stuart, B.H.; Forbes, S.L. Seasonal comparison of carrion volatiles in decomposition soil using comprehensive two-dimensional gas chromatography—Time of flight mass spectrometry. *Anal. Methods* **2014**, *7*, 690–698. [CrossRef]

20. Forbes, S.L.; Troobnikoff, A.N.; Ueland, M.; Nizio, K.D.; Perrault, K.A. Profiling the decomposition odour at the grave surface before and after probing. *Forensic Sci. Int.* **2016**, *259*, 193–199. [CrossRef] [PubMed]

21. Sampat, A.; Lopatka, M.; Sjerps, M.; Vivo-truyols, G.; Schoenmakers, P.; van Asten, A. The forensic potential of comprehensive two-dimensional gas chromatography. *TrAC Trends Anal. Chem.* **2016**, *80*, 345–363. [CrossRef]

22. Das, M.K.; Bishwal, S.C.; Das, A.; Dabral, D.; Varshney, A.; Badireddy, V.K.; Nanda, R. Investigation of gender-specific exhaled breath volatome in humans by GC×GC-TOF-MS. *Anal. Chem.* **2014**, *86*, 1229–1237. [CrossRef] [PubMed]

23. Cordero, C.; Liberto, E.; Bicchi, C.; Rubiolo, P.; Reichenbach, S.E.; Tian, X.; Tao, Q.; Giuria, V.P.; Torino, I. Targeted and non-targeted approaches for complex natural sample profiling by GC×GC-qMS. *J. Chromatogr. Sci.* **2010**, *48*, 251–262. [CrossRef] [PubMed]

24. Tranchida, P.Q.; Donato, P.; Cacciola, F.; Beccaria, M.; Dugo, P.; Mondello, L. Potential of comprehensive chromatography in food analysis. *Trends Anal. Chem.* **2013**, *52*, 186–205. [CrossRef]

25. Fullagar, R.; Hayes, E.; Stephenson, B.; Field, J.; Matheson, C.; Stern, N.; Fitzsimmons, K. Evidence for Pleistocene seed grinding at Lake Mungo, south-eastern Australia. *Archaeol. Ocean.* **2015**, *50*, 3–19. [CrossRef]

26. Fullagar, R. Residues and usewear. In *Archaeology in Practice: A student Guide to Archaeological Analyses*; Balme, J., Paterson, A., Eds.; John Wiley & Sons: New York, NY, USA, 2014; pp. 232–265.

27. Brokl, M.; Bishop, L.; Wright, C.G.; Liu, C.; McAdam, K.; Focant, J.-F. Multivariate analysis of mainstream tobacco smoke particulate phase by headspace solid-phase micro extraction coupled with comprehensive two-dimensional gas chromatography-time-of-flight mass spectrometry. *J. Chromatogr. A* **2014**, *1370*, 216–229. [CrossRef] [PubMed]

28. Stefanuto, P.-H.; Perrault, K.A.; Lloyd, R.M.; Stuart, B.H.; Rai, T.; Forbes, S.L.; Focant, J.-F. Exploring new dimensions in cadaveric decomposition odour analysis. *Anal. Method.* **2015**, *7*, 2287–2294. [CrossRef]

29. Armstrong, P.; Nizio, K.D.; Perrault, K.A.; Forbes, S.L. Establishing the volatile profile of pig carcasses as analogues for human decomposition during the early postmortem period. *Heliyon* **2016**, *2*, e00070. [CrossRef]

30. Rots, V. Towards an understanding of hafting: the macro- and microscopic evidence. *Antiquity* **2003**, *77*, 805–815. [CrossRef]

31. Barham, L. *From Hand to Handle: The First Industrial Revolution*; Oxford University Press: Oxford, UK, 2013.

32. Wynn, T. Hafted spears and the archaeology of mind. *Proc. Natl. Acad. Sci.* **2009**, *106*, 9544–9545. [CrossRef] [PubMed]

33. Wadley, L. Compound-adhesive manufacture as a behavioral proxy for complex cognition in the middle stone age. *Curr. Anthropol.* **2010**, *51*, S111–S119. [CrossRef]

34. Eggert, G. Plastiline: Another unsuspected danger in display causing black spots on bronzes. *Verb. der Restaur. zur Erhalt. von Kunst-und Kult.* **2006**, *2*, 112–116.

## 2

Separation and Determination of Some of the Main Cholesterol-Related Compounds in Blood by Gas Chromatography-Mass Spectrometry (Selected Ion Monitoring Mode)

Lucia Valverde-Som [1],*, Alegría Carrasco-Pancorbo [1] 🆔, Saleta Sierra [2], Soraya Santana [2], Cristina Ruiz-Samblás [1], Natalia Navas [1], Javier S. Burgos [2] and Luis Cuadros-Rodríguez [1]

[1] Department of Analytical Chemistry, Faculty of Science, University of Granada, C/Fuentenueva s/n, E-18071 Granada, Spain; alegriac@ugr.es (A.C.-P.); crsamblas@ugr.es (C.R.-S.); natalia@ugr.es (N.N.); lcuadros@ugr.es (L.C.-R.)

[2] Neuron Bio, P.T.S. Granada, C/Avicena 4, E-18016 Granada, Spain; ssierra@neuronbio.com (S.S.); ssantana@neuronbio.com (S.S.); jburgos@neuronbio.com (J.S.B.)

* Correspondence: luciavs@ugr.es

Abstract: Oxysterols are metabolites produced in the first step of cholesterol metabolism, which is related to neurodegenerative disorder. They can be detected by testing blood, plasma, serum, or cerebrospinal fluid. In this study, some cholesterol precursors and oxysterols were determined by gas chromatography coupled to mass spectrometry. The selected cholesterol-related compounds were desmosterol, lathosterol, lanosterol, 7α-hydroxycholesterol, 7β-hydroxycholesterol, 24(S)-hydroxycholesterol, 25-hydroxycholesterol, 7-ketocholesterol, and 27-hydroxycholesterol. A powerful method was developed and validated considering various analytical parameters, such as linearity index, detection and quantification limits, selectivity and matrix effect, precision (repeatability), and trueness (recovery factor) for each cholesterol-related compound. 7α-hydroxycholesterol, 7β-hydroxycholesterol, and desmosterol exhibited the lowest detection and quantification limits, with 0.01 and 0.03 µg/mL, respectively, in the three cases. 7-ketocholesterol and lathosterol showed matrix effect percentages between 95.5% and 104.8%, respectively (demonstrating a negligible matrix effect), and very satisfactory repeatability values (i.e., overall performance of the method). Next, the method was applied to the analysis of a very interesting selection of mouse plasma samples (9 plasma extracts of non-transgenic and transgenic mice that had been fed different diets). Although the number of samples was limited, the current study led to some biologically relevant conclusions regarding brain cholesterol metabolism.

Keywords: cholesterol-related compounds; GC-(IT)MS; selected ion monitoring mode

1. Introduction

Oxysterols play a crucial role as (a) regulators of expression of genes involved in lipid and sterol biosynthesis; (b) substrates for the formation of bile acids; and (c) mediators of reverse cholesterol transport, through which excess cholesterol is returned to the liver for excretion [1,2]. However, an overabundance of oxysterols can cause several diseases, such as cardiovascular disease, retina degeneration, inflammatory bowel disease, atherosclerosis, and neurodegenerative disorders [3–5].

Oxysterols can be found in several chemical forms and are enzymatically produced in the first steps of cholesterol (Chol) metabolism [6,7]. For example, Alzheimer's disease (AD) pathology is initiated (or accelerated) by a deregulation of the metabolism of Chol, sphingolipids, and fatty acids; in fact, hypercholesterolemia is an important risk factor for AD [8,9]. These metabolites are present in blood (serum or plasma), cerebrospinal fluid and brain tissues. Desmosterol, lanosterol, and lathosterol are the precursors of Chol synthesis. Chol is metabolized into side chain oxysterols (24(S)-hydroxycholesterol, 25-hydroxycholesterol, and 27-hydroxycholesterol) during its hepatic conversion [10] and transported into the brain by apolipoprotein E. Under normal conditions, no exchange of cholesterol occurs over the blood-brain barrier. However, Chol can be transported to the human brain in the form of a side-chain oxidized oxysterol such as 24(S)-hydroxycholesterol and 27-hydroxycholesterol [5,11,12]. As a consequence, the presence of these oxysterols has been related to neurological disorders and such oxysterols have been proposed as potential AD biomarkers [9,13].

Over the last decades, significant scientific developments and cutting-edge technologies have enabled a technological revolution in molecular biological methods. Metabolomics is a field that deals with the study of metabolites, which are the final products of cell regulation processes due to responses to genetic and environmental changes in biological systems [14].

Different approaches can be used in metabolomics [15]. Metabolite targeted analysis, for example, involves the detection, identification, and quantification of a single metabolite or a small group of specific metabolites of particular importance. The comprehensive characterization of a group of metabolites in an organism, tissue or cell implies the use of advanced analytical techniques. Oxysterols have been determined so far by liquid chromatography (LC) or gas chromatography (GC), mainly coupled to mass spectrometry (MS) and also by nuclear magnetic resonance (NMR) [6,16]. MS and NMR have a great potential to select proper biomarkers in order to diagnose associated diseases [13,17].

To determine the oxysterols present in biological fluids and tissues, there are different ways to carry out the sample preparation that obviously depend on the matrix, analyte/s and technique/s to be used. Blood plasma has been separated from blood serum by using ethylenediaminetetraacetic acid (EDTA) and further centrifugating the blood sample [18,19]. Next, sterols can be extracted from blood plasma by alkaline hydrolysis. This leads to the degradation of lipids such as phospholipids and triglycerides and avoids their potential interferences. Finally, solid phase extraction (SPE) has been utilized to separate sterols from other substances [18,20].

A derivatization reaction is usually required prior to GC analysis in order to increase the volatility of oxysterols; N,O-bis(trimethylsilyl)trifluoracetamide/trimethylchloroside (BSTFA:TMCS) is a very widely used derivatization reagent. However, N-methyl-N-(trimethylsilyl)-trifluoroacetamide-1,4-dithioerythritol-trimethyliodosilane (MSTFA:DTE:TMIS) has been proposed as one of the most efficient derivatization reagents based on its specificity/sensitivity ratio in GC-MS [21,22].

LC is recognized as a chromatographic technique that does not require a previous derivatization of the samples. Yet, in the specific case of determining oxysterols, the quantification can be more difficult if the analytes are not derivatized [22]. The use of N,N-dimethyl-glycine, picolinyl esters, and Girard P reagents has been described in LC-MS [23,24]. The derivatized compounds are more polar than the native sterols, which improves the ionization in the MS system. In fact, the number of papers describing LC-MS analytical methodologies for oxysterol determination in biological matrices has lately increased [25–27].

Traditionally, the instrumental analysis of sterols and related compounds in plasma has used GC and GC-MS [18]. Although GC and its use coupled to MS have some limitations, these techniques are widely used for sterol analysis due to their chromatographic resolving capacity, their robustness, and the relatively low cost of acquiring and operating the instruments [6,19,20,28]. GC-MS with electron ionization provides spectra rich in structural information, allowing the determination of unknown compounds [29]. By contrast, LC-MS requires Girard P derivatizations and exploring the mass-load ratio with the MS^3 mode in order to obtain structural information and identify unexpected oxysterols [27]. The use of MS in tandem (MS^n) can improve the sensitivity of the methods [30,31], but can also entail some methodological difficulties. A review conducted by Griffiths and Wang in 2009 [32] explained the biological importance of oxysterols and various relevant samples of brain and body fluids analyzed with GC-MS and LC-MS/MS. In 2016, both authors with other colleagues presented a review of MS-based methods for oxysterol analysis paying particular attention to LC-MS [27].

The aim of the present study was to develop a reliable and fit-for-purpose bioanalytical method for the determination of some Chol-related compounds (selected as potential biomarkers of AD by the biotechnology-based company Neuron Bio) in mouse plasma from six APPswe mice and three non-transgenic littermate mice of both sexes that were fed different diets from the age of 9 months to the age of 15 months. The analytical method was developed by using GC-MS in selected ion monitoring (SIM) mode. An ion trap (IT) mass analyzer was used to obtain the m/z signal of the analytes under study (7α-hydroxycholesterol (7αOHChol), desmosterol (Desm), lathosterol (Latho), 7β-hydroxycholesterol (7βOHChol), lanosterol (Lano), 24(S)-hydroxycholesterol (24OHChol), 25-hydroxycholesterol (25OHChol), 7-ketocholesterol (7KetoChol), and 27-hydroxycholesterol (27OHChol)) applying an MS method with 9 different segments. In other words, we defined specific segments or analytical windows with different MS conditions in order to selectively determine each compound. This strategy enabled us to detect and quantify Chol-related compounds with a similar structure in mouse plasma. The potential interfering effect produced by high concentrations of cholesterol was avoided by applying the SIM strategy.

2. Materials and Methods

2.1. Chemicals and Reagents

(HPLC)-grade methanol and ethyl acetate (EA) were provided by Panreac, Spain. BSTFA (*N,O*-bis (trimethylsilyl)trifluoroacetamide) with 1% TMCS (trimethylchloroside) was obtained from Sigma-Aldrich, St. Louis, MO, USA, as a derivatization reagent. Solid standards of analytes were supplied by Enzo Life Sciences, Farmingdale, NY, USA (24OHChol, >98%), MP Biomedicals, Illkirch-Graffenstaden, Strasbourg, France (25OHChol, >98%), Sigma-Aldrich, St. Louis, MO, USA (Chol, ≥99%; Latho, ≥93%; Desm, ≥84%; and 7KetoChol, ≥90%), and Avanti Polar Lipids, Alabaster, AL, USA (Lano, >99%; 7αOHChol, >99%; 7βOHChol, >99%; and 27OHChol, >99%). In addition, 98% purity betulin supplied by Avanti Polar Lipids, Alabaster, AL, USA, was used as internal standard (IS). All these chemical compounds as well their corresponding solutions were stored in the dark at −20 °C.

Betulin was chosen as internal standard because it is not a naturally occurring sterol in blood. This analyte has been also used as IS in other similar applications [33,34].

2.2. Standard Solutions

A multi-standard stock solution containing 100 μg/mL of each analyte (Chol, Desm, Lano, Latho, 7KetoOH, 7αOHChol, 7βOHChol, 24OHChol, 25OHChol, and 27OHChol) was prepared in methanol weighing the appropriate amount of the solid pure standards. A 100 μg/mL stock standard solution of betulin was also prepared in methanol.

2.3. Samples and Sample Preparation

The study was performed with two sample groups: (i) mice blood plasma as test samples and (ii) goat blood serum as quality control samples.

Each extract was prepared applying the following procedure: 200 μL of sample were poured into a glass tube along with 40 μL of internal standard and 0.5 mL of EA. The tube was shaken in a vortex mixer (1 min), sonicated (50 W) for 15 min, and centrifuged for 5 min at 10,000 rpm (approx. 8400 g). Subsequently, 100 μL of the organic phase were transferred to an Eppendorf vial and stored in the dark at −20 °C until the analysis was performed.

The final concentration of IS was 40 μg/mL in all the extracts. The solvent was evaporated with N_2 and an aliquot of 50 μL of derivatization reagent was added after that. The reaction mixture was left at room temperature for 1 h (this was previously optimized by our team and it was proved that the total derivatization of the compounds of interest was achieved). In every case, 50 μL of the derivatized sample extract were transferred to a chromatographic vial containing a 250 μL insert.

(i) *Mouse plasma as test samples*

Six 9-month-old APPswe (transgenic) mice and three 9-month-old non-transgenic (Non-Tg) littermate mice of both sexes were used. They were kept in temperature- and light-controlled rooms in Neuron Bio facilities. Three APPswe mice were fed with NST0037 (a novel statin developed by Neuron Bio (Granada, Spain) [35])—enriched diet until the age of 15 months; the rest—three APPswe and three Non-Tg littermate mice—were fed with a control diet from until the same age. In the rest of the paper, each type of mice will be denoted by using letters A, B and C: A are non-transgenic mice, Non-Tg; B are transgenic mice, APPswe; and C are transgenic mice, APPswe, fed with NST0037.

The 15-month-old mice were anesthetized with sodium pentobarbital (300 mg/kg); their blood was collected and the plasma was obtained and frozen at $-80\,^{\circ}$C.

(ii) *Goat serum as quality control samples*

The set of quality control samples was composed of 5 extracts of the blood serum of a single goat in EA. Each extract (300 μL) was prepared in the same way as the test samples, starting from 600 μL of blood sample. As indicated above, the extracts were stored in the dark at $-20\,^{\circ}$C until the analysis took place. It should be noted that the sample treatment removes the proteins; thus, extracts from matrices of plasma and serum are quite similar and serum samples can be used as representative examples of plasma samples for an effective analytical quality control.

Quality control (QC) standard solutions were prepared by adding 0.5, 1.0, 2.5, and 5.0 μg/mL of each analyte into goat serum extracts. In order to effectively monitor the analytical method, two types of QC samples were used: some were spiked before and others were spiked after the extraction.

An overall diagram showing the preparation of both sample groups (i.e., test samples and quality control samples) is shown in Figure 1.

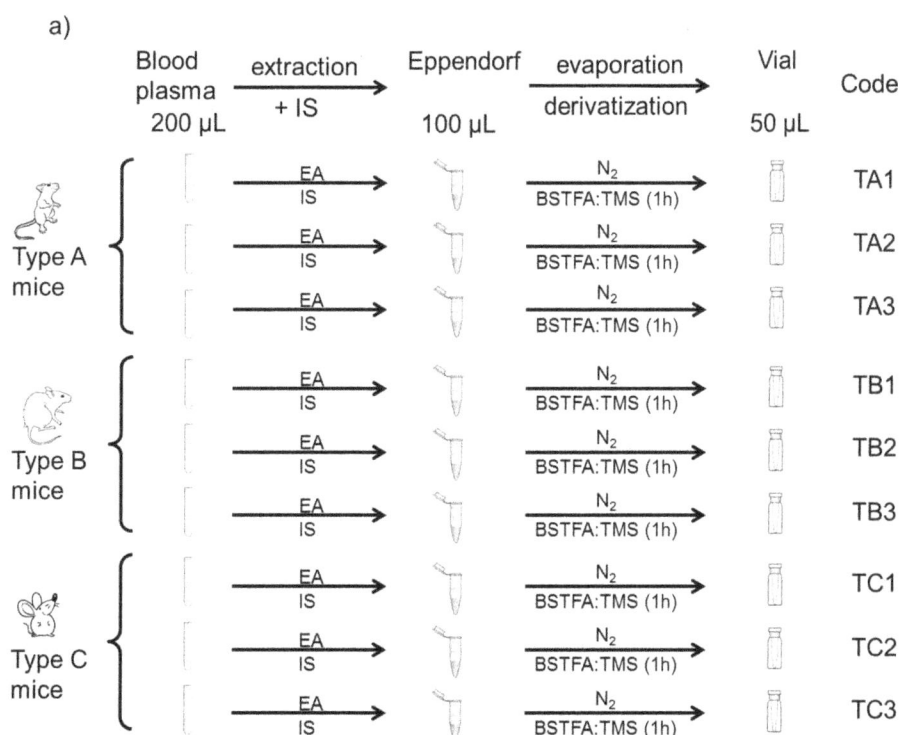

Figure 1. *Cont.*

b)

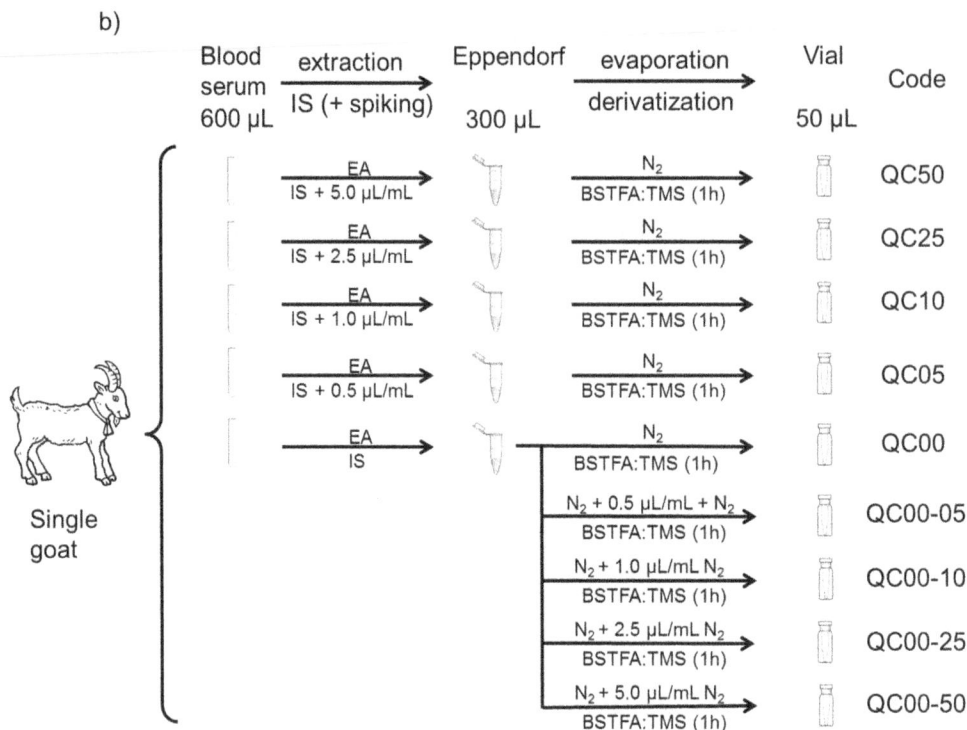

Figure 1. Diagram of the preparation of (**a**) test samples and (**b**) quality control samples. The sample code was indicated in every vial. The sample codes for the test samples (T) include information about the type of mice (A, B, or C) and the specimen number (1, 2, or 3). The codes of the quality control samples (QC) indicate the spiked concentration (50, 25, 10, 05, or 00) in µg/mL. (EA: ethyl acetate; IS: internal standard; BSTFA/TMS: silylation reagent).

2.4. GC-(EI/IT)MS Instrument and Method Conditions

All separations were performed by using a Varian GC 3800 gas chromatograph (Palo Alto, CA, USA) equipped with a split/splitless injector. The GC system was coupled to a Varian (IT)MS 4000 (ion trap) mass spectrometer (Palo Alto, CA, USA) equipped with an electron impact (EI) source. 1 µL of each sample was introduced into the GC equipment using a robotized autosampler module (CombiPal, CTC ANALYTICS, Zwingen, Switzerland). Splitless injection mode was selected and the injector temperature was held at 250 °C.

The analytical column was a capillary column coated with a 5% diphenyl-95% dimethyl polysiloxane stationary phase (DB-5MS; 30 m × 0.25 mm i.d. × 0.25 µm film thickness (Agilent Technologies J&W, Santa Clara, CA, USA). A silica deactivated pre-column (1 m × 0.25 mm i.d.) with press-fit connections (Agilent Technologies, Santa Clara, CA, USA) was used.

At the beginning of the runs, the column oven temperature was held at 200 °C for 0.5 min, then programmed to increase by 20 °C/min up to 300 °C and maintained at that value for 10 min, with a total run time of 15.50 min. Helium (99.995%) was used as the carrier gas and its flow rate was 1.2 mL/min. This method was a modification of a previously published method [21,36].

The MS conditions were as follows: the ion source temperature was held at 250 °C during the GC-MS runs. The transfer-line temperature was maintained at 280 °C. The electron energy was 70 eV and the emission current 10 µA. In full-scan mode, average spectra were acquired in the m/z range of 50–600 m/z and were recorded at a scan speed of 1.20 s. Scan control, data acquisition, and processing were performed using MS Workstation (Varian, Palo Alto, CA, USA) software data system, version 6.9.

SIM mode was selected and applied for monitoring different ions in each segment. Table 1 shows the different segments with their time range, m/z selected in each analytical window and the analyte of interest to be determined in each case. Peak identification was based on the comparison of retention times and MS signals with those obtained from the analysis of pure standards.

Table 1. Distribution of the segments used for the selective monitoring of each analyte.

Segment	Time Range (min)	m/z Selected	Analyte
1	4.00–8.98	456–458	7αOHChol
2	8.98–9.64	159, 253, 351	Desm
3	9.64–10.20	213, 255, 353, 443, 456–458	Latho, 7βOHChol
4	10.20–11.31	395–396	Lano
5	11.31–11.99	255, 323, 413, 441	24OHChol
6	11.99–12.25	131	25OHChol
7	12.25–12.70	367, 382, 472, 545	7KetoChol
8	12.70–13.60	161, 255, 417, 456, 546	27OHChol
9	13.60–15.50	189, 203	betulin (IS)

3. Results and Discussion

3.1. Optimization of Detection Conditions

The starting GC conditions were provided by Neuron Bio. Further GC-MS optimization was carried out in 5 stages to improve the starting conditions. In a first stage of the optimization, each analyte was individually injected in GC-MS (full scan mode) to determine its retention time and its characteristic m/z. The values obtained for each analyte and the IS are shown on Table 2. Regarding the m/z values, it should be noted that the MS signals appear in decreasing order of intensity.

After characterizing the individual behavior of each analyte, we analyzed a multi-standard solution (full scan mode); an example of the result is provided in Figure 2.

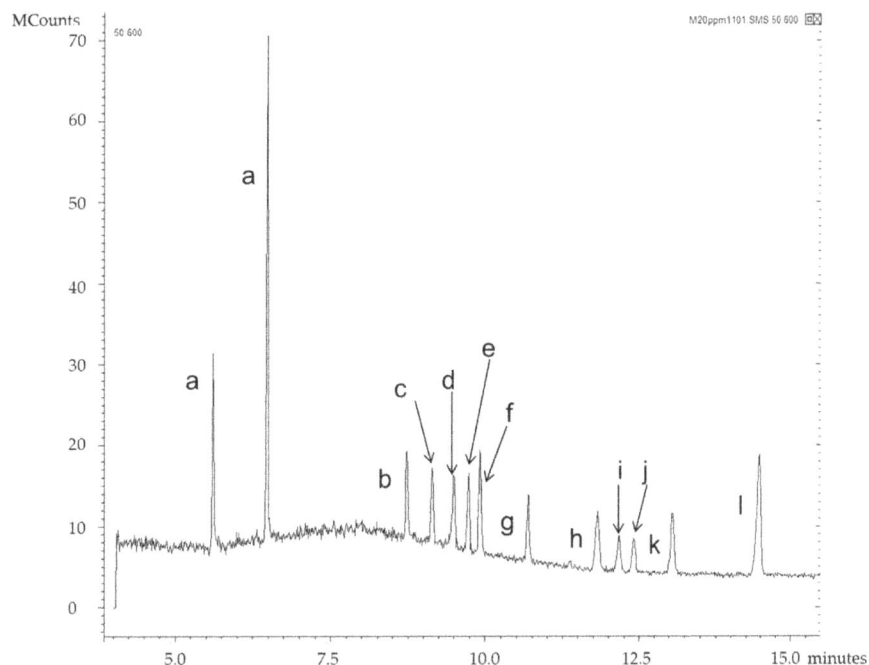

Figure 2. Chromatogram of a multi-standard solution (20 µg/mL in terms of each standard) to determine the retention time and the m/z ions of each Chol-related compound. The letters used to refer to the peaks mean the following: (a) BSTFA/TMS (silylation reagent); (b) 7α-hydroxycholesterol; (c) cholesterol; (d) desmosterol; (e) lathosterol; (f) 7β-hydroxycholesterol; (g) lanosterol; (h) 24(S)-hydroxycholesterol; (i) 25-hydroxycholesterol; (j) 7-ketocholesterol; (k) 27-hydroxycholesterol; (l) betulin (IS). The same letter codes are used in the rest of the manuscript.

Table 2. Retention time (t_R) and characteristic m/z signals of cholesterol precursors and oxysterols (in decreasing order of intensity).

Standard	m/z	t_R (min)
7α-hydroxycholesterol	456, 457	8.73
cholesterol	368, 329, 133, 353, 145	9.14
desmosterol	253, 351, 159, 456	9.49
lathosterol	255, 458, 213, 443, 353	9.74
7β-hydroxycholesterol	456, 457	9.92
lanosterol	395, 396	10.68
24(S)-hydroxycholesterol	145, 323, 413, 159, 441, 255, 546	11.82
25-hydroxycholesterol	131, 145	12.17
7-ketocholesterol	472, 367, 382, 545, 161	12.40
27-hydroxycholesterol	456, 417, 161, 546, 255, 441, 129	13.06
betulin (IS)	189, 203, 496	14.47

It is well known that the concentration of cholesterol is quite high in plasma samples; therefore, to simulate a real biological situation, we prepared a standard solution containing 2000 µg/mL of Chol (cholesterol is the most abundant sterol in human plasma with levels in the order of 1 to 3 mg/mL [20]) and 5 µg/mL of the rest of the analytes, which is enough to view the analytes with full scan mode. The result is illustrated in Figure 3, which shows that Desm (which elutes just after Chol) was properly detected even with very high concentration levels of Chol in the sample under study. However, Figure 3 also shows that when Chol was present at high concentrations, the signal of the other analytes was very low in comparison, which hampered their appropriate determination. Another issue related to Chol concentration is that the peak corresponding to 7α-hydroxycholesterol was hardly detectable. Thus, the next step of the optimization focused on improving the conditions of ion detection with MS through the use of SIM mode.

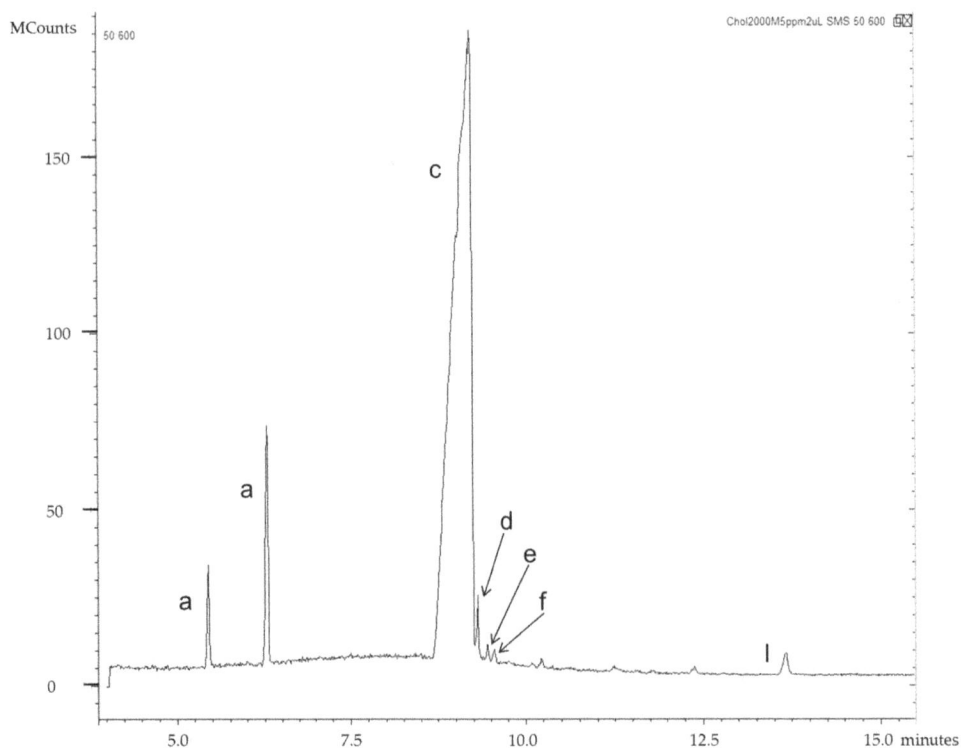

Figure 3. Chromatogram of a standard solution simulating blood cholesterol concentration (2000 µg/mL). Peak identification as in Figure 2.

We selected SIM operational mode and created specific segments or analytical windows (Table 1) in order to better detect each analyte (using specific MS parameters and looking for the characteristic *m/z* signals). A multi-standard solution containing 10 µg/mL of each analyte was injected into the GC-MS system (SIM mode). The result is shown in Figure 4.

As stated before, IT was used as mass analyzer. Even though IT can be used to perform MS/MS experiments, the use of SIM mode was selective enough to properly detect and quantify the compounds under study.

As a result, 9 analytes of interest plus the IS were detected separately by using 9 specific segments. The segments were optimized and defined taking into account the information included in Table 2 (i.e., retention times and *m/z* responses). The transfer and detection parameters were optimized aiming for the maximum sensitivity with the highest possible resolution. SIM led to very promising results in terms of improvement of the signal-to-noise ratio (S/N), sensitivity, detection and quantification limits, and some other important analytical validation characteristics.

After optimizing the GC-MS conditions, some biological samples were injected to prove the suitability of the new methodology in relatively long sequences (including biological samples, which are sometimes considerably dirty).

After 25 analyses, the need of a pre-column was quite evident to reduce the noise, which increased when a great number of consecutive injections and chromatographic runs were carried out. A silica phase with the same internal diameter as the column (0.25 mm) was selected. The pre-column had 1 m length and was connected to the column with press-fit connections. Once the pre-column was installed the S/N remained constant. The information provided in Table 2 regarding retention time was achieved with the pre-column already connected to the system.

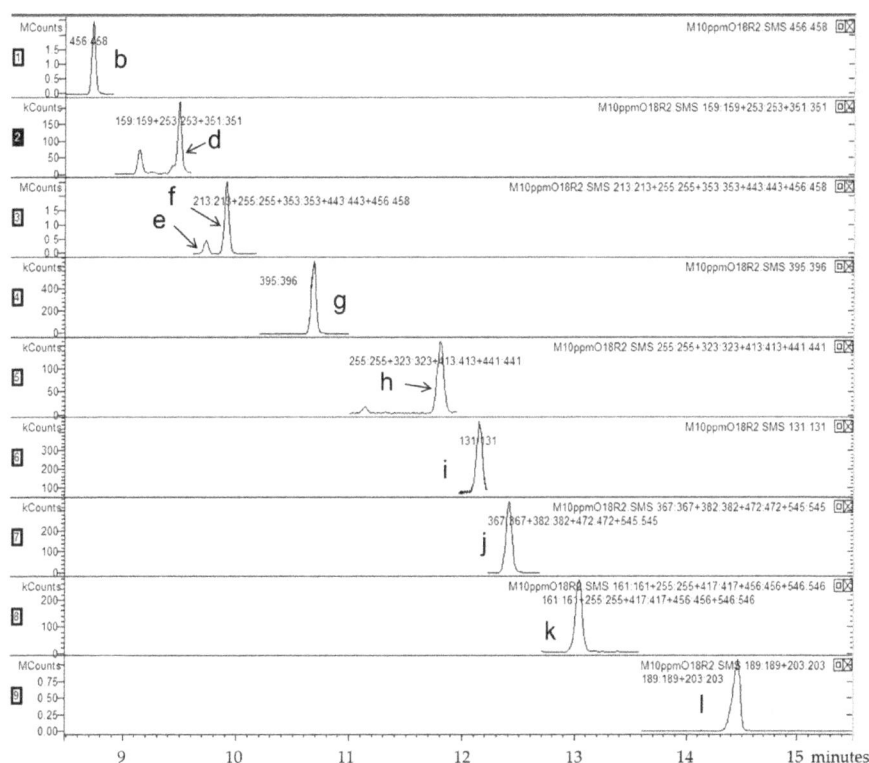

Figure 4. Result of analyzing a mix of the analytes under study (plus IS) at a concentration of 10 µg/mL using the segments created. Compounds determined in each segment (segment numbers are shown on the left side of the figure): (1) 7α-hydroxycholesterol; (2) desmosterol; (3) lathosterol and 7β-hydroxycholesterol; (4) lanosterol; (5) 24(S)-hydroxycholesterol; (6) 25-hydroxycholesterol; (7) 7-ketocholesterol; (8) 27-hydroxycholesterol; and (9) betulin (IS). Letter codes are the same as in Figure 2.

3.2. Validation of the Developed Method

Multi-standard solutions and QC samples were used to validate the method. Table 3 shows all the analytical parameters calculated during the validation of the developed method [37,38].

Multi-standard solutions were prepared at different concentration levels (0.1, 0.2, 0.5, 1.0, 2.5, and 5.0 μg/mL of each analyte) to build the calibration curves. Forty μg/mL of betulin (as for the biological samples) were added to each multi-standard solution. Fifty μL of the multi-standard solution were transferred to a vial containing a 250 μL insert. The solvent was then evaporated with N_2, after which an aliquot of 50 μL of derivatization reagent was added. The reaction mixture was left at room temperature for 1 h.

Calibration curves, one for each Chol-related compound, were fitted by least-squares regression from relative area signal (area_Chol-related analyte/area_IS) vs. Chol-related analyte concentration. The intercept was estimated as zero for all the analytes. For this reason, the relative areas were directly proportional to the concentration of each analyte. Thus, it was possible to calculate a response factor (RF) for each compound, which was constant and independent of the concentration of each analyte; values are shown in Table 3. Each point of the calibration graph corresponded to the mean value of four independent injections.

Determination coefficients and linearity indices (LIN) [39] were calculated for each calibration curve to assess goodness of fit. LIN = $1 - (SD_{slope}/slope)$ is included in Table 3 and was above 98% in all the cases, except for 25OHChol (96.5%), which had a slightly higher detection limit. Therefore, it was not possible to detect low concentrations.

The calibration curve, response factor, and linearity index were calculated considering the four replicates of each concentration level of the multi-standard solutions injected.

Table 3. Analytical parameters obtained during the validation of the method.

Compound	RF	LIN (%)	LOD (μg/mL)	LOQ (μg/mL)	Recovery Factor (%)	Matrix Effect (%)	Repeatability Consecutive Measurements	Overall Process
7αOH Chol	0.115	98.2	0.01	0.03	83.8	97.3	0.106	0.200
Desm	0.013	98.7	0.01	0.03	74.2	96.9	0.073	0.166
Latho	0.025	98.0	0.12	0.37	67.7	104.8	0.088	0.277
7βOH Chol	0.100	98.8	0.01	0.03	67.0	100.1	0.159	0.271
Lano	0.040	98.6	0.02	0.07	72.7	98.0	0.148	0.170
24OH Chol	0.0053	98.0	0.14	0.44	71.0	96.5	0.181	0.228
25OH Chol	0.022	96.5	0.35	1.06	97.8	99.4	0.146	0.294
7Keto Chol	0.020	97.9	0.07	0.20	73.2	95.5	0.114	0.197
27OH Chol	0.023	98.5	0.02	0.07	65.8	97.7	0.059	0.203

Limit of detection (LOD) and limit of quantification (LOQ) were determined considering the standard deviation calculated from six replicates of the lower sample concentration and multiplying it by 3.28 and 10, respectively. The 3.28 value is the 95%-quantile of the non-centrally t-distribution with ∞ degrees of freedom ($\alpha = \beta = 0.05$) [40]. Detection and quantification limits for 25OHChol were slightly higher than for the rest of the analytes, which can be explained because only one m/z signal was characteristic for this analyte; therefore, its determination was not as selective as that of the others. 7αOHChol, 7βOHChol, and Desm exhibited the lowest detection and quantification limits, with 0.01 and 0.03 μg/mL, respectively, in three cases.

Spiked goat serum samples were prepared to test the matrix effect and control the repeatability of the overall analytical process. To verify the matrix effect, calibration curves were prepared by spiking goat serum after the extraction with the Chol-related compounds (samples codes: QC00-05, QC00-10, QC00-25, QC00-50) at different levels: 0.5, 1.0, 2.5, and 5.0 μg/mL. The slopes of these calibration curves were compared with those achieved by using external standard calibration. The native levels of analytes present in the goat blood serum sample (QC00) were previously determined in order to perform an effective blank correction. In QC00, the following compounds were found: 7αOHChol,

7βOHChol, 24OHChol, and 7KetoChol. The other Chol-related compounds were not detected (nd; below the detection limit). The area values found for this sample were considered in all the subsequent calculations and were subtracted from the values achieved after spiking with the known concentration levels to assess the matrix effect. Such concentration levels were QC00-05 (spiked with 0.5 µg/mL of multi-standard solution), QC00-10 (spiked with 1.0 µg/mL of multi-standard solution), QC00-25 (spiked with 2.5 µg/mL of multi-standard solution), and QC00-50 (spiked with 5.0 µg/mL of multi-standard solution). As shown on Table 3, matrix effect recoveries were found between 95.5% and 104.8% for 7KetoChol and Latho, respectively (100% was the value that would be achieved if the slopes of the external standard calibration and standard addition calibration curves were identical); in other words, the matrix effect was not observed in the current study.

Moreover, to provide an estimation of overall repeatability (including sample preparation and measurement), goat serum samples spiked with the Chol-related analytes under study—before the extraction—were prepared at the same levels as shown above (samples codes: QC 05, QC 10, QC 25, and QC 50). The standard deviation (SD) was calculated from four consecutive measurements; to estimate the repeatability of the overall process, SD was calculated from three independent goat serum samples that were extracted, injected, and analyzed separately. The average SD of the determined values is provided in Table 3 and was found to be within 0.059–0.181 for repeatability of the measurements and 0.166–0.294 for the overall process.

The same preparations were useful to assess the recovery factor. The theoretical added concentration did not exactly coincide with the concentration obtained after the extraction and the corresponding measurement. Given that no matrix effect was observed in the previous step, we concluded that the difference was due to the recovery (obviously lower than 100%). Recovery values were calculated and are shown in Table 3: they ranged from 65.8% to 97.8% for 27OHChol and 25OHChol, respectively. These recovery factors should be used to calculate the real concentration of each analyte in any biological sample. They were applied to obtain the quantitative values shown in Table 4.

Table 4. Concentration levels found for each Chol-related compound in the test samples. The concentration was recalculated applying the appropriate recovery factor.

	Data	7αOH Chol	Desm	Latho	7βOH Chol	Lano	24OH Chol	25OH Chol	7Keto Chol	27OH Chol
TA1	Average (µg/mL)	0.76	0.25	d (0.13)	0.80	nd	1.28	nd	2.92	0.39
	RSD (%)	5.51	30.08	33.67	0.89	-	14.75	-	3.90	14.74
TA2	Average (µg/mL)	0.22	0.28	0.46	0.41	nd	1.06	nd	1.62	0.87
	RSD (%)	3.80	31.19	22.59	18.89	-	8.50	-	8.10	33.02
TA3	Average (µg/mL)	0.70	0.39	0.39	0.76	nd	0.99	nd	2.99	0.84
	RSD (%)	9.48	7.61	27.30	9.16	-	11.98	-	9.14	22.16
TB1	Average (µg/mL)	0.14	0.19	0.49	0.19	nd	0.76	nd	d (0.17)	0.51
	RSD (%)	19.53	30.19	21.59	21.76	-	17.01	-	27.23	26.05
TB2	Average (µg/mL)	0.42	nd	d (0.19)	0.35	nd	2.01	nd	1.70	0.42
	RSD (%)	0.24	-	30.19	10.11	-	13.94	-	11.17	10.57
TB3	Average (µg/mL)	0.11	nd	nd	0.11	nd	1.04	nd	0.53	0.35
	RSD (%)	5.24	-	-	11.97	-	1.36	-	9.00	28.14
TC1	Average (µg/mL)	0.061	nd	nd	0.04	nd	0.72	nd	d (0.16)	0.26
	RSD (%)	9.74	-	-	10.26	-	20.86	-	11.56	23.14
TC2	Average (µg/mL)	0.15	nd	d (0.20)	0.13	nd	0.85	nd	0.87	0.24
	RSD (%)	5.33	-	29.67	22.95	-	19.91	-	2.01	30.00
TC3	Average (µg/mL)	0.057	nd	d (0.34)	d (0.016)	nd	0.98	nd	0.56	0.31
	RSD (%)	1.99	-	24.09	21.72	-	19.98	-	21.57	22.01

RSD: Relative standard deviation; nd: not detected; d: detected, but below LOQ (in parentheses, when the value is <LOQ, a rough estimation is given).

3.3. Application of the Method to the Analysis of Biological Samples

Biological test samples were used for the application of the method. Each test sample was prepared and injected four times. Figure 5 shows an example of the results achieved for a mouse plasma sample (TA2) by using the SIM method with the 9 optimized segments.

In this example, all the selected analytes were found, with the exception of Lano (segment 4) and 25OHChol (segment 6); both analytes were indicated as not detected. In addition, some Chol-related compounds were detected but not quantified (they were below the LOQ). This was the case of Latho in samples TA1, TB1, TC2, and TC3 and 7KetoChol in samples TB1 and TC1. The rest of the analytes were found at concentration levels above their LOQs.

Quantitative results are summarized in Table 4, which includes the average concentration (μg/mL) of each analyte and the relative standard deviation (%RSD). The recovery factor was used to recalculate the real analyte concentration after extraction.

As stated above, there were three types of mice (A, B, and C) and three specimens in each group (1, 2, and 3). Samples belonging to the same group did not always show very similar results, what could be easily explained considering the fact that we studied different specimens.

Lano and 25OHChol were not detected in any sample. Desm was only detected in TA samples (TA1, TA2, and TA3) with concentration values ranging from 0.25 to 0.39 μg/mL, and in TB1, with a value of 0.19 μg/mL. Latho was found in the three specimens of Type A mice, with amounts of 0.39 and 0.46 μg/mL (TA3 and TA2); it was detected but not quantified in TA1. It was also found in two specimens of Type B mice (TB1 and TB2), with concentration values of 0.49 and 0.19 μg/mL respectively (the latter was a rough estimation), and in two Type C mice samples (estimated values—concentrations below LOQ—of 0.20 and 0.34 μg/mL—samples TC2 and TC3) (see Table 2). Overall, the rest of Chol-related compounds showed a decreasing trend in terms of concentration levels in mice A, B, and C.

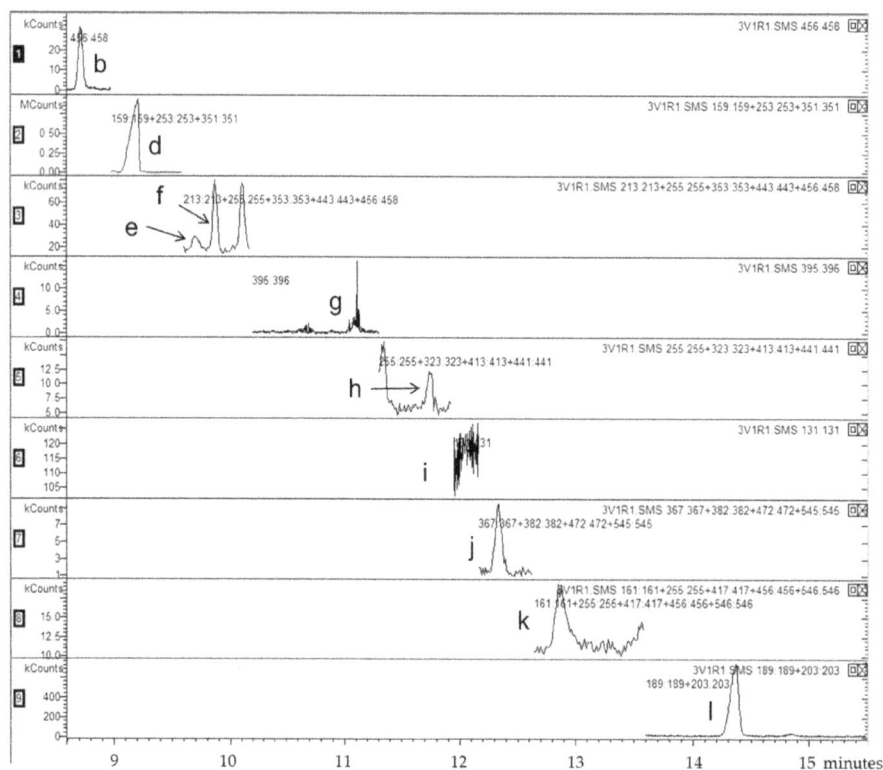

Figure 5. Results of the analysis of the test sample (TA2) using the created segments to show the application of the developed method to the analysis of biological samples. Compounds determined in each segment and numbers of segments are the same as in Figure 4.

Results showed that 7αOHChol was found at higher concentrations in Type A mice with the following values: 0.76, 0.22, and 0.70 μg/mL. In Type B samples, however, concentration values were 0.14, 0.42, and 0.11 μg/mL in the three specimens respectively. Type C mice showed the lowest values, ranging from 0.057 to 0.15 μg/mL. The situation was similar for 7βOHChol: the highest values were obtained for Type A mice (0.80, 0.41, and 0.76 μg/mL). 27OHChol and 24OHChol were found in all the samples. Their highest levels were determined in TA2 (0.87 μg/mL of 27OHChol), and TB2 (2.01 μg/mL of 24OHChol). As 24OHChol (also known as cerebrosterol) is a cholesterol catabolite produced almost exclusively in the brain, a reduction in its plasma levels after treatment with statins is regarded as an indirect marker of the inhibitory effect of these statins on brain cholesterol biosynthesis.

7KetoChol was found in all samples except for TB1 and TC1, in which it was detected but not quantified. In this case, its highest levels were determined in TA3 (2.99 μg/mL of 7KetoChol).

Overall, Type C mice showed a lower concentration of all Chol-related compounds, followed by Type B mice and Type A mice, which tended to exhibit the highest concentration values. These results are consistent with the fact that Type C mice were fed with statin. Several studies have pointed at a relationship between treatment with statins and a reduction of the risk of developing Alzheimer's disease [41–44]. This can be explained considering that statins can help to reduce oxysterol levels, which affect amyloid beta production. Our findings indicate that a statin-rich diet can affect the brain cholesterol metabolism of mice to a certain extent; further studies are still needed to clarify whether the use of statins can be a useful strategy for treating the onset of clinical dementia, but promising results can already be found in the literature.

4. Conclusions

This study proves that GC-(IT)MS, working in SIM mode, is an analytical platform suitable to detect and quantify cholesterol precursors and oxysterols in mouse plasma samples. The high cholesterol concentration present in this biological fluid may impede the proper detection of the chromatographic peaks of some Chol-related compounds that were eluting close to the cholesterol peak; however, this potential issue was avoided by using the SIM mode. The use of a GC pre-column was another methodological improvement tested in this study. We demonstrated that it can easily reduce the background noise usually found in the biological samples; this strategy also extends the lifetime of the analytical column.

Other chromatography–MS methods have been reported in the literature, showing even better performance characteristics. Yet, this study proves that the proposed method is fit for monitoring the contents of some AD biomarkers in plasma samples from mammals that are used as experimental animals. Sample selection in this study was very special as it included non-transgenic and transgenic mice that were fed different diets. Overall, Type C mice (i.e., transgenic mice fed with statin) showed a lower concentration of every Chol-related compound; this fact may be explained considering that statins can help to reduce oxysterols levels, which affect amyloid beta production. We consider the results to be very promising, however, we are aware about some limitations of our study; several experiments are already under way to increase the number of samples analyzed and ensure that the observed variations in measurements are due to the treatment and not to potential population variability.

Acknowledgments: The authors are grateful for the interesting collaboration with the biotechnology-based company Neuron Bio. The research project was funded by CEI BioTic/University of Granada.

Author Contributions: All authors contributed to the bibliographic search for the introduction; Lucia Valverde-Som prepared the multi-standard stock solution, which was designed by Lucia Valverde-Som, Alegría Carrasco-Pancorbo, Natalia Navas and Luis Cuadros-Rodríguez; Saleta Sierra, Soraya Santana and Javier S. Burgos conceived and designed the extraction treatment; Saleta Sierra and Soraya Santana prepared the sample extraction; Lucia Valverde-Som, Alegría Carrasco-Pancorbo and Cristina Ruiz-Samblás optimized the GC-(IT)MS with SIM mode method; Lucia Valverde-Som performed the experiments and analyzed the data; Alegría Carrasco-Pancorbo and Luis Cuadros-Rodríguez guided the project; Lucia Valverde-Som, Alegría Carrasco-Pancorbo and Luis Cuadros-Rodríguez wrote the paper; all the authors revised the paper.

Conflicts of Interest: The authors declare no conflict of interest.

References

1. Garenc, C.; Julien, P.; Levy, E. Oxysterols in biological systems: The gastrointestinal tract, liver, vascular wall and central nervous system. *Free Radical Res.* **2010**, *44*, 47–73. [CrossRef] [PubMed]
2. Bjökhem, I.; Heverin, M.; Leoni, V.; Meaney, S.; Diczfalusy, U. Oxysterols and Alzheimer's disease. *Acta Neurol. Scand.* **2006**, *114*, 43–49. [CrossRef] [PubMed]
3. Terao, J. Cholesterol hydroperoxides and their degradation mechanism. In *Lipid Hydroperoxide-Derived Modification of Biomolecules*; Kato, Y., Ed.; Springer: London, UK, 2014; pp. 83–91.
4. Mulder, M. Sterols in the central nervous system. *Curr. Opin. Clin. Nutr.* **2009**, *12*, 152–158. [CrossRef] [PubMed]
5. Poli, G.; Biasi, F.; Leonarduzzi, G. Oxysterol in the pathogenesis of major chronic diseases. *Redox Biol.* **2013**, *1*, 125–130. [CrossRef] [PubMed]
6. Griffiths, W.J.; Wang, Y. Analysis of oxysterol metabolomes. Biochim. *Biophys. Acta* **2011**, *1811*, 784–799.
7. Griffiths, W.J.; Abdel-Khalik, J.; Yutuc, E.; Morgan, A.H.; Gilmore, I.; Hearn, T.; Wang, Y. Cholesterolomics: An update. *Anal. Biochem.* **2017**, *524*, 56–67. [CrossRef] [PubMed]
8. Trushina, E.; Mielke, M.M. Recent advances in the application of metabolomics to Alzheimer's Disease. *Biochim. Biophys. Acta* **2014**, *1842*, 1232–1239. [CrossRef] [PubMed]
9. Leoni, V.; Caccia, C. Review. Oxysterols as biomarkers in neurodegenerative diseases. *Chem. Phys. Lipids* **2011**, *164*, 515–524. [CrossRef] [PubMed]
10. Baila-Rueda, L.; Cenarro, A.; Cofán, M.; Orera, I.; Barcelo-Batllori, S.; Pocoví, M.; Ros, E.; Civeira, F.; Domeño, C. Simultaneous determination of oxysterols, phytosterols and cholesterol precursors by high performance liquid chromatography tándem mass spectrometry in human serum. *Anal. Methods* **2013**, *5*, 2249–2257. [CrossRef]
11. Heverin, M.; Bogdanovic, N.; Lütjohann, D.; Bayer, T.; Pikuleva, I.; Bretillon, L.; Dicfalusy, U.; Winblad, B.; Bjökhem, I. Changes in the levels of cerebral and extracerebral sterols in the brain of patients with Alzheimer's disease. *J. Lipid Res.* **2004**, *45*, 186–193. [CrossRef] [PubMed]
12. Lim, W.L.F.; Martins, I.J.; Martins, R.N. The involvement of lipids in Alzheimer's disease. *J. Genet. Genom.* **2014**, *41*, 261–274. [CrossRef] [PubMed]
13. Kölsch, H.; Heun, R.; Kerksiek, A.; Bergmann, K.V.; Maier, W.; Lütjohann, D. Altered levels of plasma 24S- and 27-hydroxycholesterol in demented patients. *Neurosci. Lett.* **2004**, *368*, 303–308. [CrossRef] [PubMed]
14. De la Luz-Hdez, K. Metabolomics and mammalian cell culture. In *Metabolomics*; Roessner, U., Ed.; InTech: Rijeka, Croatia, 2012; pp. 3–18.
15. Nielsen, J. Metabolomics in functional genomics and systems biology. In *Metabolome Analysis. An Introduction*; Villas-Bôas, S.G., Roessner, U., Hansen, M.A.E., Smedsgaard, J., Nielsen, J., Eds.; John Wiley & Sons: Hoboken, NY, USA, 2007; pp. 3–14.
16. Gouveia, M.J.; Brindley, P.J.; Santos, L.L.; Correia da Costa, J.M.; Gomes, P.; Vale, N. Mass spectrometry techniques in the survey of steroid metabolites as potential disease biomarkers: A review. *Metabolis* **2013**, *62*, 1206–1217. [CrossRef] [PubMed]
17. Want, E.J.; Cravatt, B.F.; Siuzdak, G. The expanding role of mass spectrometry in metabolite profiling and characterization. *ChemBioChem* **2005**, *6*, 1941–1951. [CrossRef] [PubMed]
18. Dzeletovic, S.; Breuer, O.; Lund, E.; Diczfalusy, U. Determination of cholesterol oxidation products in human plasma by isotope dilution-mass spectrometry. *Anal. Biochem.* **1995**, *225*, 73–80. [CrossRef] [PubMed]
19. Ahmida, H.S.M.; Bertucci, P.; Franzò, L.; Massoud, R.; Cortese, C.; Lala, A.; Federici, G. Simultaneous determination of plasmatic phytosterols and cholesterol precursors using gas chromatography-mass spectrometry (GC-MS) with selective ion monitoring (SIM). *J. Chromatogr. B* **2006**, *842*, 43–47. [CrossRef] [PubMed]

20. McDonald, J.G.; Smith, D.D.; Stiles, A.R.; Rusell, D.W. A comprehensive method for extraction and quantitative analysis of sterols and steroids from human plasma. *J. Lipid Res.* **2012**, *53*, 1399–1409. [CrossRef] [PubMed]

21. Saraiva, D.; Semedo, R.; Castilho, M.C.; Silva, J.M.; Ramos, F. Selection of the derivatization reagent—The case of human blood colesterol, its precursors and phytosterols GC-MS analyses. *J. Chromatogr. B* **2011**, *879*, 3806–3811. [CrossRef] [PubMed]

22. Andrade, I.; Santos, L.; Ramos, F. Advances in analytical methods to study cholesterol metabolism: The determination of serum noncholesterol sterol. *Biomed. Chromatogr.* **2013**, *27*, 1234–1242. [CrossRef] [PubMed]

23. Griffiths, W.J.; Crick, P.J.; Wang, Y. Methods for oxysterol analysis: Past, present and future. *Biochem. Pharmacol.* **2013**, *86*, 3–14. [CrossRef] [PubMed]

24. Karuna, R.; Christen, I.; Sailer, A.W.; Bitsch, F.; Zhang, J. Detection of dihydroxycholesterol in human plasma using HPLC-ESI-MS/MS. *Steroids* **2015**, *99*, 131–138. [CrossRef] [PubMed]

25. Sugimoto, H.; Kakehi, M.; Satomi, Y.; Kamiguchi, H.; Jinno, F. Method development for the determination of 24S-hydroxycholesterol in human plasma without derivatization by high-performance liquid chromatography with tandem mass spectrometry in atmospheric pressure chemical ionization mode. *J. Sep. Sci.* **2015**, *38*, 3516–3524. [CrossRef] [PubMed]

26. Pataj, Z.; Liebisch, G.; Schmitz, G.; Matysik, S. Quantification of oxysterols in human plasma and red blood cells by liquid chromatography high-resolution tandem mass spectrometry. *J. Chromatogr. A* **2016**, *1439*, 82–88. [CrossRef] [PubMed]

27. Griffiths, W.J.; Abdel-Khalik, J.; Crick, P.J.; Yutuc, E.; Wang, Y. New methods for analysis of oxysterols and related compounds by LC-MS. *J. Steroid Biochem.* **2016**, *162*, 4–26. [CrossRef] [PubMed]

28. Menéndez-Carreño, M.; García-Herreros, C.; Astiasarán, I.; Ansorena, D. Validation of a gas chromatography-mass spectrometry method for the analysis of sterol oxidation products in serum. *J. Chromatogr. B* **2008**, *864*, 61–68. [CrossRef] [PubMed]

29. Schött, H.F.; Lütjohann, D. Validation of an isotope dilution gas chromatography-mass spectrometry method for combined analysis of oxysterols and oxyphytosterols in serum samples. *Steroids* **2015**, *99*, 139–150. [CrossRef] [PubMed]

30. Matysik, S.; Klünemann, H.H.; Schmitz, G. Gas chromatography-tandem mass spectrometry method for the simultaneous determination of oxysterols, plant sterols, and cholesterol precursors. *Clin. Chem.* **2012**, *58*, 1557–1564. [CrossRef] [PubMed]

31. Matysik, S.; Schmitz, G. Application of gas chromatography-triple quadrupole mass spectrometry to the determination of sterol components in biological samples in consideration of the ionization mode. *Biochimie* **2013**, *95*, 489–495. [CrossRef] [PubMed]

32. Griffiths, W.J.; Wang, Y. Analysis of neurosterols by GC-MS and LC-MS/MS. *J. Chromatogr. B* **2009**, *877*, 2778–2805. [CrossRef] [PubMed]

33. Narváez-Rivas, M.; Pham, A.J.; Schilling, M.W.; León-Camacho, M. A new SPE/GC-fid method for the determination of cholesterol oxidation products. Application to subcutaneous fat from Iberian dry-curred ham. *Talanta* **2014**, *122*, 58–62. [CrossRef] [PubMed]

34. Toivo, J.; Piironen, V.; Kalo, P.; Varo, P. Gas chromatographic determination of major sterols in edible oils and fast using solid-phase extraction in simple preparation. *Chromatographia* **1998**, *48*, 745–750. [CrossRef]

35. Campoy, S.; Sierra, S.; Suarez, B.; Ramos, M.C.; Velasco, J.; Burgos, J.S.; Adrio, J.L. Semisynthesis of novel monacolin J derivatives: Hypocholesterolemic and neuroprotective activities. *J. Antibiot.* **2010**, *63*, 499–505. [CrossRef] [PubMed]

36. Yamazaki, R.; Nakagawa, S.; Tanabe, A.; Ikeuchi, T.; Miida, T.; Yamato, S. Determination of 24S-hydroxycholesterol in human cerebrospinal fluid by gas chromatography/mass spectrometry. *Bunseki Kagaku* **2008**, *57*, 707–713. [CrossRef]

37. EURACHEM/CITAC Guide 2014. *The Fitness for Purpose of Analytical Methods*, 2nd ed.; 2014; Available online: www.eurachem.org (accessed on 10 September 2017).

38. Thompson, M.; Ellison, S.R.; Wood, R. *Harmonized Guidelines for Single Laboratory Validation of Methods of Analysis*; IUPAC Technical Report 2002; International Union of Pure and Applied Chemistry: Research Triangle Park, NC, USA, 2002; Volume 74, pp. 835–855.

39. Cuadros-Rodríguez, L.; García-Campaña, A.M.; Bosque-Sendra, J.M. Statistical Estimation of Linear Calibration Range. *Anal. Lett.* **1996**, *29*, 1231–1239. [CrossRef]

40. International Organization for Standardization. *Capability of Detection—Part 2: Methodology in the Linear Calibration Case*; ISO 11843-2:2000; International Organization for Standardization: Geneva, Switzerland, 2000.

41. Ostrowski, S.M.; Johnson, K.; Siefert; Shank, M.S.; Sironi, L.; Wolozin, B.; Landreth, G.E.; Ziady, A.G. Simvastain inhibits protein isoprenylation in the brain. *Neuroscience* **2016**, *329*, 264–274. [CrossRef] [PubMed]

42. Sallustio, F.; Studer, V. Targeting new pharmacological approaches for Alzheimer's disease: Potential for statins and phosphodiesterase inhibitors. *CNS Neurol. Disord.* **2016**, *15*, 647–659. [CrossRef]

43. McGuinness, B.; Craig, D.; Bullock, R.; Malouf, R.; Passmore, P. Statins for the treatment of dementia. Cochrane DB. *Syst. Rev.* **2004**, *7*, 1–74.

44. Refolo, L.M.; Pappolla, M.A.; LaFrancois, J.; Malester, B.; Schmidt, S.D.; Thomas-Bryant, T.; Tint, G.S.; Wang, R.; Mercken, M.; Pentanceska, S.S.; et al. A cholesterol-lowering drug reduces β-amyloid pathology in a transgenic mouse model of Alzheimer's disease. *Neurobiol. Dis.* **2001**, *8*, 890–899. [CrossRef] [PubMed]

## 3

Determination of Selected Aromas in Marquette and Frontenac Wine Using Headspace-SPME Coupled with GC-MS and Simultaneous Olfactometry

Somchai Rice [1,2,3] ⓘ, Nanticha Lutt [4], Jacek A. Koziel [1,2,5,*] ⓘ, Murlidhar Dharmadhikari [3,5] and Anne Fennell [6]

[1] Department of Agricultural and Biosystems Engineering, Iowa State University, Ames, IA 50011, USA; somchai@iastate.edu

[2] Interdepartmental Toxicology Graduate Program, Iowa State University, Ames, IA 50011, USA

[3] Midwest Grape and Wine Industry Institute, Iowa State University, Ames, IA 50011, USA; murli@iastate.edu

[4] Genetics and Plant Biology Program, University of California at Berkeley, Berkeley, CA 94704, USA; nantichalutt@berkeley.edu

[5] Department of Food Science and Human Nutrition, Iowa State University, Ames, IA 50011, USA

[6] Plant Science Department, South Dakota State University and BioSNTR, Brookings, SD 57007, USA; anne.fennell@sdstate.edu

* Correspondence: koziel@iastate.edu

Abstract: Understanding the aroma profile of wines made from cold climate grapes is needed to help winemakers produce quality aromatic wines. The current study aimed to add to the very limited knowledge of aroma-imparting compounds in wines made from the lesser-known Frontenac and Marquette cultivars. Headspace solid-phase microextraction (SPME) and gas chromatography-mass spectrometry (GC-MS) with simultaneous olfactometry was used to identify and quantify selected, aroma-imparting volatile organic compounds (VOC) in wines made from grapes harvested at two sugar levels (22° Brix and 24° Brix). Aroma-imparting compounds were determined by aroma dilution analysis (ADA). Odor activity values (OAV) were also used to aid the selection of aroma-imparting compounds. Principal component analysis and hierarchical clustering analysis indicated that VOCs in wines produced from both sugar levels of Marquette grapes are similar to each other, and more similar to wines produced from Frontenac grapes harvested at 24° Brix. Selected key aroma compounds in Frontenac and Marquette wines were ethyl hexanoate, ethyl isobutyrate, ethyl octanoate, and ethyl butyrate. OAVs >1000 were reported for three aroma compounds that impart fruity aromas to the wines. This study provides evidence that aroma profiles in Frontenac wines can be influenced by timing of harvesting the berries at different Brix. Future research should focus on whether this is because of berry development or accumulation of aroma precursors and sugar due to late summer dehydration. Simultaneous chemical and sensory analyses can be useful for the understanding development of aroma profile perceptions for wines produced from cold-climate grapes.

Keywords: aroma dilution analysis; wine; odor activity value; principal component analysis; solid-phase microextraction; GC-MS-Olfactometry

---

1. Introduction

The grape berry undergoes significant changes during ripening, including acid catabolism and the accumulation of sugar, anthocyanins, flavor and aroma compounds [1]. Brix measurements correspond to the percent total soluble solids (TSS), i.e., sugar, in a given weight of grape juice. Sugar content increases throughout berry ripening and is often monitored as a function of maturity.

Flavor and aroma compounds are complex. Although frequently subjected to sensory analysis, flavor and aroma compounds have limited objective measures specifically identifying the chemical and aroma links. Wines from Albillo and Muscat grape varieties (*V. vinifera*) have been shown to exhibit more "fruity" aromas and less "vegetal" and "floral" aromas at higher Brix during harvest [2]. Isobutyl methoxypyrazine, C6 alcohols, and hexyl acetate were shown to decrease in wine as Cabernet Sauvignon (*V. vinifera*) grape maturity developed [3]. The research by Bindon et al. [3] also demonstrated how higher sugar levels led to higher levels of volatile esters, dimethyl sulfide, and glycerol. Many studies have investigated the aroma profile of wines [4–7], with only a few that are focused on cold hardy grapes *V. labrusca* [8] or hybrids of *V. vinifera*, *V. labrusca*, and *V. riparia* [9,10]. In addition to TSS, pH and titratable acidity of the grapes is monitored before harvest.

Some varieties of wine have a distinct varietal aroma, and these can be attributed to a few compounds. Examples include 2-methoxy-3-isobutyl pyrazine (a green bell pepper aroma) reported in Sauvignon blanc [11] or 4-vinylguaiacol (spicy, clove aromas) found in Gewurztraminer [12]. Some varieties such as Chardonnay or Seyval do not have characteristic aromas originating from one or two specific compounds [13,14]. Understanding the volatile organic compounds (VOCs) that can contribute to the overall aroma profile of wine is important in making high quality, aromatic wines.

Use of solid-phase microextraction (SPME) and GC-MS is common for characterization of VOCs and other compounds in wine [15,16] and food grade alcohols [17,18]. VOC partitioning into headspace is suppressed with increasing concentration of ethanol in wine samples; ethanol influenced matrix-VOC partitioning more than glucose levels [19], possibly affecting SPME extraction efficiency of flavor and aroma compounds. SPME is an equilibrium extraction method and if extraction conditions such as extraction temperature, agitation, sampling time can be controlled in the laboratory quantitation using SPME is possible [20]. A previously developed method, optimizing the efficiency of SPME [21], is used in this research.

Aroma dilution analysis (ADA) is helpful in determining the major contributor to the aroma profile of the wine. ADA has been used to evaluate aroma character of Pinot noir grapes [7], Carmenere red wine [22], and Grenache rose wines [23]. It is difficult to dilute extracts while using SPME, so an ADA approach using successive sample dilutions of wine has been demonstrated [24]. Use of SPME in place of ADA has an advantage of speed due to eliminating sample preparation steps. Perceived aroma is not only based on its chemical concentration. Another consideration is the odor detection threshold of a compound (ODT), or the lowest concentration (mass/volume) needed for detection by a human nose [25]. It is critical to recognize that any odor threshold found is valid only at the conditions the test was run. For example, the odor compound will have a different threshold in water, air, and wine and is influenced by environmental conditions during the test. Odor activity value (OAV) is the ratio of the concentration of a compound to its odor detection threshold (ODT), and has been previously discussed [26–29]. It is important to recognize that for the same compound, odor threshold and taste threshold usually have very different values. Recognition thresholds and difference detection thresholds are also used in sensory evaluation of wines, but only ODT is considered in this research.

Frontenac (UMN 89 × Landot 4511) was introduced in 1996 and Marquette (MN 1094 × Ravat 262) was introduced in 2006 by University of Minnesota. These cold-climate cultivars are of importance because Marquette parentage includes Pinot Noir (grandparent), Cabernet Sauvignon and Merlot (great-great grandparents), Cabernet Franc, and Sauvignon Blanc [30]. Similar aroma profiles are expected between Marquette and Frontenac wines because the cultivars share 25% parentage from Landot Noir [31,32]. Evaluating the total aroma compounds present in headspace of Frontenac and Marquette wines is important to identify potential varietal aroma character. This information can then be used to compare the aroma of Frontenac and Marquette wines to the more recognizable vinifera wines (Figure 1).

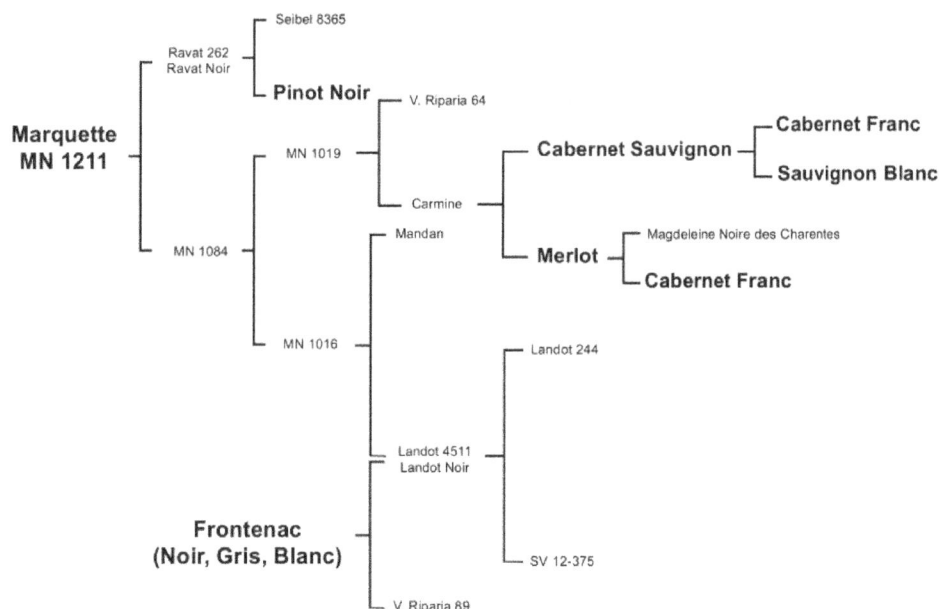

Figure 1. Pedigree of Frontenac and Marquette cultivars. Parentage of Marquette includes Pinot Noir, Cabernet Sauvignon, Merlot, Cabernet Franc, and Sauvignon Blanc. Marquette and Frontenac share common parentage (about 70%). This figure is adapted from pedigree maps from Chateau Stripmine [31,32].

The objectives of this study were to: (1) determine the relative concentrations of VOCs in the headspace of Marquette and Frontenac wines to an internal standard; (2) perform ADA on the wine samples to characterize the most odorous compounds; and (3) calculate OAV of key aroma compounds. Effects of increased berry "hang time", i.e., the time allowed to remain on the vine before harvest, on wine aroma will help enologists use viticultural practices to enhance desired winemaking styles.

2. Materials and Methods

2.1. Samples, Standards, and Matrix Blank

Marquette and Frontenac grapes from the 2014 growing season were harvested at 22° and 24° Brix at South Dakota State University (Brookings, South Dakota—44.3114° N, 93.7984° W). The vineyard was part of an NE1020 evaluation trial and was a randomized complete block design. There were four replicates for each cultivar (1 replicate per block with six vines in each replicate). Vines were grown as high cordon in standard NE1020 viticulture protocol. Three clusters were taken from each vine weekly to monitor Brix and for other sample aliquots. All berries were removed from clusters and aliquots were taken from the pool of berries for each replicate. Marquette at 22° Brix was harvested on 9 September 2013 and 24° Brix on 21 September 2013. Frontenac at 22° Brix was harvested on 13 September 2013 and 24° Brix on 21 September 2013.

A single batch of red wine was made from each harvest time point using the same winemaking protocol. Briefly, 5 gallon fermenters were used for each fermentation, 3 vines from each replicate (12 vines total). Single fermentations were used as the fruit is from the NE1020 coordinated evaluation trial, not a commercial grower.

Wines were produced according to a standard lab protocol at Tucker's Walk Vineyard and Farm Winery (Garretson, SD, USA). Red grapes were mechanically crushed/destemmed, treated with SO_2 to 25 ppm, and the must inoculated with Pasteur Red (Red star) yeast for fermentation on the skins at 70 °F (21.1 °C). After 5 days, must was dejuiced and fermentation continued in glass carboys at 70 °F (21.1 °C) until dry. Malolactic culture was added during the last third of the alcoholic fermentation, as determined by sugar content.

This wine was shipped to Iowa State University (Ames, IA, USA) for chemical and sensory analysis. Wine samples were aliquoted into 40 mL, pre-cleaned, glass amber vials with a PTFE lined screw cap. These vials were purged with helium to prevent oxidation of the wine samples and stored in a refrigerator before analysis.

A 5 mg/mL potassium bitartrate in 12.5% ethanol, pH 3.3 model wine was prepared by dissolving 5 g of potassium bitartrate (Fischer Scientific, Waltham, MA, USA) in 120 mL absolute ethanol (Fischer Scientific, Waltham, MA, USA) and q.s. to 1000 mL with deionized water in a 1000 mL volumetric flask. The solution was stirred for 10 min at room temperature, then filtered to remove any solids. The pH was adjusted with a 3.3 N hydrogen chloride. This model wine was used for all successive dilutions of wine samples and verified with the analytical method to be a suitable aroma and matrix blank.

For analysis of the undiluted sample, 4 mL of wine was pipetted into a cleaned 10 mL glass amber vial with metal screw top lid fitted with a PTFE-lined septum. The 10 mL vial also contained 2 g of sodium chloride, CAS 7440-23-5 (Sigma-Aldrich, St. Louis, MO, USA). 3-nonanone (99%), CAS 925-78-0 (Sigma-Aldrich, St. Louis, MO, USA), was used as an internal standard (IS) for semi-quantification of aroma compounds. The final concentration of IS in wine (0.205 mg/L) was achieved by adding 10 µL of an 81.9 mg/L IS in ethanol (w/v) to each 4 mL of wine. Triplicate samples were analyzed.

2.2. Aroma Dilution Analysis

A simple ADA was performed by analyzing successive dilutions of the wine sample, until the odor response from each compound or chromatographic column elution time region of interest was no longer noted at the olfactory detector. The odor dilution (OD) was assigned to the value of the sample dilution that results in odor extinction (i.e., not detected) at the olfactory detector (sniff port of GC). The higher the OD, the more significant that compound was in the overall aroma profile of the sample. Triplicate samples were analyzed, by a single trained panelist. Table 1 outlines dilution factors and weighting factors used in this research.

Table 1. Dilution and weighting factors of successive wine sample dilutions used in headspace solid phase microextraction gas chromatography-mass spectrometry-olfactometry (HS-SPME GC-MS-O) aroma dilution analysis.

Dilution Factor	Weighting Factor	Sample Volume (mL)	Model Wine Volume (mL)	Total Volume (mL)
0	1.000	4.000	0.000	4
2	0.500	2.000	2.000	4
4	0.250	1.000	3.000	4
8	0.125	0.500	3.500	4
16	0.063	0.250	3.750	4
32	0.031	0.125	3.875	4

2.3. Automated GC-MS-Olfactometry System

A CTC CombiPal™ autosampler with a heated agitator (LEAP Technologies, Inc., Carrboro, NC, USA) was used during the entire experiment. The optimized sampling parameters are described elsewhere [18]. Briefly, a 1 cm 50/30 µm divinylbenzene (DVB)/carboxen (CAR)/polydimethylsiloxane (PDMS) SPME was used for headspace sampling. Extraction time was 10 min at 50 °C, after 10 min incubation at 50 °C. Agitation speed was 500 rpm. The fiber was thermally desorbed in a 260 °C GC inlet for 2 min before exposure in sample headspace. Analytes were desorbed into the GC inlet for 2 min at 260 °C.

The analysis was performed on a 6890N GC/5973 Network mass spectrometer (Agilent Technologies, Santa Clara, CA, USA). The instrument allowed for heartcutting with a Dean's switch, cryogenic focusing, and was equipped with a FID and an olfactometry port. The GC contains two columns connected in series. The first non-polar column was a BPX-5 stationary phase with dimensions

30 m length × 0.53 mm ID × 0.5 μm film thickness (SGE, Austin, TX, USA). The second polar column was SOLGEL-Wax stationary phase with dimensions of 30 m length × 0.53 mm ID × 0.5 μm film thickness (SGE, Austin, TX, USA). A constant pressure of 5.7 psi was maintained at the midpoint between the first and second column using MultiTrax™ V.10.1 (MOCON, Round Rock, TX, USA) system automation and MSD ChemStation™ D.02.02.275 data acquisition software (Agilent, Santa Clara, CA, USA). Additional analysis was done using MassHunter Workstation (Agilent, Santa Clara, CA, USA) with NIST11 mass spectral database and BenchTop/PBM with Wiley Registry of mass spectral data, 7th edition (Palisade Corporation, New York, NY, USA). Flow from the analytical column was directed to the single quadrupole mass selective detector and the olfactometry port in an open-split interface via fixed restrictor tubing.

For this research, full heartcut was utilized from 0.05 to 35.00 min. An advantage of using two-dimensional gas chromatography is that separation is based on different and independent physico-chemical interactions (i.e., boiling point on the first column vs. polarity on the second column) for the entire chromatographic run. This serves as a fast screening method for VOCs, and can help select target VOCs for subsequent analysis. Sample flow was first directed through the non-polar column, then immediately to the polar column, therefore known retention indices for either column phase were used for identification. The following instrument parameters were used: injector, 260 °C; column, 40 °C initial, 3.0 min hold, 7 °C/min ramp, 220 °C final, 11.29 min hold; carrier gas, UHP helium (99.999%) with combination oxygen and moisture in-line gas trap. The mass detector was operated in electron ionization (EI) mode with an ionization energy of 70 eV. The mass detector ion source and quadrupole were held at 230 °C and 150 °C, respectively. Full spectrum scans were collected with the mass filter set from m/z 33 to m/z 450. The MS was auto-tuned daily before analysis. Use of full scan for data acquisition allowed for library search techniques using NIST11, and Wiley 6th edition mass spectral databases, and AMDIS with the accompanying food and flavor targeted mass spectral library.

Olfactometry data were generated using AromaTrax™ V.10.1 software (MOCON, Round Rock, TX, USA). Recorded parameters included an aroma descriptor ("note") and perceived intensity. The area under the peak of each aroma note in the aromagram is calculated as width × intensity × 100, where the width is the length of time (min) that the aroma persisted, in minutes. ADA was performed by analyzing successive dilutions of the same sample on the AromaTrax system V 10.1 (MOCON, Round Rock, TX, USA). The results files created have information about each aroma note such as elution time, intensity, and aroma descriptors. This information was combined into an ADA aromagram using a dilution and intensity weighting model. In this model, each peak in the results files that match a peak in the master file (Dilution Factor 0—Neat) is added to the ADA file based upon the intensity, multiplied by the dilution factor. For example, for a peak found in all 5 dilutions with intensities 90, 75, 45, 35, 5 and dilution factors of 1/2, 1/4, 1/8, 1/16, 1/32, the final intensity would be: (90 × 0.5) + (75 × 0.25) + (45 × 0.125) + (35 × 0.063) + (5 × 0.031) = 72. These values were then plotted in an analog manner, with % full-scale intensity vs. time.

3. Results

3.1. Semi-Quantitative Analysis of Volatiles in Wine

Ethanol was present in all samples, including the matrix blank (model wine). The chromatographic peak at RT 3.9 min is ethanol in all total ion chromatograms (TIC). Since ethanol is present in all samples, it is not included in the further discussion. The signal-to-noise ratio of total ion chromatograms of ≥10 was considered acceptable. Percent spectral match of sample spectra to library spectra was acceptable at 65% or higher.

58 unique VOCs are detected in the headspace of Marquette and Frontenac wines produced from grapes harvested at 22° and 24° Brix (Table 2). Compounds including isoamyl alcohol, ethyl octanoate, 2,4-di-tert-butylphenol, ethyl acetate, ethyl hexanoate, ethyl decanoate, isocyanatomethane, and ethyl

lactate are present in wine samples at higher concentrations relative to 0.205 ppm of 3-nonanone IS and assuming equal detector response for all analytes. A full summary of retention times, published aroma descriptors, and relative concentrations of these 58 analytes detected in all samples are given in the Supplementary Materials, Table S1.

Table 2. Compounds identified in the headspace of Marquette and Frontenac wines using AMDIS and a lab developed mass spectral library.

No.	Compound	Relative Retention [1]
1	Acetaldehyde	0.18
2	Methyl acetate	0.22
3	2-nitropropane	0.22
4	Isobutyraldehyde	0.22
5	Ethanol	0.25
6	Ethyl acetate	0.27
7	Acetic acid ethenyl ester	0.30
8	Methylbutanal	0.31
9	1,1-dimethyl-hydrazine	0.33
10	n-Propyl acetate	0.38
11	Isobutanol	0.39
12	Ethyl isobutyrate	0.42
13	1-butanol	0.46
14	Isobutyl acetate	0.46
15	Ethyl butyrate	0.51
16	Isoamyl alcohol	0.54
17	3-methylpentane	0.55
18	Isocyanatomethane	0.55
19	Ethyl methylbutyrate	0.58
20	Ethyl 3-methylbutanoate	0.59
21	Isoamyl acetate	0.65
22	Ethyl lactate	0.72
23	Styrene	0.73
24	1-hexanol	0.76
25	Acetic acid	0.79
26	1,2-dimethyl hydrazine	0.79
27	Ethyl hexanoate	0.85
28	1-Heptanol	0.92
29	2,3,4-trimethylpentane	0.94
30	Isoamyl butyrate	0.94
31	Undecane	0.97
32	Benzaldehyde	0.97
33	3-Nonanone (IS)	1.00
34	Methyl octanoate	1.06
35	1-Octanol	1.06
36	Octyl formate	1.06
37	Linalool	1.08
38	Phenylethanal	1.11
39	Ethyl octanoate	1.16
40	1-Nonanol	1.20
41	Pentanoic acid	1.25
42	Propyl octanoate	1.28
43	Ethyl nonanoate	1.30
44	Methyl salicylate	1.30
45	Methyl acetylsalicylate	1.30
46	Phenylethyl alcohol	1.34
47	Phenethyl isobutyrate	1.37
48	Phenethyl phenyl acetate	1.37
49	Ethyl decanoate	1.42
50	Octanoic Acid	1.47
51	β-damascenone	1.47
52	Isoamyl octanoate	1.49
53	Ethyl laurate	1.66
54	2,4-di-tert-butylphenol	1.76
55	Diethyl phthalate	1.85
56	Ethyl tetradecanoate	1.89
57	Ethyl hexadecanoate	2.14
58	Dibutyl phthalate	2.37

[1] Relative retention is the ratio of the compound retention time to the retention time of the internal standard 3-nonanone, in minutes.

Principal component analysis (PCA) was performed on the 58 VOCs found in all wine samples (Figure 2). Please refer to Table 2 for compound identification. Wine produced from Frontenac berries harvested at 22° Brix (F22-1, F22-2, and F22-3) is associated with higher levels of 1-octanol (compound **35**), ethyl laurate (compound **53**), methylbutanal (compound **8**), undecane (compound **31**), and ethyl tetradecanoate (compound **56**). Wine produced from Frontenac berries harvested at 24° Brix (F24-1, F24-2, and F24-3) is associated with higher levels of ethyl isobutyrate, ethyl-3-methylbutanoate, ethyl lactate, and 1-hexanol. Wine produced from Marquette berries harvested at 22° Brix (M22-1, M22-2, and M22-3) is associated with higher levels of acetaldehyde, ethyl butyrate, 1-hexanol, isoamyl butyrate, and 1-nonanal. Wine produced from Marquette berries harvested at 24° Brix (M24-1, M24-2, and M24-3) is associated with higher levels of ethyl hexanoate, propyl octanoate, and isobutyl acetate.

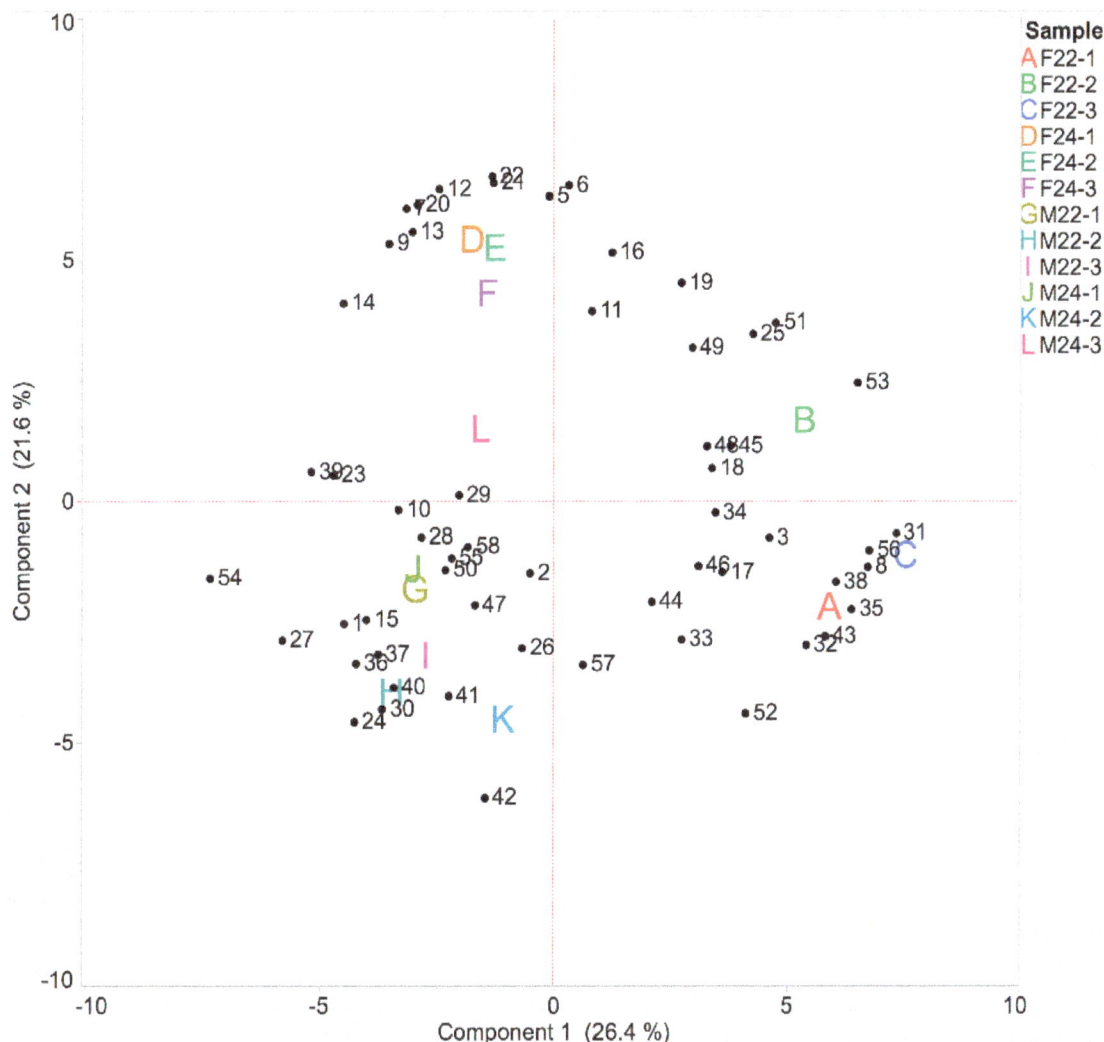

Figure 2. Results of a PCA on wines produced from Frontenac and Marquette berries harvested at 24 and 24° Brix. This biplot shows both the relationships of the wines to each other and the associations among the VOCs detected in the headspace. Vector arrows of the VOCs are not shown, but can be assumed to originate from the origin and ending at the numbered markers. The numbered markers indicate the compound identified, and are listed in Table 2. An example of sample naming convention is F22-1 used to represent wines produced from Frontenac berries, harvested at 22° Brix, analysis replicate 1.

A hierarchical cluster analysis was performed to identify the grouping of wine samples based on the degree of similarity in aroma compounds in headspace, and the constellation plot is shown in Figure 3. The cluster containing wine made from Frontenac berries harvested a 22° Brix is the most dissimilar to the other clusters representing the rest of the wines.

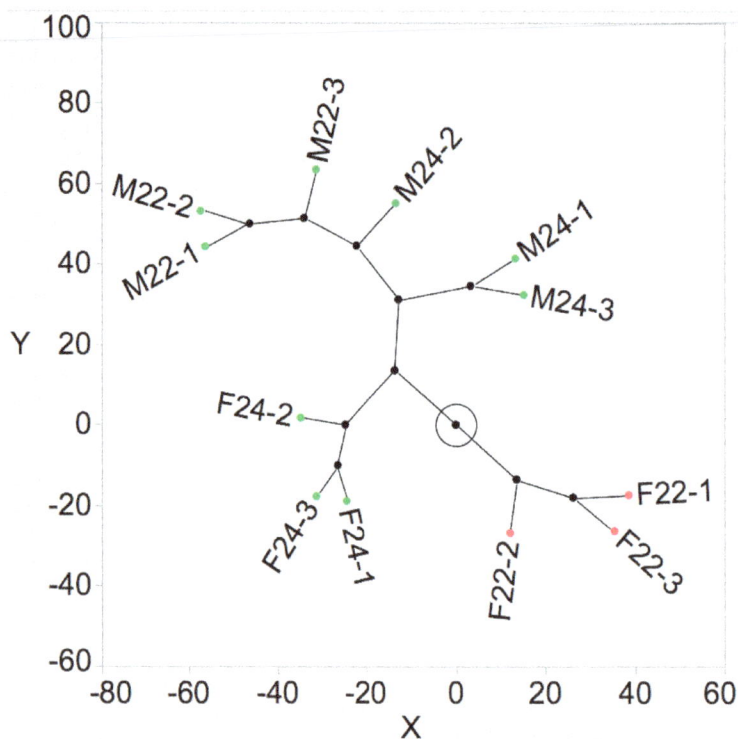

Figure 3. A constellation plot showing similarities in aroma compounds in Marquette and Frontenac wines produced from berries harvested at 22° and 24° Brix. The wine samples are arranged as endpoints and each cluster joins as a new point, with lines drawn representing membership. The longer the line, the greater the difference.

3.2. Aroma Dilution Analysis

Results of the full scan total ion chromatogram (TIC) for each 4 mL sample of wine was overlaid with the panelist generated aromagram (Supplementary Materials, Figures S1–S4). Olfactometry results including aroma descriptors, intensity, GC column retention time, and aroma event ("peak") areas were collected simultaneously with chemical analysis (Tables S1–S5). There were 15, 6, 7, and 7 aroma notes found in wines made from 22° Brix Frontenac, 24° Brix Frontenac, 22° Brix Marquette, and 24° Brix Marquette, respectively. These results were used as the master results to compare aromas from the diluted samples.

Each wine sample was diluted and analyzed again with the same methodology using dilution factors of 2, 4, 8, 16, and 32. The OD of each aroma event (detection) corresponds to the dilution of the wine in which the event was no longer present (not detected). The compound with the highest OD was contributing significantly more to the total aroma profile of the wine. After weighing the aroma notes by the dilution factor and intensities, these new values were shown in an ADA plot.

Marquette wine made from grapes harvested at 22° Brix had 14 aroma notes (excluding ethanol at 3.9 min) in the undiluted sample. After ADA (Figure 4), two events were calculated and identified as the most impactful to the total aroma profile of the wine: (1) 13.3 min, OD 8; and (2) 6.6 min, OD 2. The event at 3.9 min is ethanol, present in all samples, and therefore not included in the further discussion. These two aroma events corresponded to retention times of compounds simultaneously identified by mass spectrometer: (1) ethyl hexanoate; and (2) ethyl isobutyrate.

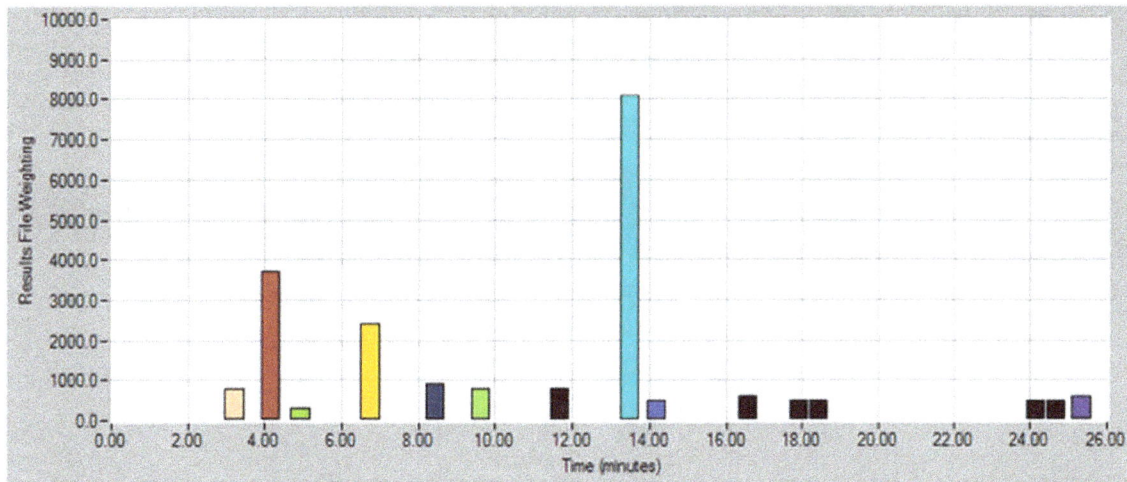

Figure 4. Aroma dilution analysis plot of wine made from Marquette grapes harvested at 22° Brix. The compound most impactful in the total aroma of this sample occurred at 13.3 min (light blue bar), assigned odor dilution number 8, and simultaneously identified using a mass spectrometer as ethyl hexanoate.

Marquette wine made from grapes harvested at 24° Brix had five aroma notes (excluding ethanol at 3.9 min) in the undiluted sample. After ADA (Figure 5), two events were calculated to be most impactful to the total aroma profile of the wine: (1) 22.0 min, OD 8 and (2) 8.3 min, OD 2. These two aroma events correspond to retention times of compounds simultaneously identified by mass spectrometer: (1) ethyl decanoate; and (2) ethyl butyrate.

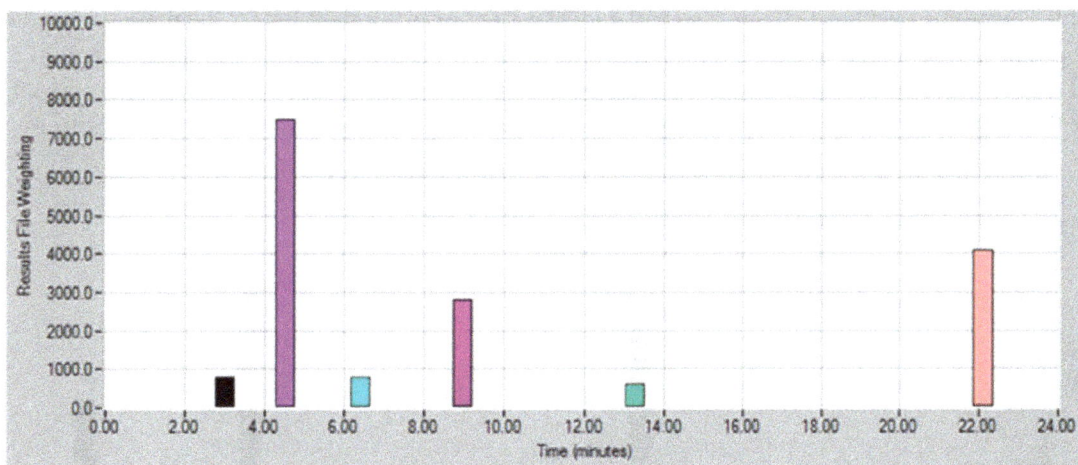

Figure 5. Aroma dilution analysis plot of wine made from Marquette grapes harvested at 24° Brix. The compounds most impactful in the total aroma of this sample occurred at: (1) 22.0 min, assigned odor dilution number 8; and (2) 8.3 min, assigned odor dilution number 2, while simultaneously identified using a mass spectrometer as: (1) ethyl decanoate; and (2) ethyl butyrate.

Frontenac wine made from grapes harvested at 22° Brix had six aroma notes (excluding ethanol at 3.9 min) in the undiluted sample. After ADA (Figure 6), two events were calculated to be most impactful to the total aroma profile of the wine: (1) 13.3 min, OD 16; and (2) 18.0 min, OD 16. These two aroma events correspond to retention times of compounds simultaneously identified by mass spectrometer: (1) ethyl hexanoate; and (2) ethyl octanoate.

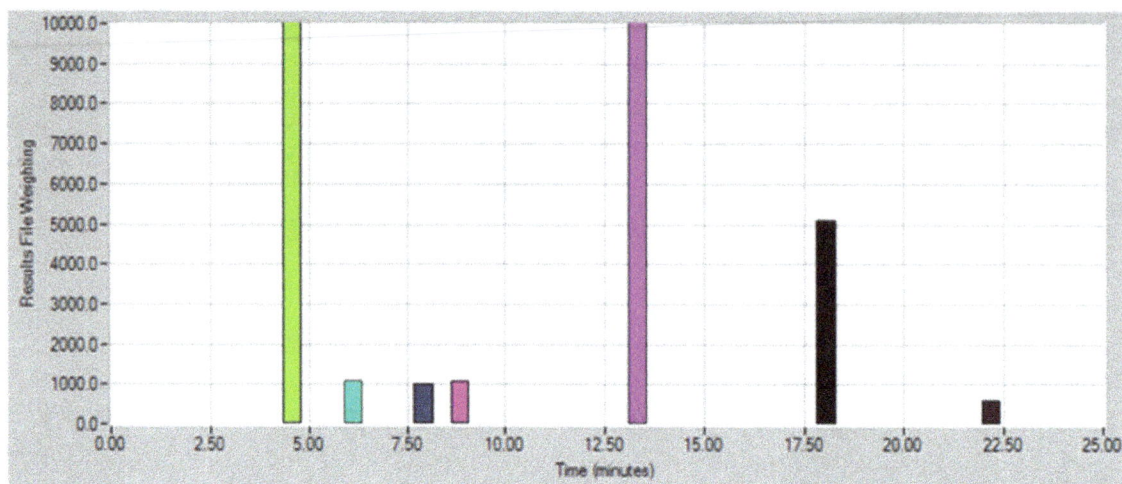

Figure 6. Aroma dilution analysis plot of wine made from Frontenac grapes harvested at 22° Brix. The compounds most impactful in the total aroma of this sample occurred at: (1) 13.3 min, assigned odor dilution number 16; and (2) 18.0 min, assigned odor dilution number 16, while simultaneously identified using a mass spectrometer as: (1) ethyl hexanoate; and (2) ethyl octanoate.

Frontenac wine made from grapes harvested at 24° Brix had six aroma notes (excluding ethanol at 3.9 min) in the undiluted sample. After ADA (Figure 7), four events were calculated to be most impactful to the total aroma profile of the wine: (1) 18.0 min, OD 8; (2) 22.3 min, OD 8; (3) 8.6 min, OD 4; and (4) 6.3 min, OD 4. These four aroma events correspond to retention times of compounds simultaneously identified by mass spectrometer: (1) ethyl octanoate; (2) ethyl decanoate; (3) 1-pentanol; and (4) ethyl isobutyrate. Ethyl butyrate was also present but not discussed because OD = 1 for this event. The aroma event at 13.0 min also had an OD = 1 and was not detected by the mass spectrometer. It is possible that not enough mass of analyte was directed to the mass spectrometer at retention time 13.0 min, via the open-split interface to sniff port, and therefore did not generate a chemical signal. It is likely that this compound has a very low ODT, and the human nose was a better detector for this compound.

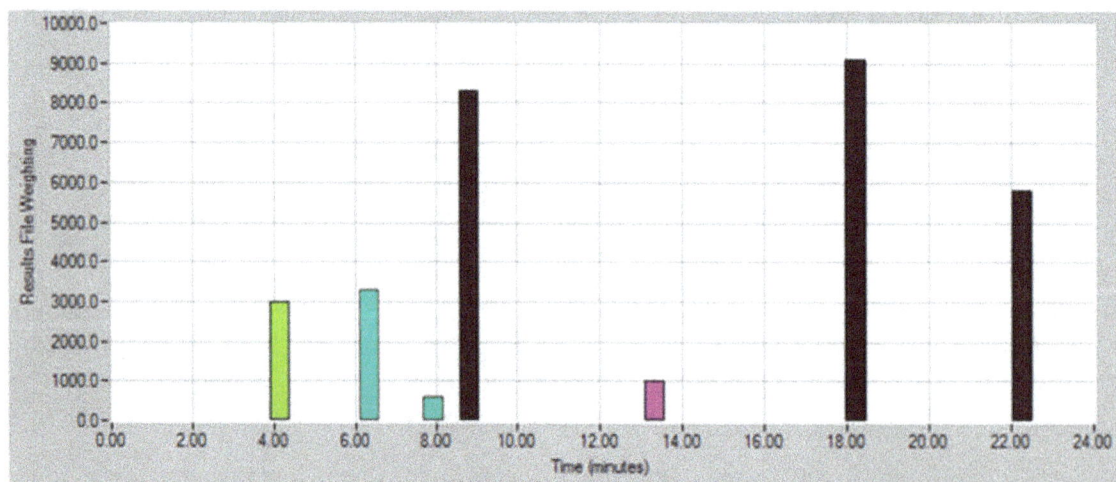

Figure 7. Aroma dilution analysis plot of wine made from Frontenac grapes harvested at 24° Brix. The compounds most impactful in the total aroma of this sample occurred at: (1) 18.0 min, assigned odor dilution number 8; (2) 22.3 min, assigned odor dilution number 8; (3) 8.6 min, assigned odor dilution number 4; and (4) 6.3 min, assigned odor dilution number 4, while simultaneously identified using a mass spectrometer as: (1) ethyl octanoate; (2) ethyl decanoate; (3) 1-pentanol; and (4) ethyl isobutyrate.

Five compounds are identified as key aromas in Marquette and Frontenac wine samples from ADA. These compounds are ethyl butyrate, ethyl decanoate, ethyl hexanoate, ethyl isobutyrate, and ethyl hexanoate. Concentrations of the five compounds in wine samples are calculated by external calibration. Quantitation range and coefficients of determination (R^2) from the linear model are given in Table 3. Ethyl butyrate and ethyl decanoate concentrations are lower than the quantitation limit for this method. Calculated concentrations of ethyl hexanoate, ethyl isobutyrate, and ethyl octanoate in Frontenac and Marquette wines produced from berries harvested at 22° and 24° Brix are summarized in Figure 8. Generally, these compounds have decreased concentrations in the headspace of wine samples as sugar content of the grapes at harvest increased. Marquette wines had higher concentrations of ethyl hexanoate (9.5 and 1.3 ppm), ethyl isobutyrate (0.7 and 0.5 ppm), and ethyl octanoate (41.9 and 17.9 ppm). Analyte concentrations in Frontenac wines were 1.3 ppm (not detected in wine produced from 24° Brix) of ethyl hexanoate, 0.5 ppm of ethyl isobutyrate, and 21.5 and 15.6 ppm of ethyl octanoate.

Table 3. Quantitation range and linear coefficients of determination for 5 key aroma compounds in wine.

Compound	Quantitation Range (ppm)	R^2
Ethyl hexanoate	2.89×10^{-2}–1.16×10^2	0.998
Ethyl isobutyrate	2.86×10^{-2}–1.15×10^2	0.999
Ethyl octanoate	2.88×10^{-2}–1.15×10^2	0.998
Ethyl decanoate	2.79×10^{-2}–1.12×10^2	0.997
Ethyl butyrate	2.93×10^{-2}–1.17×10^2	0.999

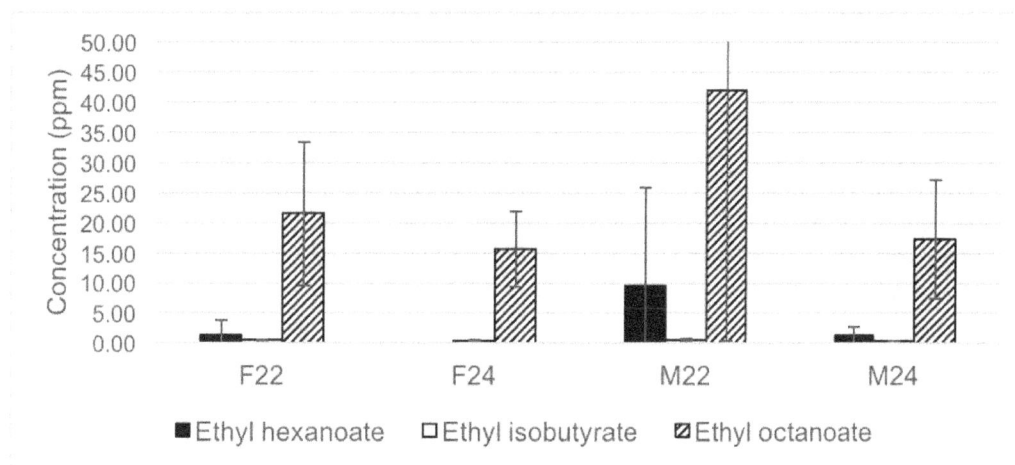

Figure 8. Concentrations of target analytes in headspace of Marquette and Frontenac wines made from berries harvested at 22° and 24° Brix. A five-point external calibration was performed using standards spiked into model wine. Data represent the mean of replicate analysis ($n = 3$) of each wine produced. F22 represents Frontenac wine produced from berries harvested at 22° Brix, F24 represents Frontenac wine produced from berries harvested at 24° Brix, etc.

3.3. Calculation of OAV

The OAV of the three quantified compounds are calculated from published odor detection thresholds and are reported as >1000 (Figure 9). The key aroma compounds impart fruity characters to the wines as determined by a single human panelist. The dominant aroma compound in all wine samples of this study is ethyl isobutyrate (OAV > 50,000), which imparts a fruity aroma.

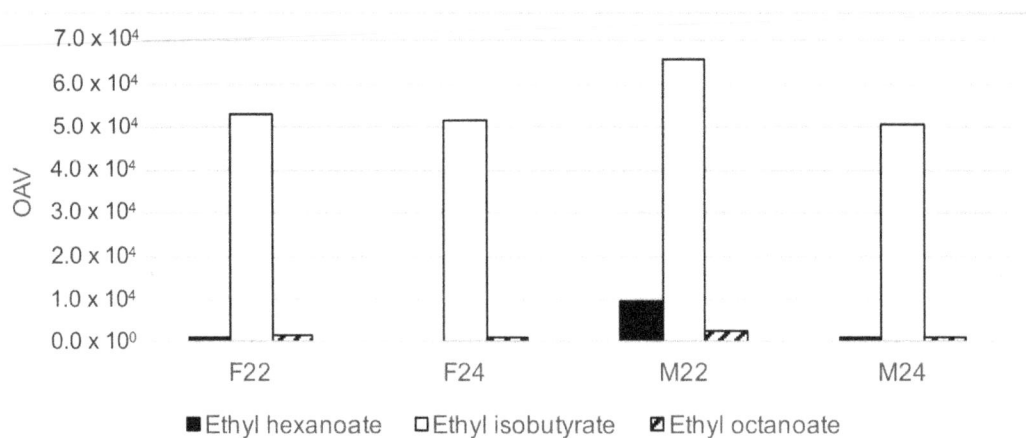

Figure 9. Odor (aroma) activity values of target analytes in headspace of Marquette and Frontenac wines made from berries harvested at 22° and 24° Brix. Odor detection thresholds for each analyte in water (and wine when available) from Leffingwell and Associates [33] were used to calculate OAV. OAV is the ratio of analyte concentration to its odor detection threshold. Ethyl octanoate is the dominate aroma compound present in headspace of Marquette and Frontenac wine samples (excluding alcohol), followed by ethyl hexanoate and ethyl octanoate. These compounds impart fruity aromas to the wine samples.

4. Discussion and Conclusions

In this research, headspace sampling using SPME, GC-MS and simultaneous olfactometry was used to investigate the key volatile aroma compounds in Frontenac and Marquette wines harvested at two maturity time points. ADA was performed to determine the key compounds present in these wine samples, and detected by human nose. Fifty-eight VOCs were detected in headspace of wines made from Frontenac and Marquette grapes harvested at 22° and 24° Brix by GC-MS, excluding ethanol and IS. In this study, the use of an internal standard was used to estimate relative chromatographic retention time and relative concentrations, assuming equal response factor across all analytes. A range of VOCs was detected by mass spectrometry including 9 alcohols, 5 aldehydes, 5 alkanes, 29 esters, 2 each of ketones, nitrogenous compounds, phenolic compounds and others. Yeast metabolism produces important wine volatiles such as higher alcohols, fatty acids, esters, and aldehydes. The reported results correspond to expected VOCs with the exception of diethy phthalate and dibutyl phthalate. Phalates in the wine could be an issue of contamination from packaging [34].

Marquette wines made from grapes harvested at 22° Brix had nine more aroma notes than grapes harvested at 24° Brix. The most impactful compounds as determined by ADA were ethyl hexanoate and ethyl isobutyrate (both from 22° Brix grapes); and ethyl decanoate and ethyl butyrate, both from 24° Brix grapes. Published aroma descriptors for ethyl hexanoate are apple peel and fruit, sweet and rubber for ethyl isobutyrate, grape for ethyl decanoate, and apple for ethyl butyrate [35]. Frontenac wines made from both harvest points had six aroma notes each. The most impactful compounds as determined by ADA were ethyl hexanoate and ethyl octanoate (both from 22° Brix grapes); and ethyl octanoate, ethyl decanoate, 1-pentanol, and ethyl isobutyrate from 24° Brix grapes. Published aroma descriptors for ethyl octanoate are fruit and fat, balsamic for 1-pentanol [35]. An external calibration was used to quantify ethyl hexanoate, ethyl isobutyrate, and ethyl octanoate, and OAVs were calculated. The most impactful aroma compound with calculated OAV >50,000 was ethyl isobutyrate contributes a fruity aroma.

A previous study reported β-damascenone, phenylethyl alcohol, acetic acid, linalool, and ethyl hexanoate quantified in Frontenac and Marquette grapes from veraision to harvest [9]. The compound ethyl hexanoate, responsible for fruity aromas increased in berry juice with increased growing degree days and differed by location in Pednault et al., 2013 [9]. In the wines in this study, no difference was found in ethyl hexanoate; however, these were taken from fruit with a greater level of ripening. However, β-damascenone, phenylethyl alcohol, acetic acid, and linalool are present in the wines from this study. Metabolic action of yeasts an influence wine aroma; specifically, precursors bound to glycosides or by decarboxylation of hydroxycinnamic acids to the equivalent vinylphenols [1]. Another study on Frontenac wine aroma reports 15 VOCs using a stir bar coated with PDMS [36]; 13 of these compounds were also detected in this study using headspace DVB/Carboxen/PDMS SPME. Similar VOCs detected were (from Table 2) compound numbers **12**, **16**, **19**, **21**, **22**, **24**, **27**, **37**, **39**, **44**, **46**, **49**, and **50**. The optimized headspace SPME method yielded more compounds detected when compared to PDMS stir bar coated sample preparation method. In a previous study of commercial Frontenac wines, using a trained panel, aroma descriptors of the most impactful characters were black currant, cherry, and cooked vegetable [37]. These aromas correspond to methoxymethylbutanethiol or mercaptomethylpentanone (black currant) [35], and acrolein, butyrophenone, or cyclohexyl cinnamate (cherry) [38]. Other significant aromas in the sensory study included black berry, jammy, cooked vegetable, fresh green, cedar, floral, geranium, tamari, and earthy [37]. There is no consensus on the mechanism for explaining olfactory perception. Mixtures of odorants, even at concentrations below the detection thresholds, act in a synergistic manner [39]. This might be the key to understanding wine aroma, where hundreds of compounds are present.

A limitation to this study is the small sample size. Only one five-gallon vinification could be completed per treatment (22° Brix and 24° Brix) because only 12 vines were available. PCA analysis, in this study, only accounted for 48% of the variance between vinifications and is more reflective of the analytical method. More replications, repeated over several years would yield more data to draw direct conclusions about the aroma of Frontenac and Marquette wines. The olfactometry portions of this study allows for a single panelist at a time. The use of more panelists, $n > 10$, could help account for variation within wine consumers due to age, sex, recognition and detection thresholds, etc.

This research adds to the growing body of knowledge about lesser-known wines from cold-hardy grapes grown in Midwest U.S. states. More research is needed to understand the aroma potential of these new cold-hardy grapes to produce quality wines that can stand on the same stage as traditional Old World wines. New cold-hardy grape cultivars that are complex hybrids of *Vitis vinifera* and native *Vitis riparia* have created a cold climate wine industry in North America. Frontenac and Marquette are vigorous, disease resistant and can grow sustainably in regions with low winter temperatures [30]. While there is significant information available for the aroma of *V. vinifera* wines, very little is known about the aroma profile of wines from cold-hardy grapes. Research is warranted to identify potential signature aromas in these new hybrid varieties as this information can advance viticultural practices, improve marketing and bolster the local economies of cold climate vineyards and wineries.

Acknowledgments: This study was funded by the USDA's Specialty Crops Research Initiative Program of the National Institute for Food and Agriculture, Project #2011-5118130850.

Author Contributions: S.R. and J.A.K. conceived and designed the experiments; N.L. and S.R. performed the experiments; S.R. analyzed the data; J.A.K., A.F. and M.D. contributed reagents/materials/analysis tools; and S.R., A.F., M.D., and J.A.K. wrote the paper.

Conflicts of Interest: The authors declare no conflict of interest. The funding sponsors had no role in the design of the study; in the collection, analyses, or interpretation of data; in the writing of the manuscript, and in the decision to publish the results.

References

1. Jackson, R.S. *Wine Science Principles and Applications*; Elsevier: Amsterdam, The Netherlands, 2008; ISBN 978-0-12-373646-8.
2. Sanchez-Palomo, E.; Diaz-Maroto, M.C.; Gonzalez-Vinas, M.A.; Soriano-Perez, A.; Perez-Coello, M.S. Aroma profile of wines from Albillo and Muscat grape varieties at different stages of ripening. *Food Control* **2005**, *18*, 398–403. [CrossRef]

3.   Bindon, K.; Varela, C.; Kennedy, J.; Holt, H.; Herderich, M. Relationships between harvest time and wine composition in Vitis vinifera L. cv. Cabernet Sauvignon 1. Grape and wine chemistry. *Food Chem.* **2013**, *138*, 1696–1705. [CrossRef] [PubMed]

4.   Gomez-Miguez, M.J.; Gomez-Miguez, M.; Vicario, I.M.; Heredia, F.J. Assessment of colour and aroma in white wines vinifications: Effects of grape maturity and soil type. *J. Food Eng.* **2007**, *79*, 758–764. [CrossRef]

5.   Vilanova, M.; Genisheva, Z.; Bescansa, L.; Masa, A.; Oliveira, J. Changes in free and bound fractions of aroma compounds of four Vitis Vinifera cultivars at the last ripening stages. *Phytochemistry* **2012**, *74*, 196–205. [CrossRef] [PubMed]

6.   Yuan, F.; Qian, M. Quantification of selected aroma-active compounds in Pinot noir wines from different grape maturities. *J. Agric. Food Chem.* **2006**, *54*, 8567–8573. [CrossRef]

7.   Yuan, F.; Qian, M.C. Aroma potential in early- and late- maturity Pinot noir grapes evaluated by aroma extract dilution analysis. *J. Agric. Food Chem.* **2016**, *64*, 443–450. [CrossRef] [PubMed]

8.   Chang, E.; Jeong, S.; Hur, Y.; Koh, S.; Choi, I. Changes in volatile compounds in vitis labrusca 'Doonuri' grapes during stages of fruit development and in wine. *Hortic. Environ. Biotechnol.* **2015**, *56*, 137–144. [CrossRef]

9.   Pedneault, K.; Dorais, M.; Angers, P. Flavor of cold-hardy grapes: Impact of berry maturity and environmental conditions. *J. Agric. Food Chem.* **2013**, *64*, 10418–10438. [CrossRef] [PubMed]

10. Slegers, A.; Angers, P.; Ouillet, E.; Truchon, T.; Pedneault, K. Volatile compounds from grape skin, juice and wine from five interspecific hybrid grape cultivars grown in Quebec (Canada) for wine production. *Molecules* **2015**, *20*, 10980–11016. [CrossRef] [PubMed]

11. Guillaumie, S.; Ilg, A.; Rety, S.; Brette, M.; Trossat-Magnin, C.; Decroocq, S.; Leon, C.; Keime, C.; Ye, T.; Baltenweck-Guyot, R.; et al. Genetic analysis of the biosynthesis of 2-methoxy-3-isobutylpyrazine, a major grape-derived aroma compound impacting wine quality. *Plant Physiol.* **2013**, *162*, 604–615. [CrossRef] [PubMed]

12. Lukic, I.; Radeka, S.; Grozaj, N.; Staver, M.; Persuric, D. Changes in physio-chemical and volatile aroma compound composition of Gewerztraminer wine as a result of late and ice harvest. *Food Chem.* **2016**, *196*, 1048–1057. [CrossRef] [PubMed]

13. Gametta, J.; Bastian, S.; Cozzolino, D.; Jeffery, D. Factors influencing the aroma composition of Chardonnay wines. *J. Agric. Food Chem.* **2014**, *62*, 6512–6534. [CrossRef] [PubMed]

14. Andrews, J.T.; Heymann, H.; Ellersieck, M. Sensory and chemical analysis of Missouri Seyval blank wines. *Am. J. Enol. Vitic.* **1990**, *41*, 116–120.

15. Canuti, V.; Conversano, M.; Calzi, M.L.; Heymann, H.; Mathews, M.A.; Eberle, S.E. Headspace solid-phase microextraction-gas chromatography-mass spectrometry for profiling the free volatile compounds in Cabernet Sauvignon grapes and wines. *J. Chromatogr. A* **2009**, *1216*, 3012–3022. [CrossRef] [PubMed]

16. Coelho, E.; Rocha, S.; Delgadillo, I.; Coimbra, M.A. Headspace-SPME applied to varietal volatile components evolution during Vitis vinifera L. cv. 'Baga' ripening. *Anal. Chim. Acta* **2006**, *563*, 204–214. [CrossRef]

17. Onuki, S.; Koziel, J.A.; Jenks, W.S.; Cai, L.; Rice, S.; van Leeuwen, J.H. Optimization of extraction parameters for quantification of fermentation volatile by-products in industrial ethanol with solid-phase microextraction and gas chromatography. *J. Inst. Brew.* **2016**, *122*, 102–109. [CrossRef]

18. Cai, L.; Rice, S.; Koziel, J.A.; Jenks, W.S.; van Leeuwen, J.H. Further purification of food-grade alcohol to make a congener-free product. *J. Inst. Brew.* **2016**, *122*, 84–92. [CrossRef]

19. Robinson, A.L.; Ebeler, S.E.; Heymann, H.; Boss, P.K.; Solomon, P.S.; Trengove, R.D. Interactions between wine volatile compounds and grape and wine matrix components influence aroma compound headspace partitioning. *J. Agric. Food Chem.* **2009**, *11*, 10313–10322. [CrossRef] [PubMed]

20. Pawliszyn, J. Solid phase microextraction in perspective. In *Handbook of Solid Phase Microextraction*; Pawliszyn, J., Ed.; Chemical Industry Press: Beijing, China, 2009; pp. 2–12. ISBN 978-7-122-04701-4.

21. Cai, L.; Rice, S.; Koziel, J.A.; Dharmadhikari, M. Development of an automated method for aroma analysis of red wines from cold-hardy grapes using simultaneous solid-phase microextraction–multidimensional gas chromatography-mass spectrometry-olfactometry. *Separations* **2017**, *4*, 24. [CrossRef]

22. Pavez, C.; Steinhaus, M.; Casaubon, G.; Schieberle, P.; Agosin, E. Identification, quantitation and sensory evaluation of methyl 2- and methyl 3- methylbutanoate in varietal red wines. *Aust. J. Grape Wine Res.* **2015**, *21*, 189–193. [CrossRef]

23. Ferreira, V.; Ortin, N.; Escudero, A.; Lopez, R.; Cacho, J. Chemical characterization of the aroma of Grenache rose wines: Aroma extract dilution analysis, quantitative determination, and sensory reconstitution studies. *J. Agric. Food Chem.* **2002**, *50*, 4048–4054. [CrossRef] [PubMed]

24. Marti, M.P.; Mestres, M.; Sala, C.; Busto, O.; Guasch, J. Solid-phase microextraction and gas chromatography olfactometry analysis of successively diluted samples. A new approach of the aroma extract dilution analysis applied to the characterization of wine aroma. *J. Agric. Food Chem.* **2003**, *51*, 7861–7865. [CrossRef] [PubMed]

25. Bi, J.; Ennis, D.M. Sensory thresholds: Concepts and methods. *J. Sens. Stud.* **1998**, *13*, 133–148. [CrossRef]

26. Rice, S.; Koziel, J.A. Characterizing the smell of marijuana by odor impact of volatile compounds: An application of simultaneous chemical and sensory analysis. *PLoS ONE* **2015**, *10*. [CrossRef] [PubMed]

27. Rice, S.; Koziel, J.A. Odor impact of volatiles emitted from marijuana, cocaine, heroin and their surrogate scents. *Data Br.* **2015**, *5*, 653–706. [CrossRef] [PubMed]

28. Rice, S.; Koziel, J.A. The relationship between chemical concentration and odor activity value explains the inconsistency in making a comprehensive surrogate scent training tool representative of illicit drugs. *Forensic Sci. Int.* **2015**, *257*, 257–270. [CrossRef] [PubMed]

29. Rice, S.; Koziel, J.A.; Dharmadhikari, M.; Fennell, A. Evaluation of tannins and anthocyanins in Marquette, Frontenac, and St. Croix cold-hardy grape cultivars. *Fermentation* **2017**, *3*, 47. [CrossRef]

30. University of Minnesota, Minnesota Hardy. Available online: http://mnhardy.umn.edu/varieties/fruit/grapes (accessed on 28 April 2016).

31. Chateau Stripmine, Frontenac Parentage. Available online: http://chateaustripmine.info/Parentage/Frontenac.gif (accessed on 10 May 2017).

32. Chateau Stripmine, Marquette Parentage. Available online: http://chateaustripmine.info/Parentage/Marquette.gif (accessed on 10 May 2017).

33. Leffingwell & Associates. Odor & Flavor Detection Thresholds in Water. Available online: http://www.leffingwell.com/odorthre.htm (accessed on 29 December 2017).

34. Chatonnet, P.; Boutou, S.; Plana, A. Contamination of wines and spirits by phthalates: Types of contaminants present, contamination sources and means of prevention. *Food Addit. Contam. A* **2014**, *31*, 1605–1615. [CrossRef] [PubMed]

35. Acree, T.; Arn, H. Flavornet and Human Odor Space. Available online: www.flavornet.org (accessed on 14 February 2018).

36. Mansfield, A.K.; Schirle-Keller, J.P.; Reineccius, G.A. Identification of odor-impact compounds in red table wines produced from Frontenac grapes. *Am. J. Enol. Vitic.* **2011**, *62*, 169–176. [CrossRef]

37. Mansfield, A.K.; Vickers, Z.M. Characterization of the aroma of red Frontenac table wines by descriptive analysis. *Am. J. Enol. Vitic.* **2009**, *60*, 435–441.

38. The Good Scents Company. Available online: www.thegoodscentscompany (accessed on 14 February 2018).

39. Patterson, M.Q.; Stevens, J.C.; Cain, W.S.; Cometto-Muniz, J.E. Detection thresholds for an olfactory mixture and its three constituent compounds. *Chem. Sens.* **1993**, *18*, 723–734. [CrossRef]

## 4

Supercritical CO_2 Extracts and Volatile Oil of Basil (*Ocimum basilicum* L.) Comparison with Conventional Methods

José Coelho [1,2,*] [iD], Jerson Veiga [1], Amin Karmali [1,3], Marisa Nicolai [4] [iD], Catarina Pinto Reis [5,6] [iD], Beatriz Nobre [2] and António Palavra [2]

[1] Chemical Engineering and Biotechnology Research Center, Instituto Superior de Engenharia de Lisboa, IPL, 1959-007 Lisboa, Portugal; jersonveiga@hotmail.com (J.V.); akarmali@deq.isel.ipl.pt (A.K.)

[2] Centro de Química Estrutural, Instituto Superior Técnico, Universidade de Lisboa, 1049-001 Lisboa, Portugal; beatriz.nobre@tecnico.ulisboa.pt (B.N.); antonio.palavra@ist.utl.pt (A.P.)

[3] CITAB-Centre for the Research and Technology of Agro-Environmental and Biological Sciences, University of Trás-os-Montes and Alto Douro, 5000-801 Vila Real, Portugal

[4] CBiOS, Research Center for Biosciences & Health Technologies, ULHT, 1749-024 Lisboa, Portugal; mhfnicolai@gmail.com

[5] iMED, ULisboa, Research Institute for Medicines, Faculty of Pharmacy, ULisboa, 1749-003 Lisboa, Portugal; catarinareis@ff.ulisboa.pt

[6] IBEB, Biophysics and Biomedical Engineering, Faculty of Sciences, ULisboa, Campo Grande, 1749-016 Lisboa, Portugal

* Correspondence: jcoelho@deq.isel.ipl.pt

Abstract: Interest in new products from aromatic plants as medical and nutritional compounds is increasing. The aim of this work was to apply different extraction methods, including the use of supercritical carbon dioxide extraction, and to test the antioxidant activity of basil (*Ocimum basilicum* L.) extracts. In vitro efficacy assessments were performed using enzymatic assays. Essential oil obtained by hydrodistillation and volatile oil obtained from supercritical fluid extraction were analyzed by gas chromatography to quantify components. The total phenolic content in the extracts ranged from 35.5 ± 2.9 to 85.3 ± 8.6 mg of gallic acid equivalents and the total flavonoid content ranged from 35.5 ± 2.9 to 93.3 ± 3.9 micromole catechin equivalents per gram of dry weight of extract. All the extracts showed an antioxidant activity with 2,2-diphenyl-1-picrylhydrazyl (DPPH), 2,2-azino-bis(3-ethylbenzthiazoline-6-sulfonic acid (ABTS), and the reducing power test. Extracts obtained from methanol had a higher antioxidant capacity per the DPPH test results ($IC_{50} = 3.05 \pm 0.36$ mg/mL) and the reducing power test assay 306.8 ± 21.8 μmol of trolox equivalents per gram of extract (TE/g) compared with ethanolic or supercritical fluid extracts. However, using the ABTS assay, the extract obtained by supercritical fluid extraction had a higher antioxidant capacity with an IC_{50} of 1.74 ± 0.05 mg/mL. Finally, the examined extracts showed practically no acetylcholinesterase (AChE) inhibitory capacity and a slight inhibitory activity against tyrosinase.

Keywords: *Ocimum basilicum*; supercritical fluid extraction; phenolic and flavonoids content; antioxidant activity; in vitro efficacy tests

1. Introduction

Basil is an aromatic plant belonging to the Lamiacecae family, used as a culinary herb and for ornamental purposes. The genus *Ocimum* contains between 50 to 150 species of plants and is found in tropical regions of Asia, Africa, and South and Central America [1–9]. Basil has been used in traditional medicine for treating problems like headaches, coughing, diarrhea, poor kidney function, and warts. Many authors have referenced basil as a medicinal plant. Basil extract and essential oil from various parts including the leaves, flowers, and roots, have been used to determine antimicrobial, antioxidant, anticancer, antidiabetic, anti-inflammatory, analgesic, sedative, and hypoglycemic activities [1,5,6,9–15], or to estimate the cost of manufacturing the extracts [16].

Consumer preference for natural products has increased. Consequently, the interest in new antioxidant sources, and aromatic herbs in particular, has also increased [17–19]. Other vegetable extracts have exhibited important activities [20–22]. Herbs can be applied as natural food preservatives to prevent the deterioration of foodstuff quality that occurs during processing and storage, mainly due to oxidative processes. Extraction techniques aim to not only extract active biocompounds from herb samples [23], but also to increase the concentration of the compound of interest. Classical extraction techniques are generally based on the extractive potential of various solvents, using heating or mixing. The disadvantages of conventional extraction techniques include the need for expensive purity solvents, long extraction times, and evaporation of significant amounts of solvents, low selectivity, and potential decomposition of thermolabile compounds [24,25]. These problems can be solved by using other extraction techniques, such as green techniques. Usually, green extraction methods use safer solvents and less toxic chemicals. Particularly, the use of supercritical fluids in the extraction of volatile oils and extracts has increased since 2000 due to the expected advantages of the supercritical extraction process. These processes have received increased attention for the production of high-value plant extracts for the pharmaceutical, cosmetics, and food industries [26–28]. Specifically, supercritical fluid extraction is considered a simple, rapid, selective, and convenient method. Supercritical fluid extraction is also a solvent-free and environmentally-friendly sample pretreatment technique. In this study, a number of factors that influence extraction yields of basil extracts were studied and a comparison with conventional extraction techniques was completed.

2. Materials and Methods

2.1. Plant Material

Ocimum basilicum L. was purchased from a commercial supplier in Portugal in March 2016. Only the aerial parts of the plant leaves and flowers were used for these experiments. Extractions with fresh plants were performed in the first three days after acquiring the plant. The remaining plants were dried in an oven (LSIS-B2V/VC111 1900 W, Cejl, Czech Republic) at 40 °C for 72 h and stored in vacuum packages at –18 °C. The flowers and leaves were ground (IKA WERKE-M20, Staufen, Germany) at low temperatures to avoid loss and thermal degradation of secondary metabolites. A set of standard sieves was used to determine the particle size distribution.

2.2. Supercritical Fluid Extraction and Conventional Extractions

To obtain the volatile oil, supercritical fluid extraction (SFE) with carbon dioxide (CO_2) and 80 g basil with a 0.6 mm mean particle size was performed for 2.5 h, at a flow rate of 1.0 kg/h CO_2, at 313 K and 9.0 MPa, in a flow apparatus using a two-stage fractional separation technique [29,30]. Separation was performed at 7.0 MPa, 263 K in the first separator, whereby waxes were mainly collected. A second separation was performed at 2.0 MPa and 273 K, from which the volatile oil (SFEO) was obtained. The amount of volatile oil obtained was assessed gravimetrically with an uncertainty of ±0.1 mg. Purity CO_2 (99.995%) was supplied by Air Liquide (Lisboa, Portugal). Due to this SFE equipment limitations (maximum pressure of 30.0 MPa) and to obtain basil extracts at higher pressures, SFE experiments were carried out in a flow apparatus from Applied Separations, SpeedTM SFE, which allows extraction to be performed at temperatures up to 393.2 K and pressures up to 60.0 MPa [31–33].

SFE was completed using 15 g samples of basil with a 0.6 mm mean particle size. The conditions of extraction were as follows: a CO_2 flow rate of 0.10 kg/h, pressure of 40.0 MPa, and a temperature of 313 K. The supercritical carbon dioxide extracts (SFEE) were collected in a separator at atmospheric pressure.

For comparison, conventional hydrodistillation was conducted in a Clevenger-type apparatus for 2.5 h, using 40 g of dry plant material, with the same particle size as that used in the supercritical extraction conditions. This essential oil (hydrodistillation with dried plants, HDD) was compared with that obtained by hydrodistillation with fresh plants (HDF). The basil extracts, by conventional extraction, were isolated by Soxhlet. A total of 20 g of basil with a particle size of 0.6 mm were extracted either with 250 mL of methanol (MeOH) (Sigma Aldrich 99.8%, Steinheim, Germany) or with ethanol (EtOH) (Panreac 99.5%, Barcelona, Spain) for 3 h at the solvent boiling point, siphoning at least five times per hour. The extract was filtered, the solvent was removed by reduced pressure evaporation in a rotary evaporator (Büchi, model R-205, Flawil, Switzerland), and the residue was dried to constant weight.

2.3. Gas Chromatography and Gas Chromatography-Mass Spectrometry Analysis

Essential oil, SFE volatile oil, and waxes were analyzed by gas chromatography (GC), for component quantification, using a Perkin Elmer Autosystem XL gas chromatograph (Perkin Elmer, Shelton, CT, USA) equipped with two columns of different polarities: a DB-1 fused-silica column (30 m × 0.25 mm i.d., film thickness 0.25 μm; J and W Scientific Inc., Rancho Cordova, CA, USA) and a DB-17HT fused-silica column (30 m × 0.25 mm i.d., film thickness 0.15 μm; J and W Scientific Inc.). For component identification, a gas chromatograph coupled to mass spectrometry (GC-MS) was used, consisting of a Perkin Elmer Autosystem XL gas chromatograph interfaced with a Perkin-Elmer Turbomass mass spectrometer (software version 4.1; Perkin Elmer, Shelton, CT, USA). The components were identified based on their comparative retention times relative to C_9–C_{27} n-alkane indices and GC-MS spectra from a lab-made library constructed based on the analyses of reference oils, laboratory-synthesized components, and commercially-available standards [34].

2.4. Determination of Total Phenolic and Flavonoid Content

Spectrophotometric (BIO-RAD Model 680 Microplate Reader, Tokyo, Japan) quantitative determination of the total phenolic content (TPC) in plant extracts was performed using the Folin-Ciocalteau (Sigma Aldrich 2M, Switzerland) micro method (Microtiter plate reader, BIO-RAD Model 680). Gallic acid (GA) (Sigma Aldrich, 98%, China) was used as the standard and the calibration curve, which was adapted from previous studies [35–37] within a concentration range of 12.5 to 600 mg/L. The absorbance calibration curve vs. the concentration of the standard was used to quantify TPC content. For the assay, 30 μL aliquots of extracts were transferred to 96-well microtiter plates (Nunc-Imuno Plate, Roskilde, Denmark), then 150 μL of Folin-Ciocalteau phenol reagent (2M) (1:10 v/v with water) were added after 4 min. Next, 120 μL of sodium carbonate solution (0.25 mg/L) was added to each well to a final volume of 300 μL. The reaction mixture was placed in the dark for 30 min at 40 °C and the absorbance was recorded at A_{665} against a blank sample. The analysis was performed at least in triplicate and the results were expressed as mg of gallic acid equivalents (GAE) per g of extract (mg GAE/g).

The total flavonoid content was measured by using an aluminum chloride colorimetric assay ($AlCl_3$-Merck, 98%), with minor modifications to the previously reported method [38,39]. The calibration curve was performed with catechin (Sigma Aldrich, 98%, China) as a standard. The flavonoid contents were measured as micromole catechin equivalents (μmol CE) per g of extract (μmol CE/g). Then, 25 μL of the dissolved extract were transferred to a microtiter (Nunc) and 7.5 μL of 5% $NaNO_2$ (w/v) (Panreac, 98%) were added to the wells. After 5 min, 7.5 μL of 10% $AlCl_3$ in ethanol were added to the mixture. Finally, after 5 min, 100 μL of 1M NaOH (Merck, P.A., Darmstadt, Germany) were added and, after 10 min, the absorbance was measured against the reagent's blank in triplicate at 750 nm using a microtiter plate reader (FLUOstar OPTIMA, Offenburg, Germany).

2.5. Antioxidant Activity Determinations

The 1,1-diphenyl-2-picrylhydrazyl (DPPH) (Sigma Aldrich, Steinheim, Germany) assay was used as described previously [37] to determine the radical scavenging activity of the extracts, with appropriate modifications. A fresh stock solution of 100 μM DPPH in methanol was prepared daily. Aliquots of 30 μL of the test extract were dissolved in methanol at concentrations ranging from 0.05 to 1.3 mg/mL and mixed in a microtiter plate (Nunc) with 270 μL of DPPH solution that was previous transferred in triplicate to a 96-well microtiter plate. The absorbance was measured at 510 nm by using a microtiter plate reader (FLUOstar OPTIMA) after 40 min of incubation at room temperature in the dark. Blank and control solutions were also prepared and measured. By using the same procedure, positive controls of trolox solution (Sigma Aldrich, 98%, China) and ascorbic acid (Panreac, 99%, Barcelona, Spain) in methanol were also completed for the same concentration range. The half maximal inhibitory concentration value (IC$_{50}$) was obtained from the linear range of the data. According to Equation (1), A is the absorbance, s is the sample, b is the blank, and c the control:

$$\text{IC50 (\%)} = \left[1 - \left(\frac{A_s - A_b}{A_c - A_b} \right) \right] \times 100 \qquad (1)$$

For the 2,2-azino-bis(3-ethylbenzthiazoline-6-sulfonic acid (ABTS) (Sigma Aldrich, ≥98%, China) assay, the experiments were performed by using the different decolorization ability of ABTS that indicates diverse scavenging activities according to a previously-described method [6,7], with some modifications. A total of 2.5 mM potassium persulfate (Acros Organics, 99%, Geel, Belgium) solution was prepared with distilled water and 10 mL of this solution were used to prepare a 7.4 mM ABTS stock solution, which was reacted for 16 h at room temperature in the dark. The stock solution was diluted with methanol until a 1.0 absorbance was attained at 730 nm in the microplate reader (BIO-RAD Model 680). To determine IC$_{50}$ values, 20 μL of the extract dissolved in methanol, at concentrations ranging from 0.05 to 1.3 mg/mL, were mixed in a microtiter plate (Nunc) with 280 μL of ABTS solution following the procedure described for DPPH. The same procedure was followed for the positive controls of the solutions using the trolox and ascorbic acid.

The reducing power test was performed by using a microtiter reader following a previously reported method [38,39] with some modifications. A total of 25 μL of different concentrations of the extracts ranging from 0.1 to 1.3 mg/mL were mixed in a microplate (Nunc) with 25 μL of sodium phosphate buffer (200 mM, pH 6.6) and 25 μL of 1% (w/v) aqueous potassium ferricyanide III (K$_3$Fe(CN)$_6$, Sigma-Aldrich, ≥97%). The mixture was incubated at 50 °C for 20 min, and 25 μL of trichloroacetic acid (10% w/v) were added. A volume of 50 μL of this solution was transferred to the microtiter plate, subsequently adding 50 μL of deionized water and 50 μL of ferric chloride (0.1% w/v). The absorbance was measured at 655 nm. The same procedure was followed by using trolox concentrations as positive controls and the final results were expressed as μmol of trolox equivalents per g of extract (μmol TE/g).

2.6. Acetylcholinesterase and Tyrosinase Activity

Acetylcholinesterase activity was assayed as described, with some modifications [40,41]. In a microtiter plate (Costar, Cambridge, MA, USA)), a mixture of 98 μL of HEPES solution (50 mM, pH 8.0) with 30 μL of extract and 7.5 μL of acetylcholinesterase (AChE) was prepared. After 15 min of incubation at 25 °C, the reaction was started with the addition of 22.5 μL of acetylthiocholine iodide (AChI, 1.20 mM) and 142 μL of 5,5'-dithiobis[2-nitrobenzoic acid] (DTNB).The microplate was then read at 405 nm every 30 s for 3 min by a microtiter plate reader (TECAN Infinito M200, Mannedorf, Switzerland). All tests were performed in triplicate at a sample and a positive control concentration of 50 μg of extract/mL, and tacrine was used as the reference standard for this procedure.

Enzyme activity was calculated as a percentage of the initial velocities compared to the assay by using buffer without any inhibitor [42]. Tyrosinase (Tyr) activity was assayed as described [43] with appropriated modifications. In a microtiter plate (Costar), a mixture of 180 μL of L-tyrosine (0.45 mM,) with 10 μL of the extract was incubated at 30 °C for 5 min (min). Tyrosinase (10 μL, 0.5 U/mL) was added and the mixture was incubated again at 30 °C for 5 min. The reaction was followed by a microtiter plate reader (TECAN Infinito M200) at 450 nm, at 2 min intervals for 12 min. Kojic acid was used as the positive control. The reaction was performed as previously described in a 50 mM phosphate buffer at pH 6.8, with 0.5 mM L-tyrosine, 10 mg/mL of kojic acid, and 5000 U/mL of mushroom tyrosinase at 30 °C [42]. Experimental conditions were also based on previous studies [44,45]. After incubation, absorbance were measured at 450 nm every 2 min, for a total of 10 min using 96-well reader [44]. The enzyme inhibition activity was calculated using Equation (2) and the rate of inhibition using Equation (3) [42]:

$$\text{Enzyme inhibition activity } (\%) = \left(\frac{\text{OD}_{\text{Control}} - \text{OD}_{\text{Sample}}}{\text{OD}_{\text{Control}}} \right) \times 100 \tag{2}$$

$$\text{Rate of inhibition} = \frac{\text{Corrected absorbance}}{\text{time (min)}} \tag{3}$$

All tests were performed in triplicate at a sample and positive control concentration of 50 μg of extract/mL and kojic acid was used as the reference standard for this procedure.

2.7. Statistical Analysis

All results are presented as mean value ± standard deviation (SD). Correlation and regression analyses were performed with the Excel software 2013 package (Microsoft Corporation, under Academic License, Microsoft of Portugal). Correlations were considered statistically significant at $p < 0.05$ according to Tukey HSD and Scheffé test.

3. Results and Discussion

3.1. Extraction Yield

The methodologies to obtain volatile oils with SFE from herbaceous matrices are well documented in the literature, specifically the selection of the working extraction conditions, such as pressure, temperature, flow rate, and particle size [27,46–48]. Additionally, to obtain an extract comparable to conventional organic solvent extraction, other operational conditions must be selected [24,49–51]. The basil extraction yields obtained by different methodologies are presented in Table 1.

Table 1. The extraction yields from basil (*Ocimum basilicum* L.) using different extraction methods.

Extraction Method	Sample Identification	Yield (%)
Hydrodistillation (Fresh plant)	HDF	0.35 ± 0.02
Hydrodistillation (Dry plant)	HDD	0.32 ± 0.02
SFE (9.0 MPa, 40 °C)	SFEO	0.39 ± 0.02
Soxhlet (Methanol)	MeOH	17.8 ± 0.9
Soxhlet (Ethanol)	EtOH	9.6 ± 0.4
SFE (40.0 MPa, 40 °C)	SFEE	2.2 ± 0.1

The extraction yield from *Ocimum basilicum* L. isolated by hydrodistillation of fresh (HDF) and dried plants (HDD) were 0.35 ± 0.02% and 0.32 ± 0.02% (*w/w*), respectively. For SFE from dried basil at 9.0 MPa and 40 °C, the volatile oil yield was 0.39 ± 0.02% (*w/w*) in the second separator. The total wax content (not presented in Table 1), recovered in the first separator of SFE, was 0.10 ± 0.01%

The operational conditions used in SFE were selected considering several experimental works [24,30,33]. The volatile oil and essential oil yields were different from the data reported in other studies [47,49,50]. Basil genotypes differed significantly with respect to their chemical composition and oil content, ranging from 0.07% to 1.92% in dry matrices [4]. The volatile oil exhibited a light yellow color with a pleasant fragrance as opposed to the essential oil that was pale yellow.

The Soxhlet extraction with the two solvents and the comparison with supercritical fluid extraction (40.0 MPa and 40 °C) showed that the highest extract yield was obtained with methanol (17.8 ± 0.9%) and the lowest was achieved with SFE, probably due to the lower polarity of CO_2.

3.2. Quantitative Analysis of the Essential Oil and the Volatile Oil

The chemical composition of the volatile compounds was analyzed by GC. The results are presented in Table 2. Methyl eugenol (29–34%), linalool (12.6–18.8%), methyl chavicol (12.6–19.9%), and cineole (5.8–11%) were found in different concentrations as the dominant volatile compounds in *Ocimum basilicum* L. Previous studies reported that the chemical composition of the oil was significantly different with diverse basil genotypes [4,52–56]. The main differences in the essential oils from fresh and dry plants can be observed in the increase of 1,8-cineole (7.6% to 11%) and methyl chavicol (13.5% to 19.9%). In contrast, the levels of linalool remained constant (18.1% to 18.8%) in HDF when compared to HDD (Table 2).

When comparing the two obtained oils, that from the dry plant showed a decrease in the levels of 1,8-cineole, linalool, and methyl chavicol, as well as a significant increase in phytol acetate 2 (0.3% to 5.4%) in HDDF with respect to SFEO. Moreover, an increase in phytol acetate 2 was observed in SFEO. No difference in wax content was observed in the oil, suggesting that extraction with supercritical CO_2 followed by two-stage separation procedure allowed the isolation of the pure basil volatile oil; whereas in the first separator, a white mass consisting of waxes was collected [56]. Paraffinic compounds ranging between C_{27} and C_{32} were the principal waxes, obtained in the first separator, and presented in Table 2.

Table 2. Percentages of compounds from the essential oil samples of fresh (HDF) and dried (HDD) basil, SFE volatile oil (SFEO), and waxes.

			Ocimum basilicum L.		
Components	**RI**	**HDF**	**HDD**	**SFEO**	**Waxes**
				2nd S	1st S
α-Pinene	930	0.3	0.6	t	t
Camphene	938	0.1	0.2	t	t
Sabinene	958	0.2	0.4	0.1	t
1-Octen-3-ol	961	0.2	0.2	0.1	t
β-Pinene	963	0.6	1.3	0.2	t
β-Myrcene	975	0.8	0.7	0.2	t
1,8-Cineole	1005	7.6	11.0	5.8	t
trans-β-Ocimene	1027	1.5	0.8	0.3	t
γ-Terpinene	1035	0.2	0.2	0.1	t
trans-Sabinene hydrate	1037	0.1	0.1	0.1	t
Terpinolene	1064	0.4	0.4	0.1	t
Linalool	1074	18.1	18.8	12.6	t
Camphor	1102	0.7	0.9	0.5	t
Borneol	1134	0.6	0.6	0.4	t
Terpinen-4-ol	1148	0.3	0.4	0.1	t
α-Terpineol	1159	0.9	1.0	0.7	t
Methyl chavicol (Estragole)	1163	13.5	19.9	12.6	t
Bornyl acetate	1265	0.4	0.5	0.3	t
Eugenol	1327	6.9	3.9	6.5	t
trans-Methyl cinnamate	1346	0.1	0.2	0.2	t
α-Copaene	1375	0.1	t	0.1	t
Methyl eugenol	1377	34.0	30.1	29.4	1.3
trans-α-Bergamotene	1434	2.8	2.4	5.6	0.1

Table 2. *Cont.*

Components	RI	HDF	HDD	SFEO	Waxes
				2nd S	1st S
α-Humulene	1447	0.5	0.3	0.6	0.4
trans-β-Farnesene	1455	1.3	1.1	3.4	1.3
Germacrene D	1474	1.4	0.3	1.7	1.6
Bicyclogermacrene	1487	0.5	0.1	0.5	0.2
γ-Cadinene	1500	0.4	0.3	0.6	0.3
β-Sesquiphellandrene	1508	0.6	0.4	0.9	0.7
Spathulenol	1551	0.2	0.1	0.2	t
T-Cadinol	1616	1.7	0.9	1.1	0.3
Phytol acetate 2	2101	0.4	0.3	5.4	3.9
n-Heptacosane	2700	t	0.1	0.2	17.8
n-Nonacosane	2900	t	t	0.2	15.4
n-Triacontane	2000	t	t	0.1	8.3
n-Hentriacontane	3100	t	t	0.1	11.6
n-Dotriacontane	3200	t	t	0.2	23.8
Identified Compounds		97.4	98.5	91.2	87.2
Grouped components					
Monoterpene hydrocarbons		4.2	4.7	1.1	t
Oxygen-containing monoterpenes		28.4	32.9	20.2	t
Sesquiterpene hydrocarbons		7.7	5.0	13.5	4.6
Oxygen-containing sesquiterpenes		2.3	1.3	6.7	4.4
Phenylpropanoids		54.4	53.9	48.5	1.3
Others		0.4	0.7	1.2	76.9

RI: In-lab calculated retention index relative to C_9–C_{27} *n*-alkanes on the DB-1 column; t: trace (<0.05%); 1° S and 2° S are the first and second separator, respectively.

3.3. Caracterization of Plant Extract

3.3.1. Total Phenolic and Flavonoid Content and Antioxidant Activity

The total phenolic and flavonoid content of the basil extracts determined by using MeOH, EtOH, and SFEE (40.0 MPa, 40 °C) are presented in Figure 1.

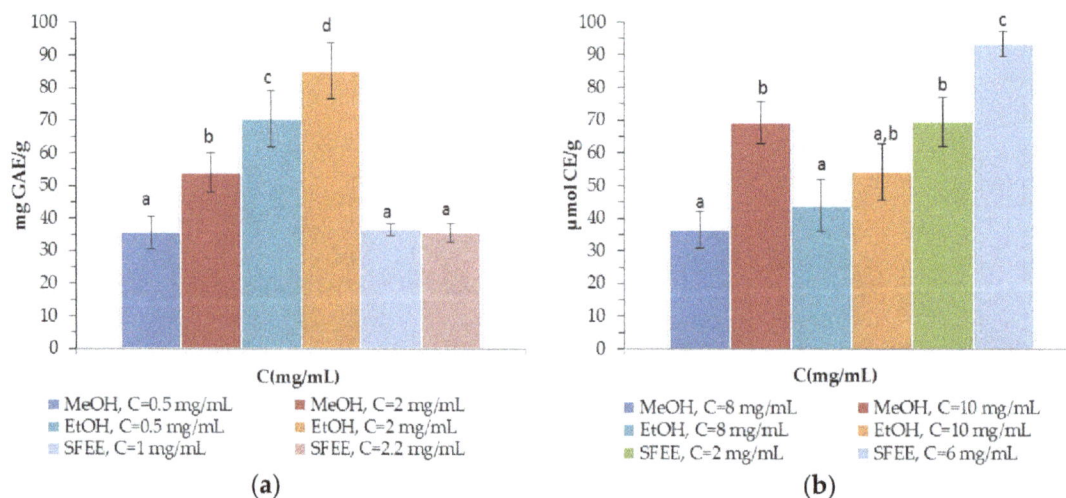

(a) (b)

Figure 1. Content of total phenols and flavonoids in basil extract by using different extraction methods. Values with different letters are significantly different ($p \leq 0.05$) according to Tukey HSD and Scheffé tests. (**a**) Total phenolic compounds in basil extract expressed as mg GAE/g extract and (**b**) total flavonoid compounds in basil extract expressed as mol CE/g extract.

The total phenolic content was calculated based on the calibration curve of gallic acid and expressed as mg of gallic acid equivalents (GAE) per g of extract (mg GAE/g). Figure 1a shows that the highest amounts, 85.3 ± 8.6 and 70.4 ± 8.6 mg GAE, were observed in the ethanolic extracts (EtOH) at concentrations of 2.0 and 0.5 mg/mL, respectively. The lower content of phenols was determined in the SFEE, at 35.5 ± 2.9 mg of GAE, from the supercritical carbon dioxide extract (SFEE). The value of 175.6 ± 2.4 mg GAE/g was determined by in a previous study [57]; however, the origin and region of the plant can influence their composition and, consequently, the extract characteristics. The total flavonoid content in basil extract was calculated as micromole catechin equivalents (μmol CE) per g of extract (μmol CE/g) as shown in Figure 1b. The highest amount of flavonoids was found in the SFE extract, and the lowest values in the ethanol and methanol extracts.

The antioxidant activities of the extracts obtained with different extraction methods are presented in Table 3. The results for the DPPH and reduction power assays revealed that the most active radical scavenger was the MeOH extract. However, in the ABTS assay, the SFEE showed higher antioxidant activity when compared with the EtOH or MeOH extracts.

These results can be explained by the compounds responsible for the antioxidant activity not necessarily being the same for each extraction method, since the extracts are complex mixtures of compounds, thus, different interactions, either synergistic or antagonistic, may occur [58]. The antioxidant activity values were greater when compared with other extracts in the literature [3,50,57], which can be explained by the environmental conditions where basil is produced, as well as their genotype.

Table 3. Antioxidant activity of 2,2-diphenyl-1-picrylhydrazyl (DPPH), 2,2-azino-bis(3-ethylbenzthiazoline-6-sulfonic acid (ABTS) (IC_{50} in mg/mL), and reducing power (μmol TE/g) of basil extracts obtained by different extraction methods.

Sample	DPPH-IC_{50} (mg/mL)	ABTS-IC_{50} (mg/mL)	Reduction Power (μmol TE/g)
MeOH	3.05 ± 0.36 [a]	4.26 ± 0.22 [a]	306.8 ± 21.8 [a]
EtOH	3.99 ± 0.55 [a]	5.38 ± 0.23 [b]	285.1 ± 18.1 [a]
SFEE	5.63 ± 0.20 [b]	1.74 ± 0.05 [c]	111.7 ± 7.3 [b]
Trolox	0.471 ± 0.088 [c]	0.425 ± 0.084 [d]	——
Ascorbic acid	0.266 ± 0.022 [d]	0.331 ± 0.050 [e]	——

Values with different letters within the columns were significantly different ($p \leq 0.05$) according to Tukey HSD and Scheffé test.

3.3.2. Acetylcholinesterase and Tyrosinase Activity

The acetylcholinesterase (AChE) and tyrosinase (Tyr) activities of basil extract using different extraction methods are shown in Figure 2. Figure 2a shows that, overall, the obtained extracts exhibited a very low inhibitory activity toward acetylcholinesterase when compared with the positive control. The SFEE extract did not present any activity, suggesting no possible inhibitory activity when compared with the values of 3% to 65% inhibitory activity in several MeOH extracts [40]. Moreover, a study reported that a 1 μg/mL hydro-methanol extract had an activity of 26.15% [59]. These results might have had some impact for cosmetic and dermatological purposes since a negative regulation of antimicrobial peptide exists in the skin through the cholinergic anti-inflammatory pathway via acetylcholine [60]. Some studies revealed that the β-adrenergic activation impairs cell motility and wound closure. The same study stated that the AChE released from keratinocytes also contributes to the regulation of local immune responses and potentially of infiltrating immune cells.

The examined extracts showed an inhibitory activity against tyrosinase (Figure 2b), although lower values were obtained when compared with the positive control (kojic acid). Tyrosinase inhibition may be related to antioxidant activity. Tyrosinase catalyzes the oxidation of phenols and its inhibition may lead to a decrease in melanin production by the rate-control step of its synthesis, for example, for potential cosmetic purposes.

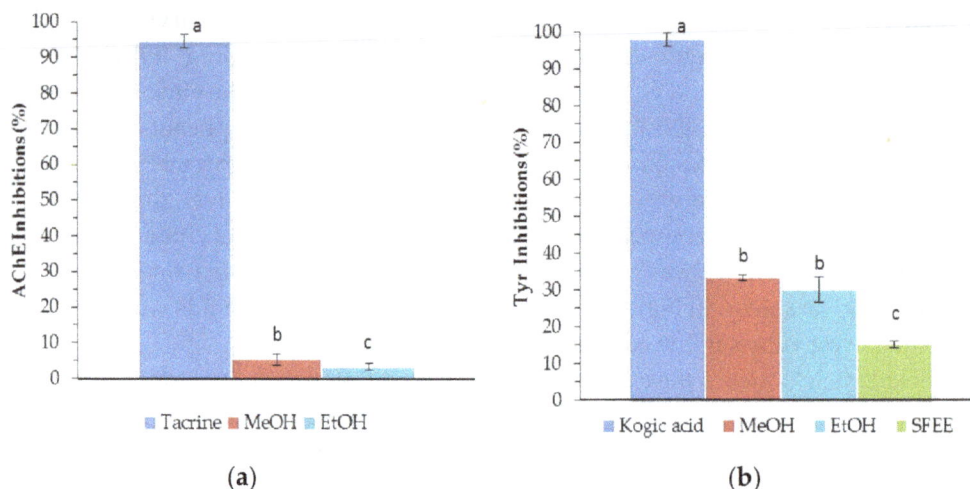

Figure 2. Acetylcholinesterase (AChE) and tyrosinase (Tyr) activity of basil extract from different extraction methods. Values with different letters were significantly different ($p \leq 0.05$) according to Tukey HSD and Scheffé tests. (**a**) AChE inhibition and (**b**) Tyr inhibition in basil extract.

4. Conclusions

The volatile oil obtained by supercritical carbon dioxide extraction of basil aerial parts was found to be similar to the essential oil obtained from hydrodistillation. The main differences in the composition were a significant increase in phytol acetate 2 in the SFEE compared with hydrodistillation oils. The composition of essential oil suggested that our *O. basilicum* chemotype is identifiable as Group 4, which is rich in estragole and linalool, by comparison with previous work [4]. The *O. basilicum* chemotype has important implications in terms of its biological properties.

From the comparison of the extracts obtained by methanol, ethanol, and supercritical CO_2, the highest yields were obtained with methanol and ethanol likely due to the higher polarity of these solvents. The methanol basil extract showed antioxidant activity, and the highest amounts of total phenolic compounds were found in the ethanol extract.

For enzyme inhibition, two enzymes (AChE and Tyr) were tested. Both enzymes are related to health problems, such as dementia and skin diseases. These extracts showed a low activity for the inhibition of the AChE enzyme. For the inhibition of TyrE, methanol and ethanol extracts of dried basil had the best activities and may be considered promising active ingredients for cosmetic purposes.

Acknowledgments: The authors are thankful for the financial support from Fundação para a Ciência e Tecnologia, Portugal, under project UID/QUI/00100/2013. The authors also gratefully acknowledge the collaboration of Ana Cristina Figueiredo (DBV-FCUL), for GC and GC-MS support.

Author Contributions: José Coelho, Beatriz Nobre and António Palavra conceived and designed the experiments; José Coelho and Jerson Veiga carried out supercritical fluid extractions; Jerson Veiga performed the analysis and with Amin Karmali carried out the antioxidant activity, the total phenolic and flavonoid content; Jerson Veiga, Catarina Pinto Reis and Marisa Nicolai performed acetylcholinesterase and tyrosinase activity experiments; José Coelho, Amin Karmali and Catarina Pinto Reis co-wrote the paper; all the authors analyzed and discuss the data.

References

1.   Avanmardi, J.; Khalighi, A.; Kashi, A.; Bais, H.P.; Vivanco, J.M. Chemical characterization of basil (*Ocimum basilicum* L.) Found in local accessions and used in traditional medicines in Iran. *J. Agric. Food Chem.* **2002**, *50*, 5878–5883. [CrossRef]

2. Ismail, M. Central properties and chemical composition of *Ocimum basilicum* essential oil. *Pharm. Biol.* **2006**, *44*, 619–626. [CrossRef]

3. Ijaz, A.; Anwar, F.; Tufail, S.; Sherazi, H.; Przybylski, R. Chemical composition, antioxidant and antimicrobial activities of basil (*Ocimum basilicum*) essential oils depends on seasonal variations. *Food Chem.* **2008**, *108*, 986–995. [CrossRef]

4. Zheljazkov, V.D.; Callahan, A.; Cantrell, C.L. Yield and oil composition of 38 basil (*Ocimum basilicum* L.) accessions grown in Mississippi. *J. Agric. Food Chem.* **2008**, *56*, 241–245. [CrossRef] [PubMed]

5. Dev, N.; Das, A.K.; Hossain, M.A.; Rahman, S.M.M. Chemical compositions of different extracts of *Ocimum basilicum* leaves. *J. Sci. Res.* **2011**, *3*, 197–206. [CrossRef]

6. Marwat, S.K.; Fazal-Ur-Rehman; Khan, M.S.; Ghulam, S.; Anwar, N.; Mustafa, G.; Usman, K. Phytochemical Constituents and pharmacological activities of sweet basil—*Ocimum basilicum* L. (Lamiaceae). *Asian J. Chem.* **2011**, *23*, 3773–3782.

7. Khair-ul-Bariyah, S.; Ahmed, D.; Ikram, M. *Ocimum basilicum*: A review on phytochemical and pharmacological studies. *Pak. J. Chem.* **2012**, *2*, 78–85. [CrossRef]

8. El-Azim, M.H.; Abdelgawad, A.A.; MohamedEl-Gerby; Ali, S.; El-Mesallamy, A. Chemical composition and antimicrobial activity of essential oil of egyptiam *Ocimum*. *Indo Am. J. Pharm. Sci.* **2015**, *2*, 837–842.

9. El-soud, N.H.A.; Deabes, M.; El-kassem, L.A.; Khalil, M. Chemical composition and antifungal activity of *Ocimum basilicum* L. essential oil. *J. Med. Sci.* **2015**, *3*, 374–379. [CrossRef] [PubMed]

10. Menaker, A.; Kravets, M.; Koel, M.; Orav, A. Identification and characterization of supercritical fluid extracts from herbs. *Comptes Rendus Chim.* **2004**, *7*, 629–633. [CrossRef]

11. Mazutti, M.; Beledelli, B.; Mossi, A.J.; Cansian, R.L.; Dariva, C.; Oliveira, V. De Caracterização química de extratos de *Ocimum basilicum* L. obtidos através de extração com CO_2 a altas pressões. *Quim. Nova* **2006**, *29*, 1198–1202. [CrossRef]

12. Zheljazkov, V.D.; Cantrell, C.L.; Evans, W.B.; Ebelhar, M.W.; Coker, C. Yield and Composition of *Ocimum basilicum* L. and *Ocimum sanctum* L. grown at four locations. *Hortic. Sci.* **2008**, *43*, 737–741.

13. Barbalho, S.M.; Maria, F.; Farinazzi, V.; Rodrigues, S.; Pereira, H.; Goulart, R.D.A. Sweet basil (*Ocimum basilicum*): Much more than a condiment. *TANG. Humanit. Med.* **2012**, *2*, 1–5. [CrossRef]

14. Barros, N.A.; Assis, A.R.; Mendes, M.F. Extração do óleo de manjericão usando fluido supercrítico: Analise experimental e matemática. *Ciência Rural* **2014**, *44*, 1499–1505. [CrossRef]

15. Sales, K.C.; Rosa, F.; Sampaio, P.N.; Fonseca, L.P.; Lopes, M.B.; Calado, C.R.C. In situ near-infrared (NIR) versus high-throughput mid-infrared (MIR) spectroscopy to monitor biopharmaceutical production. *Appl. Spectrosc.* **2015**, *69*, 760–772. [CrossRef] [PubMed]

16. Leal, P.F.; Maia, N.B.; Carmello, Q.A.C.; Catharino, R.R.; Eberlin, M.N.; Meireles, M.A.A. Sweet basil (*Ocimum basilicum*) extracts obtained by supercritical fluid extraction (SFE): Global yields, chemical composition, antioxidant activity, and estimation of the cost of manufacturing. *Food Bioprocess Technol.* **2008**, *1*, 326–338. [CrossRef]

17. Rijo, P.; Matias, D.; Fernandes, A.S.; Simões, M.F.; Nicolai, M.; Reis, C.P. Antimicrobial plant extracts encapsulated into polymeric beads for potential application on the skin. *Polymers* **2014**, *6*, 479–490. [CrossRef]

18. Rijo, P.; Falé, P.L.; Serralheiro, M.L.; Simões, M.F.; Gomes, A.; Reis, C. Optimization of medicinal plant extraction methods and their encapsulation through extrusion technology. *Meas. J. Int. Meas. Confed.* **2014**, *58*, 249–255. [CrossRef]

19. Nicolai, M.; Pereira, P.; Vitor, R.F.; Reis, C.P.; Roberto, A.; Rijo, P. Antioxidant activity and rosmarinic acid content of ultrasound-assisted ethanolic extracts of medicinal plants. *Measurement* **2016**, *89*, 328–332. [CrossRef]

20. Karmali, A. Process for Simultaneous Extraction and Purification of Fine Chemicals from Either Spent Mushroom Compost, Mushroom Stems or Partially Degraded Mushroom Fruiting Bodies. European Patent No. EP 2078755, 18 December 2009.

21. Arteiro, J.M.S.; Martins, M.R.; Salvador, C.; Candeias, M.F.; Karmali, A.; Caldeira, A.T. Protein-polysaccharides of *Trametes versicolor*: Production and biological activities. *Med. Chem. Res.* **2011**, *21*, 937–943. [CrossRef]

22. Silva, S.; Martins, S.; Karmali, A.; Rosa, E. Production, purification and characterisation of polysaccharides from *Pleurotus ostreatus* with antitumour activity. *J. Sci. Food Agric.* **2012**, *92*, 1826–1832. [CrossRef] [PubMed]

23. Pisoschi, A.M.; Pop, A.; Cimpeanu, C.; Predoi, G. Antioxidant capacity determination in plants and plant-derived products: A review. *Oxid. Med. Cell. Longev.* **2016**, *2016*, 1–36. [CrossRef] [PubMed]

24. Palavra, A.M.F.; Coelho, J.P.; Barroso, J.G.; Rauter, A.P.; Fareleira, J.M.N.A.; Mainar, A.; Urieta, J.S.; Nobre, B.P.; Gouveia, L.; Mendes, R.L.; et al. Supercritical carbon dioxide extraction of bioactive compounds from microalgae and volatile oils from aromatic plants. *J. Supercrit. Fluids* **2011**, *60*, 21–27. [CrossRef]

25. Coelho, J.P.; Palavra, A.F. Supercritical fluide of compounds from spices and herbs. *High Press. Fluid Technol. Green Food Process.* **2015**, 357–396. [CrossRef]

26. Reverchon, E.; Adami, R.; Campardelli, R.; Della Porta, G.; De Marco, I.; Scognamiglio, M. Supercritical fluids based techniques to process pharmaceutical products difficult to micronize: Palmitoylethanolamide. *J. Supercrit. Fluids* **2015**, *102*, 24–31. [CrossRef]

27. Fornari, T.; Vicente, G.; Vazquez, E.; Garcia-Risco, M.; Reglero, G. Isolation of essential oil from different plants and herbes by supercritical fluid extraction. *J. Chromatogr.* **2012**, *1250*, 34–48. [CrossRef] [PubMed]

28. Knez, Ž.; Markočič, E.; Leitgeb, M.; Primožič, M.; Hrnčič, M.K.; Škerget, M. Industrial applications of supercritical fluids: A review. *Energy* **2014**, *77*, 235–243. [CrossRef]

29. Coelho, J.A.; Grosso, C.; Pereira, A.P.; Burillo, J.; Urieta, J.S.; Figueiredo, A.C.; Barroso, J.G.; Mendes, R.L.; Palavra, A.M.F. Supercritical carbon dioxide extraction of volatiles from Satureja fruticosa Béguinot. *Flavour Fragr. J.* **2007**, *22*. [CrossRef]

30. Coelho, J.P.; Cristino, A.F.; Matos, P.G.; Rauter, A.P.; Nobre, B.P.; Mendes, R.L.; Barroso, J.G.; Mainar, A.; Urieta, J.S.; Fareleira, J.M.N.A.; et al. Extraction of volatile oil from aromatic plants with supercritical carbon dioxide: Experiments and modeling. *Molecules* **2012**, *17*, 10550–10573. [CrossRef] [PubMed]

31. Coelho, J.P.P.; Bernotaityte, K.; Miraldes, M.A.A.; Mendonca, A.F.; Stateva, R.P.P.; Mendonça, A.F.F.; Stateva, R.P.P. Solubility of ethanamide and 2-propenamide in supercritical carbon dioxide. Measurements and correlation. *J. Chem. Eng. Data* **2009**, *54*. [CrossRef]

32. Marques, A.J.V.; Coelho, J.A.P. Determination of fat contents with supercritical CO_2 extraction in two commercial powder chocolate products: Comparison with NP-1719. *J. Food Process Eng.* **2011**, *34*. [CrossRef]

33. Coelho, J.P.; Mendonça, A.F.; Palavra, A.F.; Stateva, R.P. On the solubility of three disperse anthraquinone dyes in supercritical carbon dioxide: New experimental data and correlation. *Ind. Eng. Chem. Res.* **2011**, *50*. [CrossRef]

34. Mota, L.; Figueiredo, A.C.; Pedro, L.G.; Barroso, J.G.; Ascensão, L. Glandular trichomes, histochemical localization of secretion, and essential oil composition in *Plectranthus grandidentatus* growing in Portugal. *Flavour Fragr. J.* **2013**, *28*, 393–401. [CrossRef]

35. Herald, T.J.; Gadgil, P.; Perumal, R.; Bean, S.R.; Wilson, J.D. High-throughput micro-plate HCl-vanillin assay for screening tannin content in sorghum grain. *J. Sci. Food Agric.* **2014**, *94*, 2133–2136. [CrossRef] [PubMed]

36. Attard, E. A rapid microtitre plate Folin-Ciocalteu method for the assessment of polyphenols. *Open Life Sci.* **2013**, *8*, 48–53. [CrossRef]

37. Bobo-García, G.; Davidov-Pardo, G.; Arroqui, C.; Vírseda, P.; Marín-Arroyo, M.R.; Navarro, M. Intra-laboratory validation of microplate methods for total phenolic content and antioxidant activity on polyphenolic extracts, and comparison with conventional spectrophotometric methods. *J. Sci. Food Agric.* **2015**, *95*, 204–209. [CrossRef] [PubMed]

38. Rajasekaran, M.; Kalaimagal, C. In vitro antioxidant activity of ethanolic extract of a medicinal mushroom, *Ganoderma lucidum*. *J. Pharm. Sci. Res.* **2011**, *3*, 1427–1433.

39. Reis, F.S.; Pereira, E.; Barros, L.; Sousa, M.J.; Martins, A.; Ferreira, I.C.F.R. Biomolecule profiles in inedible wild mushrooms with antioxidant value. *Molecules* **2011**, *16*, 4328–4338. [CrossRef] [PubMed]

40. Ingkaninan, K.; Temkitthawon, P.; Chuenchom, K.; Yuyaem, T.; Thongnoi, W. Screening for acetylcholinesterase inhibitory activity in plants used in Thai traditional rejuvenating and neurotonic remedies. *J. Ethnopharmacol.* **2003**, *89*, 261–264. [CrossRef] [PubMed]

41. Ferreira, A.; Proença, C.; Serralheiro, M.L.M.; Araújo, M.E.M.; Proença, C.; Serralheiro, M.L.M.; Araújo, M.E.M. The in vitro screening for acetylcholinesterase inhibition and antioxidant activity of medicinal plants from Portugal. *J. Ethnopharmacol.* **2006**, *108*, 31–37. [CrossRef] [PubMed]

42. Andrade, J.E. Unravelling New Ethnopharmacological Roles of Plectranthus Species: Biological Activity Screening. Master's Thesis, Universidade de Lisboa, Lisbon, Portugal, 2016.

43. Chang, C.; Chang, W.; Hsu, J.; Shih, Y.; Chou, S. Chemical composition and tyrosinase inhibitory activity of *Cinnamomum cassia* essential oil. *Bot. Stud.* **2013**, *54*, 1–7. [CrossRef] [PubMed]

44. Moon, J.-Y.; Yim, E.-Y.; Song, G.; Lee, N.H.; Hyun, C.-G. Screening of elastase and tyrosinase inhibitory activity from Jeju Island plants. *EurAsian J. Biosci.* **2010**, *53*, 41–53. [CrossRef]

45. Yamauchi, K.; Mitsunaga, T.; Batubara, I. Isolation, Identification and tyrosinase inhibitory activities of the extractives from *Allamanda cathartica*. *Nat. Resour.* **2011**, *2*, 167–172. [CrossRef]

46. Reverchon, E.; De Marco, I. Supercritical fluid extraction and fractionation of natural matter. *J. Supercrit. Fluids* **2006**, *38*, 146–166. [CrossRef]

47. Pourmortazavi, S.M.; Hajimirsadeghi, S.S. Supercritical fluid extraction in plant essential and volatile oil analysis. *J. Chromatogr. A* **2007**, *1163*, 2–24. [CrossRef] [PubMed]

48. Sovilj, M.N.; Nikolovski, B.G.; Spasojević, M.Ð. Critical review of supercritical fluid extraction of selected spice plant materials. *Maced. J. Chem. Chem. Eng.* **2011**, *30*, 197–220.

49. Filip, S.; Vidović, S.; Adamović, D.; Zeković, Z. Fractionation of non-polar compounds of basil (*Ocimum basilicum* L.) by supercritical fluid extraction (SFE). *J. Supercrit. Fluids* **2014**, *86*, 85–90. [CrossRef]

50. Filip, S.; Vidović, S.; Vladić, J.; Pavlić, B.; Adamović, D.; Zeković, Z. Chemical composition and antioxidant properties of *Ocimum basilicum* L. extracts obtained by supercritical carbon dioxide extraction: Drug exhausting method. *J. Supercrit. Fluids* **2016**, *109*, 20–25. [CrossRef]

51. Da Silva, R.P.; Rocha-Santos, T.A.P.; Duarte, A.C. Supercritical fluid extraction of bioactive compounds. *TrAC—Trends Anal. Chem.* **2016**, *76*, 40–51. [CrossRef]

52. Reverchon, E.; Donsi, G.; Osseo, L.S. Modeling of supercritical fluid extraction from herbaceous matrices. *Ind. Eng. Chem. Res.* **1993**, *32*, 2721–2726. [CrossRef]

53. Poletto, M.; Reverchon, E. Comparison of models for supercritical fluid extraction of seed and essential oils in relation to the mass-transfer rate. *Ind. Eng. Chem. Res.* **1996**, *5885*, 3680–3686. [CrossRef]

54. Özcan, M.; Chalchat, J.-C. Essential oil composition of *Ocimum basilicum* L. and *Ocimum minimum* L. in Turkey. *Czech J. Food Sci.* **2002**, *20*, 223–228. [CrossRef]

55. Occhipinti, A.; Capuzzo, A.; Bossi, S.; Milanesi, C.; Maffei, M.E. Comparative analysis of supercritical CO_2 extracts and essential oils from an *Ocimum basilicum* chemotype particularly rich in T-cadinol. *J. Essent. Oil Res.* **2013**, *25*, 272–277. [CrossRef]

56. Reverchon, E. Supercritical fluid extraction and fractionation of essential oils and related products. *J. Supercrit. Fluids* **1997**, *10*, 1–37. [CrossRef]

57. Vlase, L.; Benedec, D.; Hanganu, D.; Damian, G.; Csillag, I.; Sevastre, B.; Mot, A.C.; Silaghi-dumitrescu, R.; Tilea, I. Evaluation of antioxidant and antimicrobial activities and phenolic profile for *Hyssopus officinalis*, *Ocimum basilicum* and *Teucrium chamaedrys*. *Molecules* **2014**, *19*, 5490–5507. [CrossRef] [PubMed]

58. Martinez-Correa, H.A.; Magalhães, P.M.; Queiroga, C.L.; Peixoto, C.A.; Oliveira, A.L.; Cabral, F.A. Extracts from pitanga (*Eugenia uniflora* L.) leaves: Influence of extraction process on antioxidant properties and yield of phenolic compounds. *J. Supercrit. Fluids* **2011**, *55*, 998–1006. [CrossRef]

59. Kaur, S.; Singh, V.; Shri, R. In Vitro Evaluation of acetylcholinesterase inhibition by standardized extracts of selected plants. *Int. J. Biol. Pharm. Res.* **2015**, *6*, 247–250.

60. Curtis, B.J.; Radek, K.A. Cholinergic regulation of keratinocyte innate immunity and permeability barrier integrity: New perspectives in epidermal immunity and disease. *J. Investig. Dermatol.* **2012**, *132*, 28–42. [CrossRef] [PubMed]

## 5

Natural Variation of Volatile Compounds in Virgin Olive Oil Analyzed by HS-SPME/GC-MS-FID

Carlos Sanz [1],*, Angjelina Belaj [2], Araceli Sánchez-Ortiz [1] 🆔 and Ana G. Pérez [1]

[1] Department of Biochemistry and Molecular Biology of Plant Products, Instituto de la Grasa, CSIC, 41013-Seville, Spain; araceli.sanchez.ortiz@juntadeandalucia.es (A.S.-O.); agracia@cica.es (A.G.P.)

[2] IFAPA, Centro Alameda del Obispo, 14004-Cordoba, Spain; angjelina.belaj@juntadeandalucia.es

* Correspondence: carlos.sanz@ig.csic.es

Abstract: Virgin olive oil is unique among plant oils for its high levels of oleic acid, and the presence of a wide range of minor components, which are responsible for both its health-promoting properties and characteristic aroma, and only produced when olives are crushed during the industrial process used for oil production. The genetic variability of the major volatile compounds comprising the oil aroma was studied in a representative sample of olive cultivars from the World Olive Germplasm Collection (IFAPA, Cordoba, Spain), by means of the headspace solid-phase microextraction/gas chromatography–mass spectrometry–flame ionization detection (HS-SPME/GC-MS-FID). The analytical data demonstrated that a high variability is found for the content of volatile compounds in olive species, and that most of the volatile compounds found in the oils were synthesized by the enzymes included in the so-called lipoxygenase pathway. Multivariate analysis allowed the identification of cultivars that are particularly interesting, in terms of volatile composition and presumed organoleptic quality, which can be used both to identify old olive cultivars that give rise to oils with a high organoleptic quality, and in parent selection for olive breeding programs.

Keywords: *Olea europaea* L.; virgin olive oil; volatile compounds; variability; quality

1. Introduction

Virgin olive oil (VOO) is an essential component of the traditional Mediterranean diet, and attracts the interest of the scientific community for its health-promoting properties [1–6]. VOO is unique for its high levels of monounsaturated fatty acids and the presence of a wide range of minor components that are largely responsible for both their health-promoting properties and their organoleptic characteristics. Regarding the organoleptic characteristics, some of these minor components are volatile and consequently responsible for the aroma of VOO, which is characterized by a distinctive balance of green and fruity attributes.

Olías et al. [7] established that most of VOO volatiles are synthesized when enzymes and substrates join during the olive fruit milling in the oil extraction process, and that the enzymatic activities within the lipoxygenase (LOX) pathway are involved in this synthesis. Aldehydes and alcohols of six straight-chain carbons (C6), as well as their derived esters, are the major compounds in VOO aroma, either qualitatively or quantitatively [8,9]. Linoleic (LA) and linolenic (LnA) acids are the main substrates for this synthesis. LOX enzymes catalyze the production of 13-hydroperoxide derivatives in the first step of this pathway, which are later cleaved heterolytically by the hydroperoxide lyase (HPL) enzyme, forming C6 aldehydes [7,10,11]. C6 aldehydes are then reduced by alcohol dehydrogenase (ADH) enzymes to C6 alcohols [7,12], and finally converted into the corresponding esters by alcohol acyltransferase (AAT) enzymes [7,13]. The relevance of the five straight-chain carbons (C5) compounds in VOO aroma has also been suggested [9]. They seem to be produced via a parallel branch of the LOX pathway, as demonstrated in soybeans [14,15].

The synthesis of VOO volatile compounds seems to depend primarily on the availability of fatty-acid substrates to be catabolized through the LOX pathway during the VOO extraction process. In addition, LOX activity also proved to be an important limiting factor for the production of these volatile compounds, and this limitation is seemingly characteristic of each olive cultivar [16,17].

Along with its possible health benefits, its excellent organoleptic properties may well explain the continued demand for VOO. As mentioned above, volatile compounds are responsible for the aroma of this product, which is composed, when obtained from sound fruits, of a mixture of green and fruity odor notes, spiced with some other positive notes that make VOO a unique, edible oil. Odor perception and pleasantness are determined by the functional groups in the volatile compound [18,19]. To understand the relationship between volatile compounds and odor attributes in VOO, Aparicio and Morales [20] developed a statistical sensory wheel through a compilation of the data obtained from trained VOO sensory panels across Europe. As a result, odor notes with similar description and volatile compounds with a given sensory perception were grouped together into sectors. Among the most relevant to VOO aroma are green or ripe fruit odor notes. Moreover, concentration and odor threshold determine the sensory contribution of each volatile compound in the oil. The ratio between the concentration of the volatile compound and its odor threshold is the odor activity value (OAV). Those volatile compounds with OAV below one do not theoretically contribute to the aroma of VOO.

The volatile fraction of VOO is also responsible for the off-flavors present in some oils. This is of absolute importance because VOO classification into commercial categories requires a sensory analysis. This classification is officially ruled [21] and carried out by trained test panels, which assess the green and fruity odor notes typical in the aroma of virgin olive oil, when extracted from healthy, fresh olive fruits, as well as aroma defects in the oil. When VOO was obtained from infested olive fruits or those that were incorrectly processed or stored, the volatile fraction of VOO is altered and may include off-flavor compounds responsible for the sensory defects in the oils. They are mostly aliphatic carbonyls and alcohols or acids [22,23], which are produced by chemical oxidation, or by the activity of microbial enzymes present in non-sound fruits.

One of the key characteristics of the olive species is its wide genetic patrimony. The collection and conservation of olive genetic resources has received much attention. Thus, the World Olive Germplasm Collection (WOGC, Cordoba, Spain) is internationally recognized due to its large number of cultivars, and the high degree of evaluation and identification by molecular markers and agronomical traits [24]. This high genetic diversity of the olive germplasm collection could be very useful to recuperate old cultivars which produce oils with outstanding aromas, or for the selection of parents in breeding programs, in order to find new cultivars with improved olive oil aroma. In this sense, olive breeding programs are lately addressing the selection of cultivars, based on the sensory and nutritional qualities of VOO [25,26], in addition to the traditional agronomic traits [27]. For this purpose, the aim of this work was to evaluate the volatile composition and presumed aroma properties of the oils from a representative sample of cultivars from WOGC.

2. Materials and Methods

2.1. Plant Material

Sixty-one olive (*Olea europaea* L.) cultivars from the WOGC (IFAPA Alameda del Obispo in Cordoba) were studied (Table 1). Two trees per accession were grown at the experimental orchards of IFAPA Alameda del Obispo center using drip irrigation and standard culture practices. Fruits were picked by hand on two consecutive years, when they reached the turning stage, to better compare the cultivars.

Table 1. Cultivars selected from the World Olive Germplasm Collection (WOGC, Cordoba, Spain).

No.	Cultivar	No.	Cultivar
1	Abbadi Abou Gabra	32	Lastovka
2	Aggezi Shami	33	Lechín de Sevilla
3	Agouromanakolia	34	Lentisca
4	Amygdalolia Nana	35	Levantinka
5	Arbosana	36	Maurino
6	Argudell	37	Merhavia
7	Azapa o Arauco	38	Negrinha
8	Blanqueta	39	Ouslati
9	Blanqueta-48	40	Palomar
10	Borriolenca	41	Patronet
11	Bosana	42	Pecoso
12	Canetera	43	Pequeña de Casas Ibañez
13	Caninese	44	Picholine
14	Chorreao de Montefrío	45	Picholine Marroquí
15	Cipresino	46	Rapasayo
16	Coratina	47	Sabatera
17	Corbella	48	Shami
18	Cornicabra	49	Torcio de Cabra
19	Curivell	50	Ulliri i Kuq
20	Elmacik	51	Vallesa
21	Figueretes	52	Vaneta
22	Galega Vulgar	53	Vera
23	Haouzia	54	Verde Verdelho
24	Istarska Bjelica	55	Verdial de Badajoz
25	Jaropo	56	Verdial de Vélez-Málaga
26	Joanenca	57	Vinyols
27	Kaesi	58	Wardan
28	Kalokerida	59	Yun Gelebi
29	Kan Celebi	60	Zaity
30	Klon-1081	61	Zalmati
31	Kolybada		

2.2. Olive Oil Extraction

Olive oil was extracted from olive fruits (2–3 kg) using an Abencor system (Comercial Abengoa, S.A., Seville, Spain) that mimics, at laboratory scale, the industrial process of VOO extraction [28]. Fruits were milled using a stainless steel hammer mill operating at 3000 rpm and with a 5 mm sieve, malaxation performed for 30 min at 28 °C, and centrifugation of the paste carried out in a basket centrifuge at 3500 rpm for 1 min. The oils were decanted, filtered through paper, and stored under nitrogen at −20 °C until analysis.

2.3. Analysis of Volatile Compounds

Analysis of volatile compounds was carried out according to Pérez et al. [29]. Olive oil samples (0.5 g) were prepared in 10-mL vials and placed in a vial heater at 40 °C for a 10 min equilibration time. Volatile compounds from the headspace were adsorbed onto a headspace solid-phase microextraction (SPME) fiber divinylbenzene/carboxen/polydimethylsiloxane (DVB/CAR/PDMS) 50/30 μm (Supelco Co., Bellefonte, PA, USA). Sampling time was carried out at 40 °C for 50 min. Desorption of volatile compounds was carried out directly in the GC injector. Volatile compounds were identified on a 7820A/GC-5975/MSD system (Agilent Technologies, Santa Clara, CA, USA), including a DB-Wax capillary column (60 m × 0.25 mm i.d., film thickness, 0.25 μm: J&W Scientific, Folsom, CA, USA) under the following conditions: Injection port operated in splitless mode at 250 °C; He as the carrier gas at a flow rate of 1 mL/min; column held for 6 min at 40 °C, then programmed at 2 °C min^{-1} to 168 °C; the mass detector was operated in the electronic impact mode at 70 eV; source temperature set at 230 °C; and the mass spectra scanned at 2.86 scans/s in the m/z 40–550 amu range. Compounds were identified by mass spectrum, comparing to the Wiley/NBS and NIST libraries, retention index in agreement with literature, and co-elution with chemical standard when available. Quantification was performed in triplicate on a HP-6890 GC apparatus (Agilent Technologies, Santa Clara, CA, USA), which was equipped with a similar column and operated under the following operating conditions in order to obtain the same retention times for volatile compounds such as those obtained with the 7820A/GC-5975/MSD system: N_2 as the carrier gas at 17 psi; injector and detector at 250 °C; oven held for 6 min at 40 °C and then programmed at 2 °C min^{-1} to 168 °C. For the quantification of the different compounds, calibration curves obtained by adding known amounts of the different compounds to re-deodorized, high-oleic sunflower oil were used. Limits of quantification and concentration ranges were established according to typical contents in virgin olive oil volatile fractions. Linear regression curves were found for all the compounds, with regression coefficients higher than 0.99. At the beginning, and regularly during the sample analyses, a blank containing no oil and a mixture of volatile standards were run as controls.

2.4. Statistical Analysis

The data were analyzed using STATISTICA (Statsoft Inc., Tulsa, OK, USA). Factor Analysis and Principal Component Analysis (PCA) were used to assess the associations among the volatile compounds from the WOGC cultivar sample.

3. Results and Discussion

The extensive genetic patrimony of the olive tree is currently used in different olive breeding programs, but only recently includes the sensory properties of VOO as main traits, in addition to the agronomic traits [25,27,30]. Regarding these sensory properties, the natural variation of aroma compounds in VOO was studied using the WOGC olive cultivar collection. For this purpose, around 15% of this germplasm collection was considered (Table 1) [24]. Only healthy fruits were picked at the turning stage, and mild operating conditions were employed for the oil extraction process, to minimize the production of compounds that cause sensory defects in the oils. This subset of the olive germplasm collection showed a high degree of variability, in terms of the content of volatile compounds (Figure 1 and Table 2). As mentioned, most of the volatile compounds found in the oils were synthetized by the enzymes included in the so-called LOX pathway [7], and can be grouped according to the length of the chain, the fatty acid substrate (C6/LnA, C6/LA, C5/LnA, C5/LA), and the origin of esters (LOX esters). Moreover, compounds derived from amino acid metabolism, which has a branched-chain chemical structure (BC), and a terpene also contributed quantitatively to the volatile fraction of VOO. C6 compounds derived from linolenic acid (C6/LnA) represented, on average, about 70% of total volatile fraction in the oils (Figure 1). The C6/LnA compounds varied from 5.53 to 67.94 µg/g oil, a variability higher than that found for oils produced from a collection of 39 olive cultivars, cultivated under the same edaphoclimatic conditions (2.52–18.11 µg/g oil) [31]. The C6/LnA aldehyde group was the most abundant, of which (E)-hex-2-enal was the main compound (91% of total C6/LnA compounds on average, Figure 1). The mean content of (E)-hex-2-enal in the oils was 27.19 µg/g oil, ranging from 4.70 to 65.03 µg/g oil. The large amount, and the relatively low odor threshold [32], of this C6/LnA compound make it one of the main contributors to the aromas of the oils, which seems to be typical of oils from the olive species [29,31].

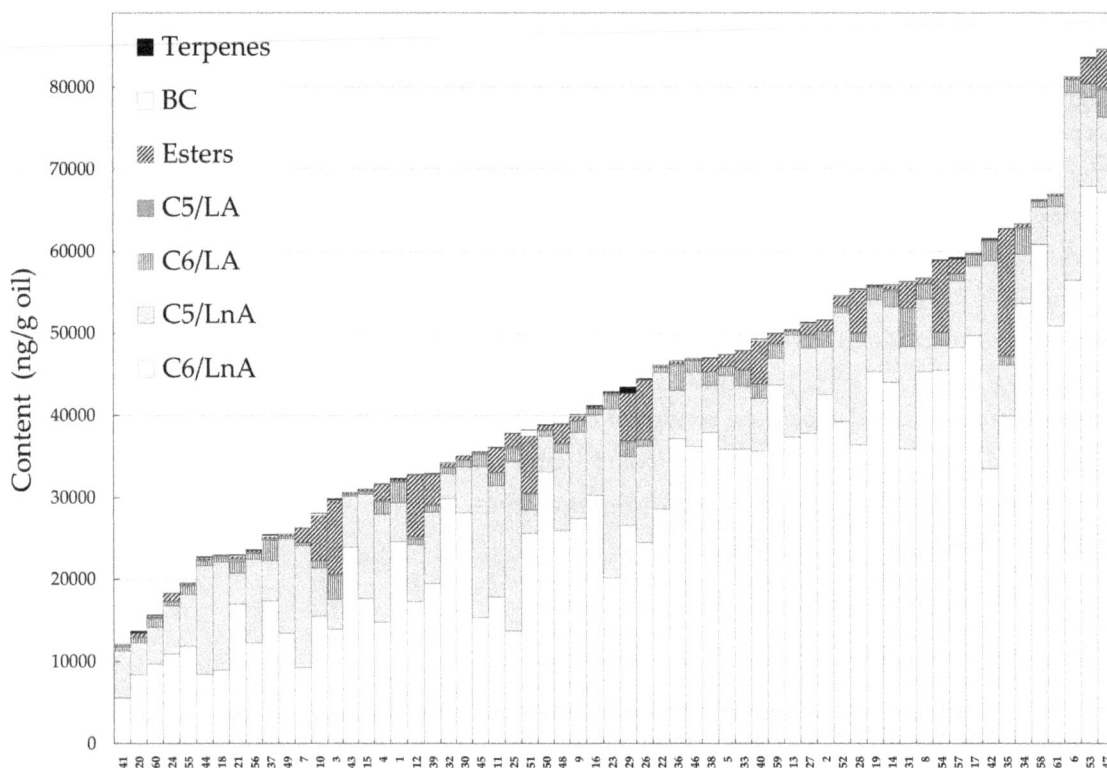

Figure 1. Content (ng/g oil) of the main groups of volatile compounds in the oils from the WOGC cultivar subset. Numbers correspond to the WOGC cultivars displayed in Table 1.

C5/LnA compounds showed contents close to those of the C6/LnA compounds. Quantitatively, the pentene dimers are the major compounds in this group. They are thought to be produced in the same branch of the LOX pathway as the rest of the C5 compounds [33]. The pentene dimers represented 82% of the C5/LnA content on average (Figure 1), but seem to have a quite low or negligible sensory contribution to the VOO aroma, according to the estimated odor thresholds for these compounds [29]. The rest of the C5/LnA compounds seem to contribute to VOO aroma according to their OAVs (Table 2). Pent-1-en-3-one and the pent-2-en-1-ols are especially remarkable. All the cultivars showed to have contents of pent-1-en-3-one above its odor threshold (0.73 ng/g oil). The odor of this compound is considered unpleasant [9], while the aroma of pent-2-en-1-ol is pleasant, green-fruity. Most of the olive cultivars (90%) presented (E)-pent-2-en-1-ol contents above its odor threshold [29] (Table 2). On the contrary, (Z)-pent-2-en-1-ol contents suggest that this compound is of little relevance to VOO aroma, as it was present below its threshold concentration in the oils.

Fruity odor notes are positive attributes of the oil aroma. Volatile esters are the main compounds responsible for these odor notes, especially LOX esters. Their content in the olive collection subset was 1844 ng/g oil on average, and is present in a range of 14–15,369 ng/g oil, much higher than those values found in a progeny of the cross of cultivars Picual and Arbequina [29]. Only a few accessions (3%) had (E)-hex-2-en-1-yl acetate contents that contribute to their aroma (OAV > 1, Table 2). However, (Z)-hex-3-en-1-yl acetate contents suggest this compound to be an important contributor to VOO aroma in more than half of the olive collection subset (Table 2).

Table 2. Content of the volatile compounds (ng/g oil) in the oils from the WOGC cultivar subset and percentaje of cultivars with odor activity values (OAV) higher than 1 for each volatile compound.

	Volatile Compound	Code	Min	Max	Mean	% cv OAV > 1
C6/LnA aldehydes	(E)-hex-3-enal	6C-1	116	924	371	25
	(Z)-hex-3-enal	6C-2	91	4598	884	100
	(Z)-hex-2-enal	6C-3	121	1290	646	80
	(E)-hex-2-enal	6C-4	4705	65,029	27,190	100
C6/LnA alcohols	(E)-hex-3-enol	6C-5	0	836	35	0
	(Z)-hex-3-enol	6C-6	45	3577	581	11
	(E)-hex-2-enol	6C-7	43	3293	252	0
C6/LA aldehyde	hexanal	6C-8	227	4451	1037	89
C6/LA alcohol	hexan-1-ol	6C-9	13	1463	314	26
C5/LnA carbonyls	pent-1-en-3-one	5C-1	126	1615	644	100
	(Z)-pent-2-enal	5C-2	8	175	57	0
	(E)-pent-2-enal	5C-3	45	226	105	0
C5/LnA alcohols	pent-1-en-3-ol	5C-4	80	634	257	10
	(Z)-pent-2-en-1-ol	5C-5	15	190	54	0
	(E)-pent-2-en-1-ol	5C-6	187	1564	510	90
Pentene dimers	pentene dimer-1	5C-7	77	1195	457	0
	pentene dimer-2	5C-8	65	1036	360	0
	pentene dimer-3	5C-9	346	4840	1812	0
	pentene dimer-4	5C-10	240	6363	2011	0
	pentene dimer-5	5C-11	0	4513	1282	0
	pentene dimer-6	5C-12	121	3909	1260	0
	pentene dimer-7	5C-13	0	2847	628	0
C5/LA carbonyls	pentan-3-one	5C-14	17	270	91	0
	pentanal	5C-15	12	233	48	0
C5/LA alcohol	pentan-1-ol	5C-16	0	26	5	0
LOX esters	hexyl acetate	E-1	5	6632	639	16
	(E)-hex-2-en-1-yl acetate	E-2	8	1620	66	3
	(Z)-hex-3-en-1-yl acetate	E-3	0	8696	1139	52
non-LOX esters	methyl acetate	E-4	6	42	19	0
	ethyl acetate	E-5	5	354	24	0
	methyl hexanoate	E-6	6	128	29	0
	ethyl hexanoate	E-7	0	359	45	0
BC aldehydes	3-methyl-butanal	BC-1	6	351	43	100
	2-methyl-butanal	BC-2	5	342	27	100
BC alcohol	2-methyl-butan-1-ol	BC-3	6	159	36	0
Terpene	limonene	T-1	0	743	59	2

The levels of the branched-chain (BC) compounds in the different accessions of the olive collection subset are also important for the aroma of the oils, despite their low concentrations in the oils (average 106 ng/g oil, range of 20–740 ng/g oil). The 2- and 3-methyl-butanal contents suggest that these BC aldehydes contribute decisively to the VOO aroma, since they have OAV values higher than one in all the cultivars. This aroma significance for VOO is especially notable because they are located in the 'ripe fruit' sector of the statistical sensory wheel [20]. However, 2-methyl-butan-1-ol, which is responsible for the fusty off-flavor of the oil [34], was found below its odor threshold in all the cultivars assessed (Table 2), so it should not contribute to the aroma of the oils. Finally, only 2% of the cultivars of the olive collection subset had an OAV above one for limonene (Table 2), which suggests that this terpene is not an important contributor to VOO.

Factor analysis allowed the correlations within the different volatile compounds in the olive oils to be explained. Figure 2 displays the factor analysis, considering only those factors with eigenvalues higher than one and using the normalized Varimax method. Those factors explained 77.97% of the total variance. The first factor explained 20.95% of the variance, while the second factor explained 12.66%. As shown, most of the C6 compounds spread along Factor 2 (dark gray ellipse), while most of the C5 compounds distributed along Factor 1. Three different groupings may be distinguished for the latter (Figure 2, light gray ellipses), which correspond to the C5/LnA alcohols, carbonyls, and pentene dimers, respectively. The group of C5/LnA alcohols is the closest to the C6 compounds, which would indicate a greater metabolic proximity and would support the hypothesis that C5/LnA alcohols would be the first products formed from the homolytic branch of the LOX pathway, which involves the activity of a LOX protein forming an alkoxy radical from a polyunsaturated fatty acid [33], parallel to the pentene dimers formation but in different ways, and finally oxidized enzymatically to C5/LnA carbonyls. Moreover, it was found previously that the pentene dimer content is not related to the contents of the C5/LA carbonyls (5C-14, 5C-15) or alcohol (5C-16), suggesting that pentene dimers are only synthesized from LnA [35]. Accordingly, those compounds are located in Figure 2 far from the group of pentene dimers.

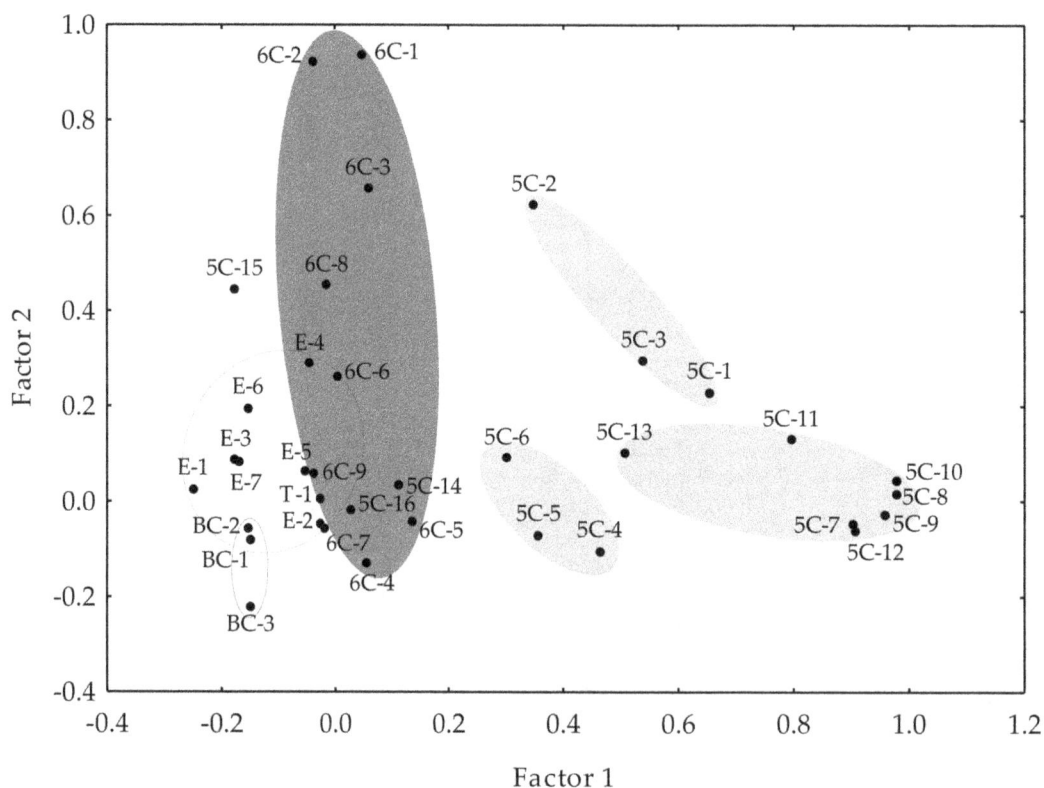

Figure 2. Factor analysis. Position of the volatile compounds in the oils from the WOGC cultivar subset on the first two factors using the normalized Varimax method.

On the other hand, the esters grouped together, overlapping the space of the C6 compounds, where the main precursors of esters are located [(hexan-1-ol (6C-9) and (Z)-hex-3-enol (6C-7)]. This partial disconnection from the LOX pathway mainstream could be due to the limitation of alcohol production during the oil extraction process, mainly due to the inactivation of ADH activity during VOO production, as previously demonstrated [36].

The PCA approach explains the associations between the different classes of volatile compounds in the olive collection subset, and allows us to distinguish those olive cultivars that are especially rich in some compounds (Figure 3). The first factor explains 19.87% of the variance, whereas the second factor explains 16.54%. Similar values were found when evaluating the content of the main volatile compounds in the oils of the Core-36 nuclear collection from the WOGC [35], and the progeny of the cross of cultivars Picual and Arbequina [29]. Cultivars producing oils with high contents of C6/LnA aldehydes are situated in the fourth quadrant of the plot, while those producing oils with high contents of C5/LnA are located opposite in the second quadrant. This distribution allows the selection in the fourth quadrant of those cultivars producing oils characterized by high contents of C6/LnA aldehydes, LOX esters, as well as BC aldehydes, which might act synergistically to provide green fruity odor notes. Most of the compounds included in those volatile classes are situated in sectors 'green fruit' and 'ripe fruit' of the statistical sensory wheel [20]. Therefore, it is possible to select cultivars from the germplasm collection subset whose oils have a prevailing green aroma, such as those cultivars located in the bisector of the first quadrant.

Table 2 shows those volatile compounds which present contents in the cultivar oils that make them major contributors to the oil aroma (OAV > 1). However, other volatile compounds contribute to the oil aroma of only a certain number of cultivars. A new PCA was carried out considering as variables those volatiles with OAV higher than one in more than 5% of the cultivars, thus truly contributing to the aroma of the oil (Figure 4). Most of them are considered desirable for VOO aroma, except for hexan-1-ol (6C-9), pent-1-en-3-one (5C-1), and pent-1-en-3-ol (5C-4). The latter give unpleasant sensations [9,20,34]. Factors 1 and 2 explain 43% of the data variation, quite similar to what was found for the Core-36 nuclear collection from the WOGC [35], and the progeny of the cross of cultivars Picual and Arbequina [29]. When using as variables those volatile compounds with OAV higher than one in more than 5% of the cultivars, the cultivars are distributed mainly along the Factor 1 axis (Figure 4b). When using the sums of the OAV of the compounds as supplementary variables in the PCA, which according to the literature are desirable (Pleasant) or undesirable (Unpleasant) to the oil aroma, the area in which the cultivars had desirable odor notes (Figure 4b, third quadrant) is clearly differentiated from that in which those cultivars characterized by having undesirable odor notes are located (Figure 4b, fourth quadrant).

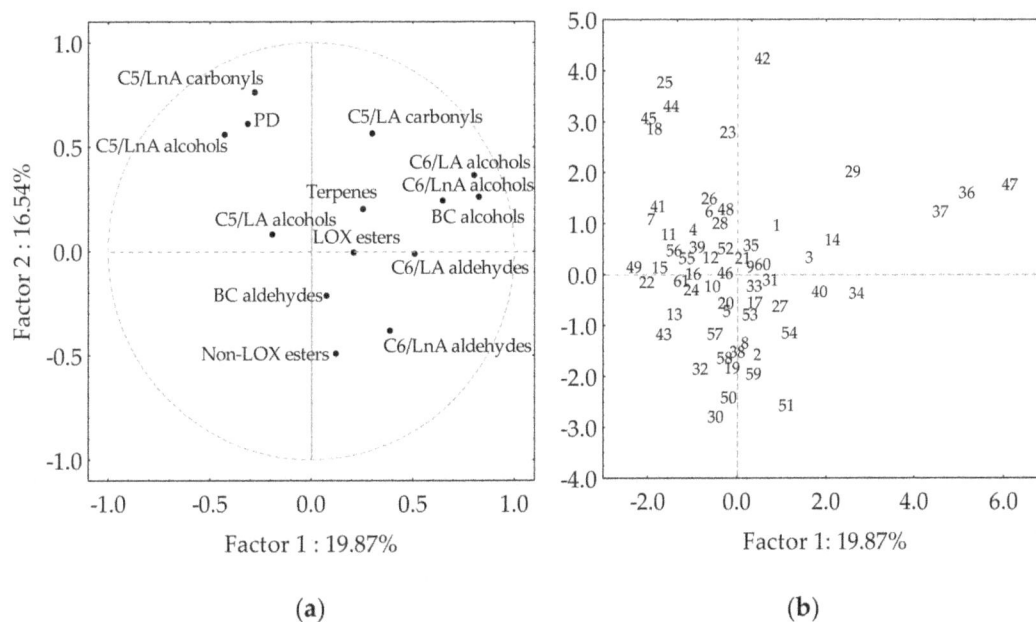

Figure 3. Main groups of volatile compounds in the oils from the WOGC cultivar subset. (**a**) Vector distribution of the volatile groups. (**b**) Distribution of the cultivars.

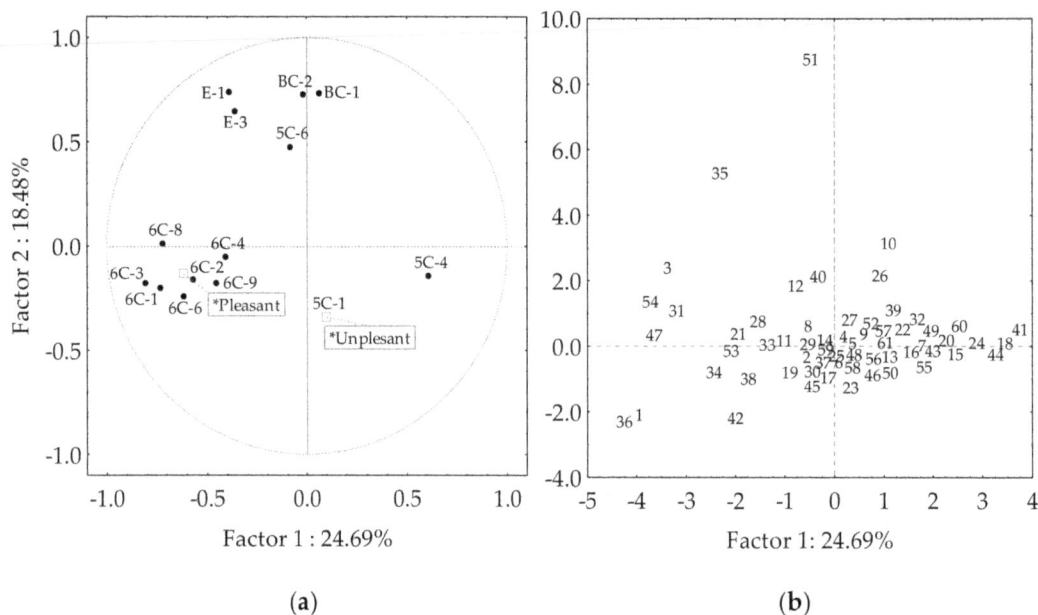

Figure 4. Selected volatile compound contributors to the aroma (OAV > 1) of the oils from the WOGC cultivar subset. (**a**) Vector distribution of the selected volatiles. (**b**) Distribution of the cultivars.

4. Conclusions

Olive species present a high variability for the VOO volatile compounds and, likewise, of the aroma quality. This natural variation of the aroma quality, and the high genetic diversity of the cultivar germplasm collection, suggest that it is possible both to pinpoint old olive cultivars that give rise to oils with a high organoleptic quality, and to select parents useful for breeding programs in order to develop olive cultivars with an enriched oil aroma. Multivariate analysis seems be a useful tool for this purpose.

Acknowledgments: Funding for this research came from the OLEAGEN project of Genoma España and the project AGL2011-24442 from the Programa Nacional de Recursos y Tecnologías Agroalimentarias, both financed by the Spanish Government.

Author Contributions: C.S. and A.G.P. conceived and designed the experiments; A.S.-O., A.G.P., and C.S. performed the experiments; A.G.P. and C.S. analyzed the data; A.B. contributed materials; C.S. collected data, performed statistical analysis, and wrote the first draft of the manuscript.

Conflicts of Interest: The authors declare no conflict of interest.

References

1.   Lucas, L.; Russell, A.; Keast, R. Molecular mechanisms of inflammation. Anti-inflammatory benefits of virgin olive oil and the phenolic compound oleocanthal. *Curr. Pharm. Des.* **2011**, *17*, 754–768. [CrossRef] [PubMed]

2.   Estruch, R.; Ros, E.; Salas-Salvadó, J.; Covas, M.I.; Corella, D.; Arós, F.; Gómez-Gracia, E.; Ruiz-Gutiérrez, V.; Fiol, M.; Lapetra, J.; et al. Primary prevention of cardiovascular disease with a Mediterranean diet. *N. Engl. J. Med.* **2013**, *368*, 1279–1290. [CrossRef] [PubMed]

3.   Visioli, F.; Bernardini, E. Extra virgin olive oil's polyphenols: Biological activities. *Curr. Pharm. Des.* **2011**, *17*, 786–804. [CrossRef] [PubMed]

4.   Konstantinidou, V.; Covas, M.I.; Muñoz-Aguayo, D.; Khymenets, O.; de La Torre, R.; Saez, G.; del Carmen Tormos, M.; Toledo, E.; Marti, A.; Ruiz-Gutiérrez, V.; et al. In vivo nutrigenomic effects of VOO polyphenols within the frame of the Mediterranean diet: A randomized trial. *FASEB J.* **2010**, *24*, 2546–2557. [CrossRef] [PubMed]

5.   Pérez-Jiménez, F.; Ruano, J.; Pérez-Martínez, P.; López-Segura, F.; López-Miranda, J. The influence of olive oil on human health: Not a question of fat alone. *Mol. Nutr. Food Res.* **2007**, *51*, 1199–1208. [CrossRef] [PubMed]

6. Psaltopoulou, T.; Kosti, R.I.; Haidopoulos, D.; Dimopoulos, M.; Panagiotakos, D.B. Olive oil intake is inversely related to cancer prevalence: A systematic review and a meta-analysis of 13800 patients and 23340 controls in 19 observational studies. *Lipids Health Dis.* **2011**, *10*, 127. [CrossRef] [PubMed]

7. Olias, J.M.; Perez, A.G.; Rios, J.J.; Sanz, C. Aroma of virgin olive oil: Biogenesis of the green odor notes. *J. Agric. Food Chem.* **1993**, *41*, 2368–2373. [CrossRef]

8. Morales, M.T.; Aparicio, R.; Ríos, J.J. Dynamic headspace gas chromatographic method for determining volatiles in virgin olive oil. *J. Chromatogr. A* **1994**, *68*, 455–462. [CrossRef]

9. Angerosa, F.; Mostallino, R.; Basti, C.; Vito, R. Virgin olive oil odour notes: Their relationships with the volatile compound from the lipoxygenase pathway and secoiridoid compounds. *Food Chem.* **2000**, *68*, 283–287. [CrossRef]

10. Salas, J.; Williams, M.; Harwood, J.L.; Sanchez, J. Lipoxygenase activity in olive (*Olea europaea* L.) fruit. *J. Am. Oil Chem. Soc.* **1999**, *76*, 1163–1169. [CrossRef]

11. Salas, J.; Sanchez, J. Hydroperoxide lyase from olive (*Olea europaea* L.) fruits. *Plant Sci.* **1999**, *143*, 19–26. [CrossRef]

12. Salas, J.J.; Sanchez, J. Alcohol dehydrogenases from olive (*Olea europaea*) fruit. *Phytochemistry* **1998**, *48*, 35–40. [CrossRef]

13. Salas, J.J. Characterization of alcohol acyltransferase from olive fruit. *J. Agric. Food Chem.* **2004**, *52*, 3155–3158. [CrossRef] [PubMed]

14. Gardner, H.W.; Grove, M.J.; Salch, Y.P. Enzymic pathway to ethyl vinyl ketone and 2-pentenal in soybean preparations. *J. Agric. Food Chem.* **1996**, *44*, 882–886. [CrossRef]

15. Fisher, A.J.; Grimes, H.D.; Fall, R. The biochemical origin of pentenol emissions from wounded leaves. *Phytochemistry* **2003**, *62*, 159–163. [CrossRef]

16. Sánchez-Ortiz, A.; Pérez, A.G.; Sanz, C. Cultivar differences on nonesterified polyunsaturated fatty acid as a limiting factor for biogenesis of virgin olive oil aroma. *J. Agric. Food Chem.* **2007**, *55*, 7869–7873. [CrossRef] [PubMed]

17. Sánchez-Ortiz, A.; Romero-Segura, C.; Sanz, C.; Pérez, A.G. Synthesis of volatile compounds of virgin olive oil is limited by the lipoxygenase activity load during the oil extraction process. *J. Agric. Food Chem.* **2012**, *60*, 812–822. [CrossRef] [PubMed]

18. Bedoukian, P.Z. The seven primary hexenols and their olfactory characteristics. *J. Agric. Food Chem.* **1971**, *19*, 1111–1114. [CrossRef]

19. Hatanaka, A.; Kajiwara, T.; Horino, H.; Inokuchi, K.I. Odor-structure relationships in n-hexanols and n-hexenales. *Z. Naturforsch.* **1992**, *47*, 183–189.

20. Aparicio, R.; Morales, M.T. Sensory wheels: A statistical technique for comparing QDA panels Application to virgin olive oil. *J. Sci. Food Agric.* **1995**, *67*, 247–257. [CrossRef]

21. European Commission. Commission Regulation 640/2008/EC. *Off. J. Eur. Union* **2008**, *L178*, 11–16.

22. Angerosa, F. Influence of volatile compounds on virgin olive oil quality evaluated by analytical approaches and sensor panels. *Eur. J. Lipid Technol.* **2002**, *104*, 639–660. [CrossRef]

23. Morales, M.T.; Luna, G.; Aparicio, R. Comparative study of virgin olive oil sensory defects. *Food Chem.* **2005**, *91*, 293–301. [CrossRef]

24. Belaj, A.; Dominguez-Garcia, M.C.; Atienza, S.G.; Martin-Urdiroz, N.; de la Rosa, R.; Satovic, Z.; Martin, A.; Kilian, A.; Trujillo, I.; Valpuesta, C.; et al. Developing a core collection of olive (*Olea europaea* L.) based on molecular markers (DArTs, SSRs, SNPs) and agronomic traits. *Tree Genet. Genomes* **2012**, *8*, 365–378. [CrossRef]

25. León, L.; Beltrán, G.; Aguilera, M.P.; Rallo, L.; Barranco, D.; de la Rosa, R. Oil composition of advanced selections from an olive breeding program. *Eur. J. Lipid Sci. Technol.* **2011**, *113*, 870–875. [CrossRef]

26. Rjiba, I.; Gazzah, N.; Dabbou, S.; Hammami, M. Evaluation of virgin olive oil minor compounds in progenies of controlled crosses. *J. Food Biochem.* **2011**, *35*, 1413–1423. [CrossRef]

27. Lavee, S. Aims, methods and advances in breeding of new olive (*Olea europaea* L.) cultivars. *Acta Hortic.* **1989**, *286*, 23–40. [CrossRef]

28. Martinez, J.M.; Munoz, E.; Alba, J.; Lanzon, A. Report about the use of the 'Abencor' analyser. *Grasas Aceites* **1975**, *26*, 379–385.

29. Perez, A.G.; de la Rosa, R.; Pascual, M.; Sanchez-Ortiz, A.; Romero-Segura, C.; Leon, L.; Sanz, C. Assessment of volatile compound profiles and the deduced sensory significance of virgin olive oils from the progeny of Picual × Arbequina cultivars. *J. Chromatogr. A* **2016**, *1428*, 305–315. [CrossRef] [PubMed]

30. El Riachy, M.; Priego-Capote, F.; Rallo, L.; Luque de Castro, M.D.; León, L. Phenolic profile of virgin olive oil from advanced breeding selections. *Span. J. Agric. Res.* **2012**, *10*, 443–453. [CrossRef]

31. Luna, G.; Morales, M.T.; Aparicio, R. Characterisation of 39 varietal virgin olive oils by their volatile composition. *Food Chem.* **2006**, *98*, 243–252. [CrossRef]

32. Reiners, J.; Grosch, W. Odorants of virgin olive oils with different flavor profiles. *J. Agric. Food Chem.* **1998**, *46*, 2754–2763. [CrossRef]

33. Angerosa, F.; Camera, L.; d'Alessandro, N.; Mellerio, G. Characterization of seven new hydrocarbon compounds present in the aroma of virgin olive oils. *J. Agric. Food Chem.* **1998**, *46*, 648–653. [CrossRef] [PubMed]

34. Angerosa, F.; Lanza, B.; Marsilio, V. Biogenesis of "fusty" defect in virgin olive oils. *Grasas Aceites* **1996**, *47*, 142–150. [CrossRef]

35. García-Vico, L.; Belaj, A.; Sanchez-Ortiz, A.; Martínez-Rivas, J.M.; Pérez, A.G.; Sanz, C. Volatile compound profiling by HS-SPME/GC-MS-FID of a core olive cultivar collection as a tool for aroma improvement of virgin olive oil. *Molecules* **2017**, *22*, 141. [CrossRef] [PubMed]

36. Sanchez-Ortiz, A.; Romero-Segura, C.; Gazda, V.E.; Graham, I.A.; Sanz, C.; Perez, A.G. Factors limiting the synthesis of virgin olive oil volatile esters. *J. Agric. Food Chem.* **2012**, *60*, 1300–1307. [CrossRef] [PubMed]

## 6

Methodology to Remove Strong Outliers of Non-Climacteric Melon Fruit Aroma at Harvest Obtained by HS-SPME GC-MS Analysis

Juan Pablo Fernández-Trujillo [1,*] (iD)**, Mohamed Zarid** [1] **and María Carmen Bueso** [2] (iD)

[1]   Department of Agricultural and Food Engineering, Regional Campus of International Excellence "Campus Mare Nostrum" (CMN), Technical University of Cartagena (UPCT), Paseo Alfonso XIII, 48, ETSIA, E-30203 Cartagena, Murcia, Spain; m.zarid@gmail.com

[2]   Department of Applied Mathematics and Statistics, CMN, UPCT, Doctor Fleming s/n, ETSII, E-30202 Cartagena, Murcia, Spain; mcarmen.bueso@upct.es

*   Correspondence: juanp.fdez@upct.es

Abstract: A methodology for making consistent studies of outliers of non-climacteric melon volatile organic compounds at harvest is reported. The juice was squeezed from the fruit of the 'Piel de sapo' cultivar harvested during two consecutive seasons and the aroma volatiles were extracted by headspace solid phase microextraction and measured by gas chromatography coupled to mass-spectrometry. A deconvolution analysis was performed to obtain volatile organic compounds. For multivariate the reliable identification of outliers, compound classes were studied as a percentage of total area counts of the melon compounds identified in the chromatogram by principal component analysis and partial least-squares discriminant analysis, and then verified by correlation analysis, box-whisker plot, and formal tests for univariate outliers. Principal component analysis was the key methodology for selecting outliers in variables that mostly did not follow a normal distribution. The presence of an excess in terms of relative percentage of area and the diversity of minor compounds such as alcohols, terpenes, acids, among others, are usually a sign of anomalous data that can be considered outliers in the aroma of this non-climacteric cultivar. This multivariate approach removed outliers, but kept the variability of aroma among the samples of every cultivar.

Keywords: *Cucumis melo* L.; exploratory data analysis; gas-chromatography mass spectrometry; multivariate analysis; solid phase microextraction; volatile organic compounds

1. Introduction

Variability among fruits in the analysis of volatile aroma compounds (VOCs) of non-climacteric melons is usually a problem due to the differential effect of fruit maturity, and the complex interaction between genotype and environment found for other quality attributes [1]. There is even variability among the years in which the fruits of each cultivar or breeding line are harvested because of difficulties in harvesting melons in the same stage of maturity, particularly in non-climacteric types. These visual indices, such as the skin color in contact with the soil, are affected by differences in the environment depending on the season.

Studies of the aroma of the parental line 'Piel de sapo' have demonstrated the inter-season variability and the lower aromatic potential compared with other climacteric varieties and, to a lesser extent, compared with almost non-climacteric or hybrid isogenic lines of the same type [2–7]. This pattern is typical of the *inodorus* cultivars studied by other authors [8,9]. Although certain data considered anomalous for this reference parental are usually discarded, no one methodology has been described for making the decision to remove outliers in a consistent way across years or seasons, although statistical tools, such as Grubbs' test for a single outlier based on the assumption of normality, are available [10]. However, this type of test is designed to detect the presence of only one outlier, which is why some authors have proposed tests for multiple outliers (e.g., Iglewicz and Hoaglin's two sided and robust test), or outlier high dimensional data analysis [11,12], the latter also applicable to fruit aromas which are composed of many individual chemical compounds.

In previous studies we tested different multivariate and univariate statistical methods for the data analysis of VOCs and other quality traits [2,3,5,6,13], also considering "zero" type data, when the VOC variable should be considered consistent [5], or even the use of deconvolution to reduce this problem [6,7]. The groups of different compound classes are of interest for the potential discrimination of physiological behavior or for quantitative trait loci (QTL) mapping of VOCs [5,6]. In particular, principal component analysis (PCA) and partial least-squares discriminant analysis (PLS-DA) were applied to data for unsupervised and supervised dimension reduction, respectively. The projection of data in a lower dimension space is a useful technique for identifying outlier points based on the distance between the rest of the points.

The box-whisker plot, originally introduced by Tukey [14] as a tool in exploratory data analysis, has become one of the most widely used statistical graphs for the detection of univariate outliers. Outliers in a box-whisker plot are the data with distance to the box of at least 1.5 times the interquartile range (IQR) while, in the case of extreme outliers, the distance is at least three times the IQR.

Data classified as outliers using different criteria have been considered as strong outliers. The objective of the present work was to develop a methodology for the systematic observation of strong outliers in VOC analysis of samples in an integrated manner that allows some potentially anomalous data to be excluded in the determination of fruit aromas. This is particularly critical for QTL mapping and in the determination of QTL × environment effects because the variability could be higher than in other cases, particularly in plots in open fields for breeding purposes, and/or in fruit where the exact degree of maturity is difficult to assess due to biological variance, e.g., melon [15]. For this, the study was based on the same parental studied in two different seasons.

2. Materials and Methods

2.1. Plant Material and Experimental Design

The Spanish melon parental (*Cucumis melo* L., cv. T111, *inodorus* group) of the 'Piel de sapo' type (PS) was grown in Torre Pacheco (Murcia, Spain) in typical Mediterranean conditions [6] during two consecutive seasons (S1 and S2). The experimental plot consisted of six plants per replicate (three plants located in two adjacent rows) with n = 21 and n = 9 replicates in S1 and S2, respectively. Two to three fruits per replicate were harvested between 07:00 and 10:00 h in one week in each season according to previously-reported minimum and optimum harvest maturity indices for PS [6,15]. The most critical parameters for externally identifying full maturity at harvest in the field were a developed stem scar, followed by peduncle lignification, and the onset of a light yellow color in the skin surrounding the peduncle, a dull dark green skin color, and a minimum light yellow color in the skin in contact with the soil. The total soluble solids (TSS) content, which was regularly monitored, showed a maximum that depended on the season, and fruit with less than 10 °Brix were discarded. Flesh volatiles were analyzed in both seasons.

2.2. Flesh and Juice Sampling for Volatile AnalyFTWsis

Flesh cylinder samples (20 mm length × 15 mm diameter) were obtained from the equator of individual fruit using an apple core borer following the previously reported methodology [2,6]. Each replicate consisted of a single analysis of fruit (usually two different melons) per field replicate. Juice was squeezed from the cylinders with a Simplex Super cast aluminum manual juicer and filtered through a four-layer cheesecloth. After 3 min at 23 °C, the mixture (20 mL juice and 8 mL $CaCl_2$ saturated solution, i.e., 71.4% *v/v* of juice) was poured into two 5 mL and one 15 mL sterile polypropylene vials, which were stored at −80 °C until analysis (after around two months). For the analysis, samples were thawed (about 15 min at 20 °C) and 2.32 mL of the mixture was poured into 10-mL vials. After adding 0.02 mL of the internal standard (a solution of 4.27 mg L^{-1} of 1-phenylethanol, CAS 98-85-1, in dichloromethane), the vials were capped hermetically (SU860101 silicone/PTFE 18 mm, 35 shore A, screw cap, Supelco, Bellefonte, PA, USA), and put into a thermostatted tray at 13 °C. Around 15 samples per day were analyzed in a random order.

2.3. Volatile Analysis: Headspace (HS) Formation, Solid-Phase Micro-Extraction (SPME), and Gas Chromatography-Mass Spectrometry (GC-MS) Analysis

Solid-phase microextraction was conducted after each season according to the indications of Amaro et al. [16] with slight modifications. A retention time-locking gas-chromatography mass-spectrometry method, using n-pentadecane (Sigma-Aldrich, Saint Louis, MO, USA; Merck KGaA, Darmstadt, Germany) as reference to block the method (at a retention time of 24.022 min), was used. Therefore, the column operated in constant pressure mode that slightly differed each year in order to keep the retention times constant. This methodology practically avoids the need for alignment [17], because differences in retention times among chromatograms were negligible.

Volatiles were extracted without stirring during 120 min at 35 °C in the heating tray of the gas chromatograph (6890N, Agilent Technologies, Wilmington, DE, USA) by using static HS-SPME. The fiber was 1 cm long and coated with 50/30 μm divinylbenzene/carboxen on polydimethylsiloxane (DVB/CAR/PDMS) (57329-U DVB/CarboxenTM/PDMS Stable FlexTM, Supelco, Bellefonte, PA, USA). This fiber is recommended for automatic holders and volatile and semi-volatile (C3–C20) analyses. The fiber (20 mm length and 20 gauge needle size) entered 22 mm into the vial headspace and remained for 30 min absorbing volatiles at 35 °C. The volatiles were desorbed from the SPME fiber into the GC injection port (mass spectrum detector transfer line heater) with a bake-out step of 3 min at 280 °C. The analyses were carried out with a MPS2 Gerstel Multipurpose sampler coupled to the GC-MS. The injection port was operated at 260 °C in splitless mode and subjected to a pressure of 13.4 psi adjustable for the retention time of n-pentadecane to be the same. The purge flow rate of the inlet was 40 mL·min^{-1} and the purge time was 2 min. The post-run column conditions were 10 min at 220 °C with a pressure of 26.2 psi. The fiber bake-out parameters were 15 min pre-bake-out time, 44 mm of injector penetration, and 1 min post-bake-out time. A solvent delay of 0.5 min was applied. The inlet liner used was a 2637505 SPME/direct (Supelco), 78.5 mm × 6.5 mm × 0.75 mm. Volatiles were separated on a 30 m × 0.25 mm i.d. × 0.25 μm thick capillary column (HP-5MS ultra inert, Agilent Technologies, Wilmington, NC, USA) that contained 5% phenyl-methyl silicone as a stationary phase. The carrier gas was helium (purity > 99.999%) with a flow rate of 1.5 mL·min^{-1}. The initial oven temperature was 35 °C, followed by a ramp of 2 °C·min^{-1} up to 75 °C, and then at 50 °C·min^{-1} to reach a final temperature of 250 °C, which was held for 5 min. Oven equilibration time was 0.5 min. Mass spectra were obtained by electron ionization (EI) at 70 eV. The detector worked at 230 °C and in full-scan with data acquisition and ion mass captured between 40 and 450 amu (4 scan·s^{-1}) during the run time (27.5 min).

2.4. Identification of Volatile Organic Compound and Data Calculations

The peaks were identified by a mass spectrometer (5973 Network Mass Selective Detector, Agilent Technologies) coupled to the GC, comparing the experimental spectra with those of NIST 11 MS library (National Institute of Standards and Technology, Gaithersburg, MD, USA). The chromatographic analysis data for the volatile compounds were automatically integrated using MSD ChemStation software (F.01 001 1903 2013, Agilent Technologies). To reduce the number of additional peaks that would need to be manually introduced in the subsequent deconvolution process, the automatic Chemstation parameters were set as follows: initial area reject 0; initial peak width 0.045; shoulder detection off; initial threshold 14.0. These preliminary data only served as a basis for manual integration, deconvolution, and individual reverification of the peaks based on deconvolution analysis using NIST 11 MS library and the automated mass spectral deconvolution and identification system AMDIS (version 2.7, 2011, National Institute of Standards and Technology, Gaithersburg, MD, USA). Data for the best deconvolution models were selected from those offered by automatic mass spectral deconvolution and identification system (AMDIS) according to the higher match quality index for the target compound (usually 2–3 models, and always less than five models per peak). After several trials with different options, the deconvolution settings of AMDIS were: component width equal to 16, two adjacent peak subtractions, low resolution, very high sensitivity, and low shape requirements. These parameters can be modified, especially to obtain the results for some flat peaks (i.e., some alkanes or poorly detectable compounds) that cannot always be confirmed with the above parameters: for example, by increasing the resolution to high.

Chemical compounds were verified by a comparison of linear retention indices (LRI) calculated with a homologous series of n-alkanes (C6–C20; Sigma-Aldrich) for the HP-5 MS UI column [18] and analyzed under the same conditions. Propane, butane, and pentane retention times were estimated based on previous analysis of several seasons. LRI were compared with the literature results compiled by NIST [19], although sometimes the LRI data were estimated from the non-polar retention index (*n*-alkane scale) offered by NIST MS Search 2.0. We also used a profile of melon VOC LRIs (more than 500 compounds) obtained in hundreds of previous analyses using cultivars of different types and the same methodology, and others with HP-5 MS columns [20] (Fernández-Trujillo et al., 2013). A coefficient of variation for *LRI* was calculated as follows:

$$\text{CV}_{\text{LRI}} = \frac{(LRI \text{ calculated } - \ LRI \text{ literature})}{LRI \text{ literature}} \times 100 \qquad (1)$$

The compounds accepted always showed $\text{CV}_{\text{LRI}} < 1\%$ (usually less than 0.5%). The concentration of the individual aroma volatile compounds identified (expressed as $\text{ng}\cdot\text{mL}^{-1}$ of juice) was quantified from the relation between their areas and that of the internal standard (1-phenylethanol), obtained from the total chromatograms, using a response factor of 1 [6].

The IUPAC names of individual aroma volatiles were verified with the CAS number of the NIST database [19] in accordance with several databases [21,22].

The chemical volatile compounds were classified into eleven groups, according to Dos-Santos et al. (2013) [5]: volatile acids (ACD); acetate esters (ACE); alcohols (ALC); aldehydes (ALD); alkanes (AHA); ketones (KET); non-acetate esters (NAE); sulfur-derived compounds (SDC); terpenes (TER); others (OTH). Unidentified (NID) compounds which did not demonstrate a clear comparison with the NIST11 database were not included in the main analysis. Each fruit was classified by the number of replicates (21 in season 1, and nine in season 2) and a letter (A, B, C or D), and sometimes by adding the letter "r", indicating re-analysis of the same individual fruit sample.

We tested the following six variables calculated automatically by using a Microsoft® Excel 2013 worksheet programmed for this purpose and the calculations of the *LRI* and CV_{LRI}:

— Total area counts (match quality or similarity index MQ > 50).
— Percentage of total area counts of identified compound classes (only classified compounds excluding unidentified and exogenous, with MQ > 50).
— Percentage of total area counts of the whole chromatogram (including compound classes classified as exogenous, with MQ > 50, but not unidentified compounds).
— Total area counts (without MQ criteria).
— Percentage of total area counts of compound classes, including exogenous and unidentified compounds.
— Percentage of total area counts (total area of the chromatogram, without MQ criteria; including exogenous and unidentified ones).

The first and fourth variables expressed in area counts were also examined after normalization of the compound class variables to the response of the internal standard (1-phenylethanol) by multiplying the areas by the corresponding factor calculated for each season, as follows:

$$F_{IE}ij = \frac{\text{Area of internal standard of the sample } j \text{ of season } i}{\text{Average area of Internal standard of season } i} \qquad (2)$$

The two former percentage variables showed similar levels. Additionally, the last two percentage variables calculated without MQ criteria showed similar results, and lower levels than the first two percentage variables calculated with the constraint of MQ.

Data of individual fruit aroma analysis computed in Microsoft® Excel were aligned using software specifically developed for this purpose in one of our projects (unpublished). The former software aligned compounds based on names of chemical compounds generated by the NIST library, and CAS numbers were used for further verification and, particularly, to avoid mistakes. Unidentified compounds (NID) were also aligned based on names automatically generated using the concatenate function of Microsoft® Excel with its respective retention time of two decimals (Unidentified (x)). A pseudo CAS number of unidentified compounds was also generated per compounds using its respective x RT and its y linear retention index (LRI) as follows: (NIDxLRIy). This information served for further manual verification in order to identify potential target unidentified compounds for future research. Two Dell® Precision WorkStations (M6800 and M4400, both with 16 Gb of RAM) were required to perform these processes.

2.5. Total Soluble Solids

Total soluble solids content were determined using a digital refractometer (Pocket, Atago PAL-1, Tokyo, Japan). The mean of each replicate plus SE are reported.

2.6. Statistical Analysis and Outlier Identification

Exploratory data analysis consisted of a normal probability plot and box-whisker plot analysis of the six variables obtained for the compound classes of aroma volatiles of both seasons separately, followed by principal component analysis and partial least-squares discriminant analysis applied to the scaled and mean-centered data. For each univariate variable, the Grubbs' test for outlier identification was performed when normality assumption was not rejected according to the Shapiro-Wilk test ($p \leq 0.05$) [23]. Correlation analysis was also carried out for each pair of variables within each type of the former six. Classical and robust 97.5% confidence ellipses of the data were calculated. The robust method used 50% of the observations for MCD (minimum covariance determinant) estimations.

All statistical analyses were performed using the free R software (R Core Team 2017), the FactoMineR library for the principal component analysis [24], and the *caret* library [25] for the PLS-DA analysis. Grubbs' test and correlation analysis form part of the outliers and mvoutlier packages of R, respectively [26,27].

The first step in the procedure to detect strong outliers consisted of checking PCA and PLS-DA output and marking the potential candidates, then verifying the candidates by correlation analysis, box-whisker plots, and Grubb's test when available. Additionally, dispersion graphs of different quality traits (e.g., TSS vs. the variables of the different VOC compound classes in percentage), served for the final decision on outlier classification and to study their potential origin.

3. Results

In a previous examination of the data by PCA, there were several mistakes because of a malfunction in our alignment program (two samples) and a sample with many acetate esters and sulfur-derived compounds that were also unique in fruit VOC of both seasons (Supplementary Materials Figure S1).

In general, the Grubbs' test only allowed detection of a few outliers because the normality assumption was rejected for most of the variables according to the Shapiro-Wilk test at a significance level $p \leq 0.05$ (data not shown).

3.1. Chemical Compounds Classes Based on Total Areas

All these data were studied visually taking into account in each fruit sample the other samples of the same or different season. The results for the three groups of variables representing the total areas of the chromatograms (with or without the MQ criteria, with or without normalizing the areas to the internal standard) were similar, and judged by PCA. Finally, the outliers considered with this type of variables were the fruits 2B and 6A (S1), and, to a lesser extent 3B (S2) (Figure 1; Supplementary Materials Figure S2A).

Total areas

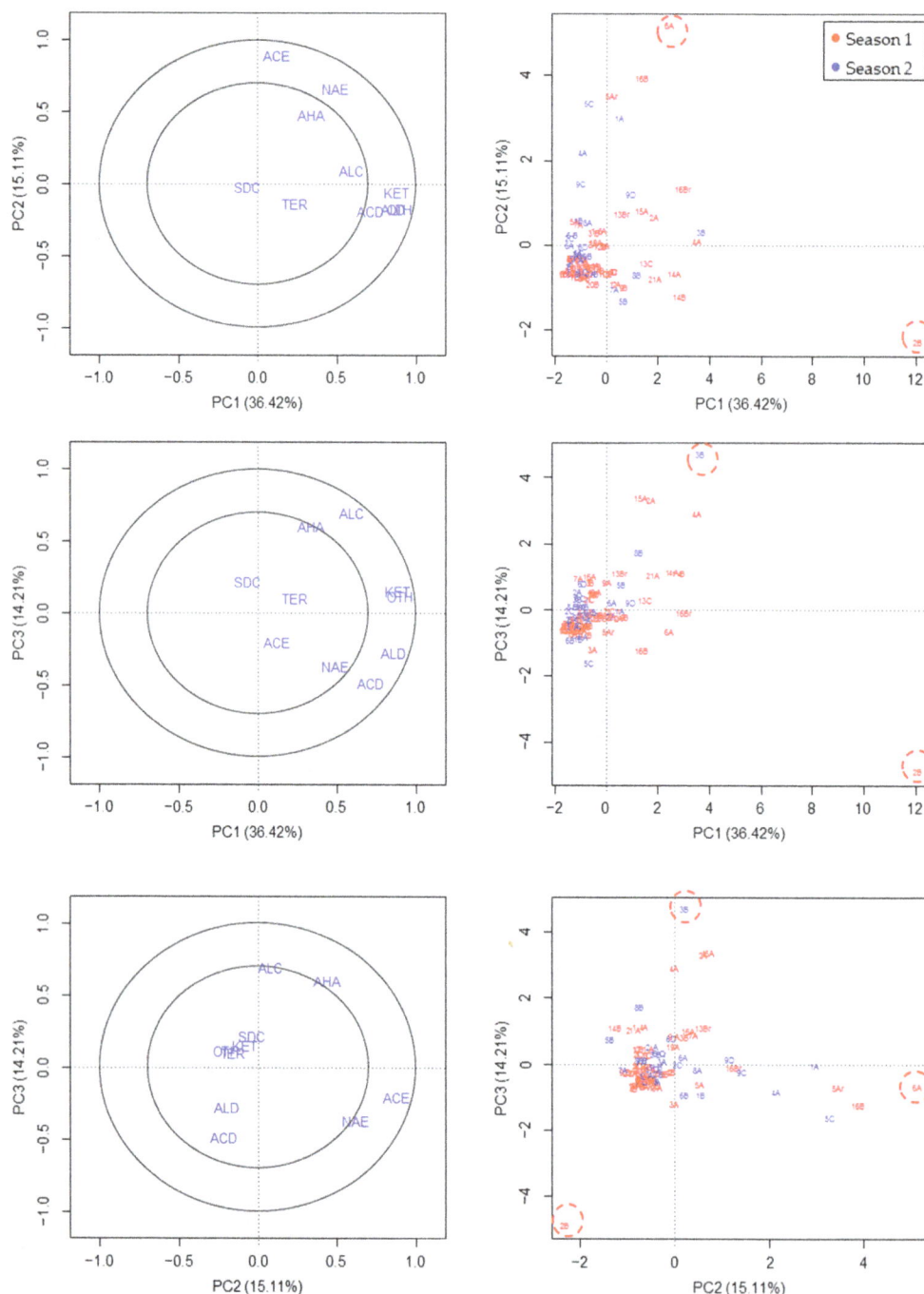

Figure 1. Correlation (**left**) and score (**right**) plots of the first three components of the PCA (with 67.82% of variance explained) applied to compound classes of aroma variable based on total areas (only compounds with match quality >50 without exogenous compounds), obtained in two seasons. The areas of compound classes were normalized to the response of the internal standard (1-phenylethanol) for each season. ACE, acetate ester; NAE, non-acetate ester; ALD, aldehyde; ALC, alcohols, KET, ketones; SDC, sulfur derived-compounds; TER, terpenes; AHA, alkanes and aliphatic compounds; OTH, other compounds. Circles represent $r^2 = 50\%$ and 100% variability explained by the components. Potential outliers are marked by dashed red line.

The outliers observed by using PLS-DA were the same and also 16B (S1) and 2A (S2), and, to a lesser extent, 5B or 6D (S2), because these two were only identified in the case of one of the variables of the total area considered without applying the normalization of areas to the internal standard (Supplementary Materials Figure S2).

From an observation of the figures and association outlier-centroid of each group of VOC compound class (centroids on the left of each figure for PCA and on the right for PLS-DA; Figure 1; Supplementary Materials Figure S2), the main reasons identifying the outliers in the first season was the excess of acetate esters (6A (S1), 16B (S1)) and aldehydes, ketones, and others in 2B (S1). In the second season, fruit 2A (S2) and 6D (S2) were considered outliers due to the excess of sulfur-derived compounds, and 5B (S2) due to the unusually high relative abundance and diversity of terpenes.

The main problem with the total area variables was that they did not consider the differences among different fruit analyses that are usually corrected when percentages or total areas are used, sometimes by using internal standards, and this is the reason for verifying the proposed outliers with the variables in percentages.

3.2. Chemical Compound Classes Based on Percentages of Sum of Areas

The results of the four groups of variables representing percentages of different compound classes based on different total areas of the chromatograms were similar. The reason is that MQ values increased when deconvolution was used, and so the percentages proposed with MQ are not altered much by this limitation.

Essentially, one outlier was detected in the VOCs of fruit 15A (S1) by using PCA and PLS-DA, due to the high proportion of alcohols, while no outliers were detected in S2 (Figure 2). Additionally, fruit 4B (S1) could be an outlier as judged by PLS-DA, due to the greater proportion of terpenes and/or acids than in other fruit (Supplementary Materials Figure S2). Certain differences between seasons were detected by PCA or the PLS-DA of percentages (e.g., in the graph of the first two axes; Figure 2 and Supplementary Materials Figure S3). The fruit classified as outliers in the previous section by reference to the area in absolute values were not classified as outliers using PCA, but using variables of compound classes based on percentages of the sum of the areas.

In the last of the percentage variables (without MQ criteria and including total area), fruit 5A (S1) was also considered as outlier (only by PLS-DA) due to a slightly higher proportion of acids, acetate esters and/or alkanes (data not shown). This datum was not finally considered as a strong outlier.

Percentage of areas

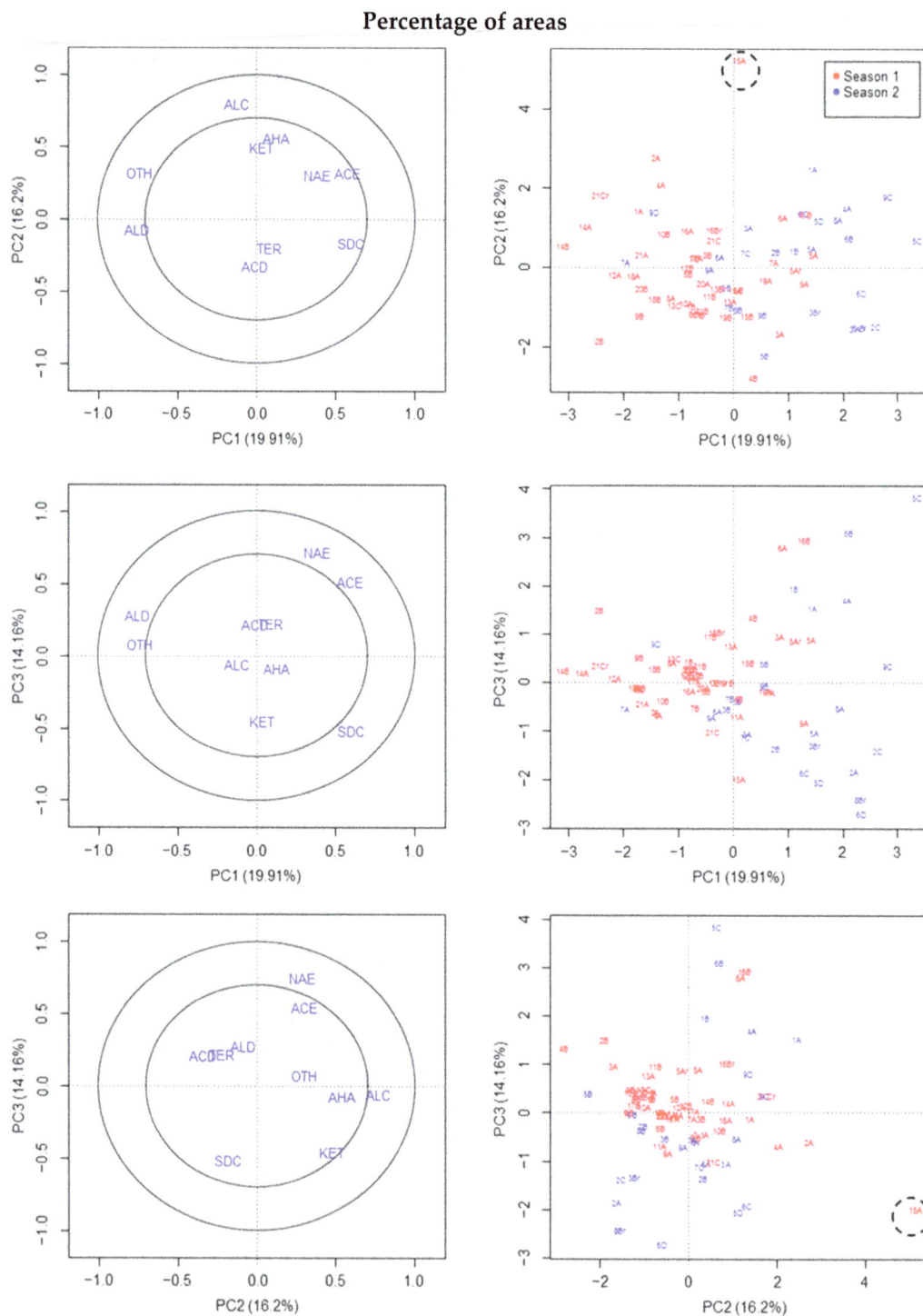

Figure 2. Correlation (**left**) and score (**right**) plots of the first three components of the PCA (with 50.27% of explained variance) applied to compound classes of aroma variable of percentages of different compound classes based on total area counts of such compounds with match quality >50, and without exogenous compounds, obtained in two seasons. ACE, acetate ester; NAE, non-acetate ester; ALD, aldehyde; ALC, alcohols, KET, ketones; SDC, sulfur derived-compounds; TER, terpenes; AHA, alkanes and aliphatic compounds; OTH, other compounds. Circles represent $r^2 = 50\%$ and 100% variability explained by the components. Potential outliers are marked by dashed black line.

3.3. Box-Whisker Plots of Both Types of Variables

A high number of outliers were detected by box-whisker plots based on the lack of normality in most of the variables of the compound classes tested, irrespective of the type of variable considered. The two fruits proposed as strong outlier candidates (15A and 4B of S1) were reassessed by means of box-whisker plots. The outlier of sample 15A (S1) was due to its high relative levels of alcohols, and, after removing this fruit, both variables showed similar behavior in the box-whisker plot (Figure 3A). This was also the case for acids and/or terpenes when the outlier of fruit 4B (S1) was removed (Figure 3B,C). Both outliers were relatively isolated from the rest of S1 and S2 fruits according to the above multivariate analyses and, therefore, can be classified as strong outliers.

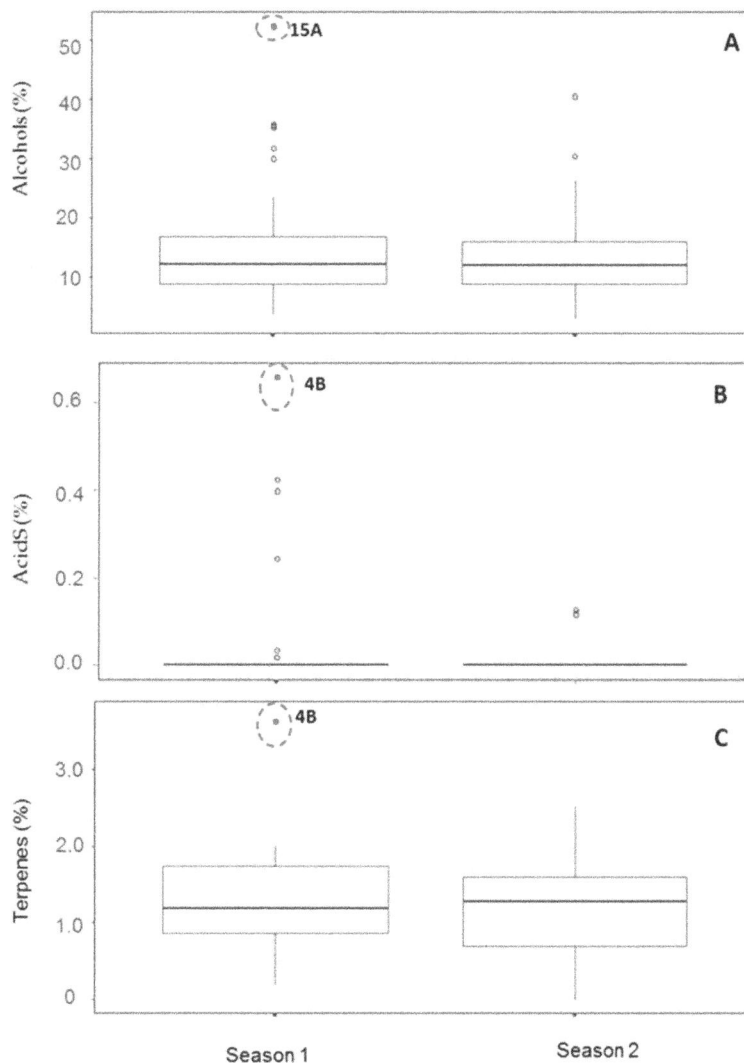

Figure 3. Box whisker-plot of three variables of compound classes (in percentage of area counts of the melon compound identified with match quality >50 and without exogenous compounds). (**A**) Alcohols; (**B**) acids; and (**C**) terpenes. Bold red circles indicate the selected outliers (with the fruit code beside this circle).

3.4. Correlation Analysis among Variables

This type of analysis produced many apparent outliers in each correlation pair (Supplementary Materials Figure S4), but the strong outliers (15A and 4B of S1) also appeared in graphs when alcohols and terpene variables appeared as percentages (Figure 4; Supplementary Materials Figure S4A,C,D).

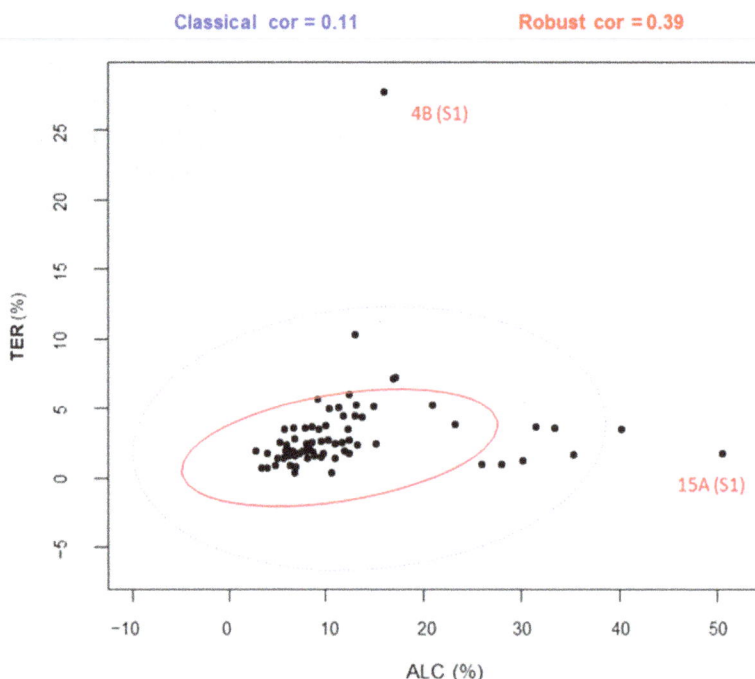

Figure 4. Correlation analysis of alcohol (ALC) and terpene (TER) percentages based on total areas of the melon compounds identified with match quality > 50, and without exogenous compounds). Classical and robust 97.5% confidence ellipses of the data (blue and red, respectively). The robust method used 50% of the observations for minimum covariance determinant (MCD) estimations. The strong outliers labeled are the fruits 4B and 15A from season 1.

3.5. Total Soluble Solids

Considering all the fruit after discarding outliers, the TSS means $\pm$ SD were significantly higher (Tukey test, $p = 0.001$) in season 2 (12.0 $\pm$ 0.2 °Brix; $n = 9$) compared with season 1 (9.7 $\pm$ 0.3 °Brix; $n = 21$), and the same was true (and with similar values) when they were not removed (data not shown). The TSS did not correlate with any of the variables referring to the total levels of compounds (in area or percentage), although this parameter could be useful for identifying some outliers. The same lack of correlation was found with other quality traits, such as the ratio between TSS and titratable acidity, among others (data not shown).

3.6. PCA of VOCs as Percentage after Removal of Outliers

The removal of strong outliers did not influence the loss of replicates of the experiment because at least one or two extra fruit can be used to calculate the average per replicate. According to all the PCAs (Figure 5) and PLS-DA (Supplementary Materials Figure S5), they were able to discriminate between seasons. However, fruit 2B (S1) appeared to be an extra outlier (Figure 5).

Though the outliers reported are not associated with aldehydes, 2B (S1), a potential additional outlier showed a high relative concentration of hexanal or pentanal, and, to a lesser extent, nonanal, together with an unusual diversity of other aldehydes within one sample, but in lower proportions (butanal, heptanal, octanal, decanal, undecanal, dodecanal, tetradecanal, (2E,4E)-hepta-2,4-dienal, (2E,4E)-nona-2,4-dienal, (E)-hept-2-enal, etc.). Additionally, some of these aldehydes were only detected by deconvolution analysis.

Percentage of areas

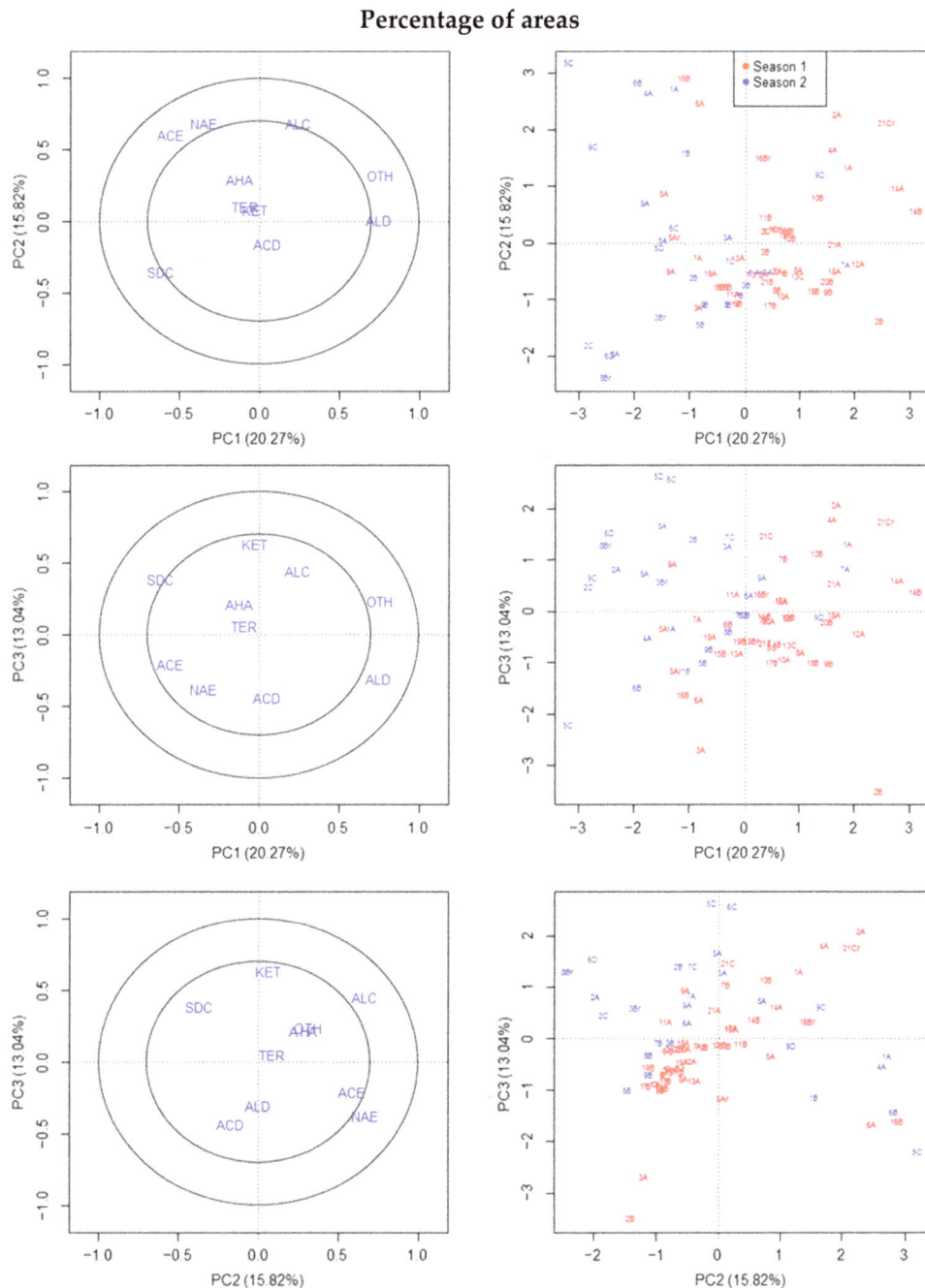

Figure 5. Correlation (**left**) and score (**right**) plots of the first three components of the PCA (with 49.13% of explained variance) applied to variables of different volatile organic compound classes based on percentages of total areas (only compounds with match quality >50) and without exogenous compounds included, obtained in two seasons, after removing the two fruit with outliers of season 1. ACE, acetate ester; NAE, non-acetate ester; ALD, aldehyde; ALC, alcohols, KET, ketones; SDC, sulfur derived-compounds; TER, terpenes; AHA, alkanes and aliphatic compounds; OTH, other compounds. Circles represent r^2 = 50% and 100% variability explained by the components.

3.7. Individual Aroma Volatiles

The individual aroma volatiles tentatively identified in more than 10% of the fruit analyzed were 126 compounds, irrespective of the season (data not shown). However, only 54 of them appeared in more than 50% of the fruit, and only 21 in more than in 90% of fruits with a similarity index above ninety (Table 1). Of the former, most were aldehydes (particularly hexanal), alcohols, sulfur-derived compounds and, to a lesser extent, other groups.

Table 1. Main volatile organic compounds extracted by headspace solid phase microextraction, measured by gas-chromatography coupled to mass spectrometry and tentatively identified by deconvolution analysis versus NIST11 database. Chemical compound were identified in more than 90% of the melon fruit samples analyzed (outliers excluded from the calculations) in order of retention time. The column used was an HP-5MS UI.

RT Mean	Chemical Abstract Service	Volatile Organic Compound	Linear Retention Index Calculated	Match Quality Mean
(Min)	Name	(IUPAC Name)	(ud)	(%)
1.279	000074-93-1	Methanethiol	<400	92
3.572	000624-92-0	Methyldisulfanylmethane	737	94
4.143	000071-41-0	Pentan-1-ol	763	88
4.949	000066-25-1	Hexanal	799	92
7.581	000111-27-3	Hexan-1-ol	867	90
8.974	000111-71-7	Heptanal	902	91
11.951	000100-52-7	Benzaldehyde	954	83
12.427	003658-80-8	Methylsulfanyldisulfanylmethane	962	88
14.814	000124-13-0	Octanal	1004	89
16.607	000104-76-7	2-ethylhexan-1-ol	1033	86
18.709	000098-86-2	1-phenylethanone	1067	92
20.146	000617-94-7	2-phenylpropan-2-ol	1091	81
20.814	000124-19-6	Nonanal	1107	94
20.916	004621-04-9	4-Isopropylcyclohexanol (isomer)	1114	75
21.433	000464-49-3	1,7,7-trimethylbicyclo[2.2.1]heptan-2-one	1151	95
21.629	010340-23-5	(Z)-non-3-en-1-ol	1165	81
22.015	000098-19-1	1-tert-butyl-3,5-dimethylbenzene	1193	72
22.112	000112-40-3	Dodecane	1200	82
22.177	000112-31-2	Decanal	1208	92
22.911	71186-24-8	(6Z)-6-[(E)-but-2-enylidene]-1,5,5-trimethylcyclohexene	1305	59
23.463	000629-59-4	Tetradecane	1400	65

The levels of alcohols of the outlier 15A (S1) was 52.4% (percent of identified compounds with MQ > 50), mostly due to the high relative concentration of hexan-1-ol followed by pentan-1-ol (70% of the area counts of alcohols identified), but also to the presence of other alcohols.

The outlier of fruit 4B (S1) has terpene compounds such as alpha-pinene (4,7,7-trimethylbicyclo [3.1.1]hept-3-ene), or beta-pinene (7,7-dimethyl-4-methylidenebicyclo[3.1.1]heptane) or camphene (6,6-dimethyl-5-methylidenebicyclo[2.2.1]heptane), apart from the terpenoid camphor (1,7,7-trimethylbicyclo[2.2.1]heptan-2-one). The acid detected in 4B (S1) was tentatively identified as 2-phenylpropanoic acid.

Other fruit also at the limit of distribution have high relative concentrations of ketones due to 1-phenylethanone, or some ester (isobutyl acetate or phenylmethyl acetate, and, to a lesser extent, isopropyl acetate) or sulfur-derived esters (data not shown).

4. Discussion

Overall, the outliers detected from the total areas did not match with those detected by the percentages because of the potential variability in the analysis and intensity of aroma among fruits. Transformation of the percentages into logarithms of the variables of compound classes based on percentages is not advisable except for classification or prediction purposes. The PLS-DA of the total area (with or without normalization of each compound class to the response of the internal standard) gave a higher number of outliers compared with PCA (Supplementary Materials Figure S2 vs. Figure 1), while, in percentages, these differences were minimized. Outliers detected by box-whisker plots were also more frequent than in this case because the variables frequently could not be analyzed by assuming normality, as demonstrated by the Shapiro-Wilk test. This problem also meant that Grubb's test could not be used in our case.

The main problem with applying Grubb's test is that individual variables can only be used if they follow a normal distribution, thus losing the perspective of a multivariate analysis. Additionally, this test can only be applied to the two fruits with extreme distribution data [10]. However, more than one fruit could be considered as an outlier in one extreme, particularly if the variable does not follow a normal distribution.

An important point when two seasons are being compared is that the potential anomaly detected by the box-whisker plot should be in the same range (Figure 3), which probably explains why more outliers were not found in the multivariate statistical analysis compared with box-whisker plots.

The outliers detected by correlation analysis (Figure 4) were more than those finally considered as strong outliers. Removing all of them from certain variables of compound classes using this methodology or box-whisker plot would be problematic for considering the whole aroma of a fruit, and also incorrect from a multivariate point of view.

Variability in melon fruit VOCs could be associated with differences in physiological maturity, but also fruit sampling, particularly the start of the ripening process in the placental tissue containing the seeds [28,29] because fruit ripening enhances the presence of alcohols. Some of the alcohols, such as hexan-1-ol, are good substrates for alcohol acetyltransferases or AAT [30], and potential accumulation in some fruits can be a sign of a lack of activity of the enzyme compared with aromatic cultivars [31], or a fruit harvested in a stage when the activity of the enzyme is still low [30]. On the other hand, the presence of terpenes or acids is sometimes associated with skin tissue [32], though terpenes have been found at higher levels in PS than in other cultivars [33]. According to our experience and other authors, acids are rarely detected in pulp samples by the HS-SPME test [5,33], but, by using stir-bar sorptive extraction (SBSE), it is possible to detect them as a precursor of aromas in whole PS melons, and also in the pulp of climacteric cultivars [20,34] and in other juices [35].

Previous publications have applied PCA and PLS-DA to the classification of and comparison between VOCs of melon near-isogenic lines and the corresponding parental [5,6], the PLS-DA providing a better classification with fewer variables involved. For a metabolomics approach, PCA combined with multiblock hierarchical PCA has been used [36].

Another possibility in the formation of outliers is the onset of analytical artifacts during the process, for example, due to septum bleeding, or the saline solution used (always stored at 2 °C and prepared before each season). No artifacts were observed after a blank SPME of the saturated saline solution or water used to prepare the saline solution alone, in agreement with general recommendations [8,33].

From a physiological point of view, the higher the relative acetate ester concentration in flesh tissue, the more senescent the PS melon flesh [37]. This was the case with some samples when area counts were observed, but, in the percentage of total area counts, the outliers apparently disappeared. The esters, particularly isobutyl acetate or phenylmethyl acetate, and, to a lesser extent isopropyl acetate, are typical of climacteric cultivars [6]. Other compounds, such as methyl 2-methylpropanoate, or the sulfur-derived esters, which are very specific to some cultivars and more abundant in climacteric ones [38,39], can also help detect outliers. A few esters can develop with certain abundance in some non-climacteric cultivars synthetized by ethylene-independent pathways, [40], particularly in certain seasons [6]. The high level of these esters dramatically reduced the proportion of the aldehydes typical of non-climacteric cultivars and accessions based on total area counts [8,9,37].

Another factor of importance is whether esters were found in the sampling procedure, particularly in large experiments involving climacteric aromatic and non-climacteric non-aromatic cultivars or lines for breeding purposes [39]. In the case of non-climacteric cultivars, placental tissue or local over-mature tissue should be avoided [3,28,36].

Overall, variables in the percentages of total area counts, such as total level of acids, terpenes, or alcohols, increased during PS melon fruit ripening, while aldehydes decreased, although non-acetate esters, alcohols and aldehydes also slightly increased close to senescence [34,37]. The detection of the excess of some of these compounds within a sample of different fruit of one parental may help to provide more reliable control data for QTL mapping of this type of compound at harvest using near-isogenic lines [4,5].

Several strategies are feasible to reduce fruit sampling outliers, although improving the signal to noise ratio in GC-MS analysis is probably the best by using techniques such as dynamic headspace purge and trap methodology for extraction [41]. The additional validation of the results by SBSE GC-MS is another interesting strategy [34], but deconvolution would be essential for integration [35], particularly if full-scan analysis, instead of using quantification, monitoring one or two ions per VOC [38], was used.

Another suggestion for studying outliers using the proposed methodology is to group the compounds according to their metabolic pathways, taking into account their precursors [42]. Finally, the methodology proposed can be helpful in association studies of VOCs with aroma obtained by sensorial evaluation, particularly with small datasets in which outliers may or may not control what relationships are found, and some variables show collinearity [43].

5. Conclusions

We propose a method for detecting outliers using PCA, double-checked with box-whisker plots, using the VOC variables obtained as percentages of chemical compound classes based, for example, on the total area of identified compounds in a chromatogram with an MQ above 50 (without exogenous compounds). The differences with other variables based on percentages were irrelevant. An excess of compounds, such as alcohols, terpenes, acids, or others in fruit samples was typical of some outliers.

Author Contributions: J.P.F.-T. conceived, designed, and performed the experiments; M.Z. contributed to the materials analysis tools; M.Z. and J.P.F.-T. integrated the chromatograms; M.C.B. statistically analyzed the data; and J.P.F.-T. and M.C.B. wrote the paper. All the authors discussed the data and corrected the paper.

Funding: Funds were provided by Fundación Séneca de la Región de Murcia-Spain (grant 11784/PI/09) and UE-FEDER and MINECO in Spain (AGL2010-20858).

Acknowledgments: We are grateful to IRTA-CRAG for the seeds, and to Plácido Varó and co-workers (CIFEA-Torre Pacheco, Consejería de Agricultura, Murcia) for crop management. We thank Laura Llanos, Yineth Piñeros, Paola Jiménez, Libia A. Chaparro, Javier M. Obando-Ulloa, Mohammad-Mahdi Jowkar, Mercedes Gutiérrez, Estanislao Cuadros, and Noelia Dos-Santos for sampling and VOC analysis, and María José Roca (SAIT-UPCT) and Pablo Castillo (Proquimur, Biomaster Group) for GC-MS assistance. Mohamed Zarid is indebted to the UE Erasmus Mundus program for a grant.

Conflicts of Interest: The authors declare no conflict of interest.

References

1. Eduardo, I.; Arús, P.; Monforte, A.J.; Obando, J.; Fernández-Trujillo, J.P.; Martínez, J.A.; Alarcón, A.L.; Alvarez, J.M.; van der Knaap, E. Estimating the genetic architecture of fruit quality traits in melon using a genomic library of near- isogenic lines. *J. Am. Soc. Hortic. Sci.* **2007**, *132*, 80–89.

2. Obando-Ulloa, J.M.; Moreno, E.; García-Mas, J.; Nicolai, B.; Lammertyn, J.; Monforte, A.J.; Fernández-Trujillo, J.P. Climacteric or non-climacteric behavior in melon fruit 1. Aroma volatiles. *Postharvest Biol. Technol.* **2008**, *49*, 27–37. [CrossRef]

3. Obando-Ulloa, J.M.; Nicolai, B.; Lammertyn, J.; Bueso, M.C.; Monforte, A.J.; Fernández-Trujillo, J.P. Aroma volatiles associated with the senescence of climacteric or non-climacteric melon fruit. *Postharvest Biol. Technol.* **2009**, *52*, 146–155. [CrossRef]

4. Obando-Ulloa, J.M.; Ruiz, J.; Monforte, A.J.; Fernández-Trujillo, J.P. Aroma profile of a collection of near-isogenic lines of melon. *Food Chem.* **2010**, *118*, 815–822. [CrossRef]

5. Dos-Santos, N.; Bueso, M.C.; Fernández-Trujillo, J.P. Aroma volatiles as biomarkers of textural differences at harvest in non-climacteric near-isogenic lines of melon. *Food Res. Int.* **2013**, *54*, 1801–1812. [CrossRef]

6. Chaparro-Torres, L.A.; Bueso, M.C.; Fernández-Trujillo, J.P. Aroma volatiles at harvest obtained by HSSPME/GC-MS and INDEX/MS-E-nose fingerprint discriminate climacteric behavior in melon fruit. *J. Sci. Food Agric.* **2016**, *96*, 2352–2365. [CrossRef] [PubMed]

7. Canales, I.; Fernández-Trujillo, J.P.; Bueso, M.C.; Zarid, M. Volatile changes in non-climacteric melons with introgression in linkage group X at three stages of maturity. *Acta Hortic. (ISHS)* **2018**, *1194*, 351–356. [CrossRef]

8. Verzera, A.; Dima, G.; Tripodi, G.; Ziino, M.; Lanza, C.M.; Mazzaglia, A. Fast quantitative determination of aroma volatile constituents in melon fruits by headspace solid–phase microextraction and gas chromatography mass spectrometry. *Food Anal. Meth.* **2011**, *4*, 141–149. [CrossRef]

9. Verzera, A.; Dima, G.; Tripodi, G.; Condurso, C.; Crinò, P.; Romano, D.; Mazzaglia, A.; Lanza, C.M.; Restuccia, C.; Paratore, A. Aroma and sensory quality of honeydew melon fruits (*Cucumis melo* L. subsp. melo var. inodorus H. Jacq.) in relation to different rootstocks. *Sci. Hortic.* **2014**, *169*, 118–124. [CrossRef]

10. Grubbs, F. Procedures for detecting outlying observations in samples. *Technometrics* **1969**, *11*, 1–21. [CrossRef]

11. Aggarwal, C.C.; Yu, P.S. Outlier detection for high dimensional data. *ACM Sigmoid Rec.* **2001**, *30*, 37–46. [CrossRef]

12. Ro, K.; Zou, C.; Wang, Z.; Yin, G. Outlier detection for high dimensional data. *Biometrika* **2015**, *102*, 589–599. [CrossRef]

13. Fernández-Trujillo, J.P.; Obando-Ulloa, J.M.; Monforte, A.J.; Sanmartín, P.; Kessler, M.; Bueso, M.C. Métodos estadísticos multivariantes aplicables a estudios de calidad postcosecha del fruto de melón. In Proceedings of the Fifth Virtual Iberoamerican Congress on Quality Management in Laboratories, Congress V IBEROLAB, Madrid, Spain, 5 April 2009; Available online: http://www.iberolab.org/opencms/export/sites/IberolabV/comunicaciones/Comunicaciones/documentos_comunicaciones/Requisitos_Tecnicos_0003.pdf (accessed on 28 March 2018).

14. Tukey, J.W. *Exploratory Data Analysis*; Addison-Wesley Pub. Co.: Boston, MA, USA, 1977; ISBN 0201076160.

15. Tijskens, L.M.M.; Dos-Santos, N.; Jowkar, M.M.; Obando, J.; Moreno, E.; Schouten, R.E.; Monforte, A.J.; Fernández Trujillo, J.P. Postharvest firmness behaviour of near-isogenic lines of melon. *Postharvest Biol. Technol.* **2009**, *51*, 320–326. [CrossRef]

16. Amaro, A.L.; Fundo, J.F.; Oliveira, A.; Beaulieu, J.C.; Fernández-Trujillo, J.P.; Almeida, D.P.F. 1-Methylcyclopropene effects on temporal changes of aroma volatiles and phytochemicals of fresh-cut cantaloupe. *J. Sci. Food Agric.* **2013**, *93*, 828–837. [CrossRef] [PubMed]

17. Agilent Technologies, Retention Time Locking with the MSD Productivity ChemStation. Technical Overview 2008. Available online: http://www.chem.agilent.com/Library/technicaloverviews/Public/5989-8574EN.pdf (accessed on 28 March 2018).

18. Van den Dool, H.; Kratz, P.D. A generalization of the retention index system including linear temperature programmed gas–liquid partition chromatography. *J. Chromatogr. A* **1963**, *11*, 463–471. [CrossRef]

19. National Institute of Standards and Technology. U.S. Department of Commerce. NIST Chemistry WebBook, SRD 69. 2017. Available online: http://webbook.nist.gov/chemistry/cas-ser/ (accessed on 28 March 2018).

20. Fernández-Trujillo, J.P.; Dos-Santos, N.; Martínez-Alcaraz, R.; Le Bleis, I. Non-destructive assessment of aroma volatiles from a climacteric near-isogenic line of melon obtained by headspace sorptive bar extraction. *Foods* **2013**, *2*, 401–414. [CrossRef] [PubMed]

21. Chemindustry. Chemical Information Search. Available online: http://www.chemindustry.com/apps/chemicals (accessed on 28 March 2018).

22. The Good Scent Co. Available online: http://www.thegoodscentscompany.com/search2.html (accessed on 28 March 2018).

23. Shapiro, S.S.; Wilk, M.B. An analysis of variance test for complete samples. *Biometrika* **1965**, *52*, 591–611. [CrossRef]

24. Le, S.; Josse, J.; Husson, F. FactoMineR: An R package for multivariate analysis. *J. Stat. Softw.* **2008**, *25*, 1–18. [CrossRef]

25. Kuhn, M.; Wing, J.; Weston, S.; Williams, A.; Keefer, C.; Engelhardt, A.; Cooper, T.; Mayer, Z.; Kenkel, B.; The R Core Team; et al. Package 'Caret'. Classification and Regression Training. R Package Version 6.0–78. 2016. Available online: http://CRAN.R-project.org/package=caret (accessed on 28 March 2018).

26. Filzmoser, P.; Gschwandtner, M. Mvoutlier: Multivariate Outlier Detection Based on Robust Methods. R Package Version 2.0.8. 2017. Available online: https://CRAN.R-project.org/package=mvoutlier (accessed on 29 March 2018).

27. Komsta, L. Outliers: Tests for Outliers. R Package Version 0.14. 2011. Available online: https://CRAN.R-project.org/package=outliers (accessed on 28 March 2018).

28. Moing, A.; Aharoni, A.; Biais, B.; Rogachev, I.; Meir, S.; Brodsky, L.; Allwood, J.W.; Erban, A.; Dunn, W.B.; Kay, L.; et al. Extensive metabolic cross-talk in melon fruit revealed by spatial and developmental combinatorial metabolomics. *New Phytol.* **2011**, *190*, 683–696. [CrossRef] [PubMed]

29. Nattaporn, W.; Pranee, A. Effect of pectinase on volatile and functional bioactive compounds in the flesh and placenta of 'Sunlady' cantaloupe. *Intl. Food Res. J.* **2011**, *18*, 819–827.

30. Shalit, M.; Katzir, N.; Tadmor, Y.; Larkov, O.; Burger, Y.; Shalekhet, F.; Lastochkin, E.; Ravid, U.; Amar, O.; Edelstein, M.; et al. Acetyl-CoA: Alcohol acetyltransferase activity and aroma formation in ripening melon fruits. *J. Agric. Food Chem.* **2001**, *49*, 794–799. [CrossRef] [PubMed]

31. Gonda, I.; Bar, E.; Portnoy, V.; Lev, S.; Burger, J.; Schaffer, A.A.; Tadmor, Y.; Gepstein, S.; Giovannoni, J.; Katzir, N.; et al. Branched-chain and aromatic amino acid catabolism into aroma volatiles in *Cucumis melo* L. fruit. *J. Exp. Bot.* **2010**, *61*, 1111–1123. [CrossRef] [PubMed]

32. Portnoy, V.; Benyamini, Y.; Bar, E.; Harel-Beja, R.; Gepstein, S.; Giovannoni, J.J.; Schaffer, A.A.; Burger, J.; Tadmor, Y.; Lewinsohn, E.; et al. The molecular and biochemical basis for varietal variation in sesquiterpene content in melon (*Cucumis melo* L.) rinds. *Plant Mol. Biol.* **2008**, *66*, 647–661. [CrossRef] [PubMed]

33. Condurso, C.; Verzera, A.; Dima, G.; Tripodi, G.; Crinò, P.; Paratore, A.; Romano, D. Effects of different rootstocks on aroma volatile compounds and carotenoid content of melon fruits. *Sci. Hortic.* **2012**, *148*, 9–16. [CrossRef]

34. Fernández-Trujillo, J.P.; Fernández-Talavera, M.; Ruiz-León, M.T.; Roca, M.J.; Dos-Santos, N. Aroma volatiles during whole melon ripening in a climacteric near-isogenic line and its inbred non-climacteric parents. *Acta Hortic. (ISHS)* **2012**, *934*, 951–958. [CrossRef]

35. Barba, C.; Thomas-Danguin, T.; Guichard, E. Comparison of stir bar sorptive extraction in the liquid and vapour phases, solvent-assisted flavour evaporation and headspace solid-phase microextraction for the (non)-targeted analysis of volatiles in fruit juice. *LWT-Food Sci. Technol.* **2017**, *85*(Part B), 334–344. [CrossRef]

36. Biais, B.; Allwood, J.W.; Deborde, C.; Xu, Y.; Maucourt, M.; Beauvoit, B.; Dunn, W.B.; Jacob, D.; Goodacre, R.; Rolin, D.; et al. 1H NMR, GC-EI-TOFMS, and data set correlation for fruit metabolomics: Application to spatial metabolite analysis in melon. *Anal. Chem.* **2009**, *81*, 2884–2894. [CrossRef] [PubMed]

37. Escudero, A.A.; Zarid, M.; Bueso, M.C.; Fernández-Trujillo, J.P. Aroma volatiles during non-climacteric melon ripening and potential association with flesh firmness. *Acta Hortic. (ISHS)* **2018**, *1194*, 363–366. [CrossRef]

38. Gonda, I.; Lev, S.; Bar, E.; Sikron, N.; Portnoy, V.; Davidovich-Rikanati, R.; Burger, J.; Schaffer, A.A.; Tadmor, Y.; Giovannonni, J.J.; et al. Catabolism of L–methionine in the formation of sulfur and other volatiles in melon (*Cucumis melo* L.) fruit. *Plant J.* **2013**, *74*, 458–472. [CrossRef] [PubMed]

39. Esteras, C.; Rambla, J.L.; Sánchez, G.; López-Gresa, M.P.; González-Mas, M.C.; Fernández-Trujillo, J.P.; Bellés, J.M.; Granell, A.; Picó, M.B. Fruit flesh volatile and carotenoid profile analysis within the *Cucumis melo* L. species reveals unexploited variability for future genetic breeding. *J. Sci. Food Agric.* **2018**, *98*, in press. [CrossRef] [PubMed]

40. Oh, S.; Lim, B.S.; Hong, S.J.; Lee, S.K. Aroma volatile changes of netted muskmelon (*Cucumis melo* L.) fruit during developmental stages. *Hortic. Environ. Biotechnol.* **2011**, *52*, 590–595. [CrossRef]

41. Fredes, A.; Sales, C.; Barreda, M.; Valcárcel, M.; Roselló, S.; Beltrán, J. Quantification of prominent volatile compounds responsible for muskmelon and watermelon aroma by purge and trap extraction followed by gas chromatography-mass spectrometry determination. *Food Chem.* **2016**, *190*, 689–700. [CrossRef] [PubMed]

## 7

Optimization and Application of a GC-MS Method for the Determination of Endocrine Disruptor Compounds in Natural Water

José Gustavo Ronderos-Lara [1], Hugo Saldarriaga-Noreña [1,*] (iD), Mario Alfonso Murillo-Tovar [2] (iD) and Josefina Vergara-Sánchez [3] (iD)

1   Centro de Investigaciones Químicas, Instituto de Investigación en Ciencias Básicas y Aplicadas, Universidad Autónoma del Estado de Morelos, Av. Universidad 1001, 62209 Cuernavaca, Mexico; ronderos92@gmail.com

2   Cátedras, Consejo Nacional de Ciencia y Tecnología, Av. Insurgentes Sur 1582, Colonia Crédito Constructor, Del. Benito Juárez, 03940 Ciudad de México, Mexico; mario.murillo@uaem.mx

3   Laboratorio de Análisis y Sustentabilidad Ambiental, Escuela de Estudios Superiores de Xalostoc, Universidad Autónoma del Estado de Morelos, 62715 Ayala, Morelos, Mexico; vergara@uaem.mx

*   Correspondence: hsaldarriaga@uaem.mx

Abstract: Bisphenol A (BPA), 4-nonylphenol (4NP), estradiol (E_2), and ethinylestradiol (EE_2) are considered as endocrine disruptors or mutagens. These compounds are commonly called endocrine disrupter chemicals (EDCs). BPA and 4NP are widely used as plastic additives, lacquers, resins, or surfactants, while E_2 is one of the predominant female sex hormones during the reproductive years, and EE_2 is an estrogen derived from estradiol, used in the production of contraceptive pills. All of these can be usually found in wastewater. In Mexico, it is common for water from rivers, lakes, and canyons to be reused for different purposes. Unfortunately, there is little information on the concentration of many of the pollutants present in such bodies of water. To determine the presence of these compounds in samples of wastewater in the Apatlaco River, an accurate and reproducible method was developed by coupling gas chromatography to mass spectrometry (GC-MS). A solid-phase extraction with Chromabond RP-18 cartridges was carried out, and the elution was performed with an acetone/methanol mixture. After isolation, the solvent was removed and a silylation step was carried out using N,O-bis(trimethylsilyl)trifluoroacetamide (BSTFA). Recoveries for spiked samples were between 71.8% and 111.0%. The instrumental limits of detection (IDL) ranged between 24.7 and 37.0 ng mL^{-1}. In total, 16 samples were taken in 2015 at the microbasin of the Apatlaco River, located in the state of Morelos. The maximum concentrations found were 4NP (85.5 ng mL^{-1}), BPA (174.6 ng mL^{-1}), E_2 103.6 (ng mL^{-1}), and EE_2 (624.3 ng mL^{-1}).

Keywords: endocrine disruptors; surface water; environmental risk; GC-MS

1. Introduction

The presence of endocrine disrupter chemicals (EDCs) in water is harmful to the development of biota and to human health. Once in the environment, these compounds can be transported by the aquatic current, deposited on the bottom of bodies of surface water, stored in sediments, and/or bioaccumulate [1]. In Mexico, this situation constitutes a high risk, since rivers and lakes are frequently used in the irrigation of crops, fish farming, and recreational activities. Also, there is a lack of information on the content of EDCs in surface water, in part because most of the studies have focused mainly on the determination of heavy metals [2–5].

In the state of Morelos, there are several bodies of surface water which have suffered significant deterioration because some of them are the destination of wastewater discharges generated from daily activities. The sector of agriculture in Morelos uses approximately 800 thousand m^3/year of surface water, while altogether, domestic, urban, industrial, and recreational activities use approximately 60 thousand m^3/year of surface water [6]. A specific case is the microbasin of the Apatlaco River, which emerges as a channel in the Chapultepec spring in Cuernavaca, Morelos, Mexico. It flows into the Yautepec River, which in turn, flows into the Amacuzac River, and discharges its waters into the Balsas River, to end its path in the Pacific Ocean. This body of water is important for the development of daily life in some places in the state of Morelos, and its waters are used mainly to feed agricultural irrigation systems, contributing approximately 186 thousand m^3/year [5], in addition to having a great diversity of aquatic species.

The basin is located northwest of the state of Morelos and covers approximately 746 km^2 of the total surface of state of Morelos [7]. The formation of the Apatlaco River is mainly due to the runoff of water that flows through the soil. This river presents a decline that goes from 3690 to 880 m above sea level [6]. During this path, it receives the direct discharge of residual waters that come from diverse activities, residential, agricultural, industrial, and hospital zones.

For this reason, this study aims to establish environmental levels of four endocrine disrupting compounds (17α-ethinylestradiol, 17β-estradiol, 4-nonylphenol, and bisphenol-A), all them considered markers of anthropic activities, using gas chromatography and mass spectrometry (GC-MS). Also, the health risk derived from the presence of these substances in the surface water that flows through the microbasin of the Apatlaco River will be evaluated.

2. Material and Methods

2.1. Reagents and Materials

All used reagents were analytical grade. 4-Nonylphenol ($\geq$99.8%), 17α-ethynilestradiol ($\geq$98.0%), and 17β-estradiol ($\geq$98.0%) were obtained from Sigma-Aldrich (St. Louis, MO, USA) and Bisphenol A (99.0%) from Supelco (St. Louis, MO, USA). Meanwhile N,O-bis(trimethylsilyl)-trifluoroacetamide + trimethyl-chlorosilane, HPLC water, and pyridine ($\geq$99%) were obtained from Sigma Aldrich. Acetone (99.9%) was purchased from Meyer (México City, México) and methanol (99.7%) from Fermont (Monterrey, N.L., México). The Millipore nylon filters from Pall Corporation (Ann Arbor, MI, USA) and solid phase extraction cartridges C_{18} (500 mg/6 mL) were bought from Chromabond (Düren, Germany).

2.2. Standard Preparation

Individually, approximately 1 mg of standard was dissolved in acetone. Then, a series of dilutions was made to obtain a calibration curve with five different concentration levels (5, 25, 125, 250, and 500 ng mL^{-1}). To evaluate extraction efficiency, repeatability, and reproducibility, a synthetic sample was prepared (80 ng mL^{-1}).

2.3. Extraction and Elution

A synthetic sample of 500 mL was extracted in C_{18} cartridges previously conditioned using 6 mL of methanol/acetone (3:2), followed by 6 mL of methanol, and finally, 6 mL of reagent water. Spiked samples were passed through the cartridge applying vacuum at a flow rate of 6.0 mL min^{-1}. The elution of retained compounds was performed using 10 mL of methanol/acetone (3:2). Eluates were reduced in a rotary evaporator until the volume was approximately 1 mL, and the remaining solvent was completely eliminated by applying a soft stream of nitrogen (99.99%).

2.4. Derivatization

The dry extract was resuspended in 50 μL of pyridine and 50 μL of derivatizing agent BSTFA + TMCS (99:1). Then, to complete the reaction of derivatization, vials were submerged in a water bath (60 °C) for 60 min.

2.5. Gas Chromatography in Tandem with Mass Spectrometry

The derivatized samples were analyzed in Agilent Technologies gas chromatograph (GC), model 6890 coupled to a mass spectrometer (MS) 5973. The separation of compounds was performed in an HP-5MS 30 m × 0.25 mm capillary column, and 0.25 μm film thickness (Agilent, Santa Clara, CA, USA). Carrier gas was helium (99.998%) at a flow rate of 1 mL min^{-1}. The injector temperature was 300 °C in splitless mode using an injection volume of 1.0 μL. The oven program started at 150 °C and was maintained for 2 min, then the temperature was incremented at 15 °C per minute to reach 250 °C, then immediately incremented at 5 °C per minute until it reached 280 °C, where it was held for 15 min. The MS was operated in electron impact mode (70 eV), holding the temperatures of the ion source and quadrupole filter at 230 °C and 150 °C, respectively (Table 1).

Table 1. Monitored ions (m/z) in mass spectrometry for each of the trimethylsilyl derivated (TMS) compounds of interest.

Compound	Molecular Weight (g/mol)	Trimethylsilyl Derivated Compound	Molecular Weight (g/mol)	Ion Quantitation	Ion Confirmation
4NP	220.35	TMS-4NP	292.54	292	207, 277
BPA	228.29	TMS-BPA	372.65	357	372, 207, 73
E$_2$	272.38	TMS-E$_2$	416.75	416	285, 232, 129
EE$_2$	296.40	TMS-EE$_2$	440.77	425	440, 300, 285

2.6. Application on Natural Water Samples

The samples were collected according the to the United States Environment Protection Agency method 1698 [8], briefly, the samples were collected in amber glass bottles, previously conditioned and maintained at 4 °C while transporting to a laboratory. Before extraction, samples were filtered through a nylon filter (0.45 μm). In total, 16 samples were collected during November and December 2015 (after the rainy season) in the basin of Apatlaco in three different municipalities: Cuernavaca, Temixco, and Jiutepec (Figure 1).

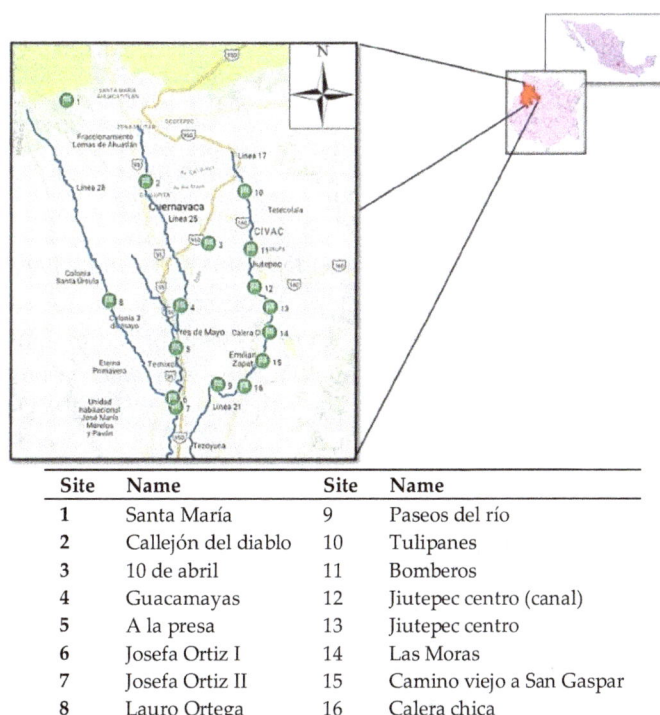

Site	Name	Site	Name
1	Santa María	9	Paseos del río
2	Callejón del diablo	10	Tulipanes
3	10 de abril	11	Bomberos
4	Guacamayas	12	Jiutepec centro (canal)
5	A la presa	13	Jiutepec centro
6	Josefa Ortiz I	14	Las Moras
7	Josefa Ortiz II	15	Camino viejo a San Gaspar
8	Lauro Ortega	16	Calera chica

Figure 1. Location of the sampling sites.

3. Results and Discussion

3.1. Optimization of the Analytical Conditions

Selectivity was determined by the complete separation of the four compounds by gas chromatography, and the complete absence of interference signals. Linearity was calculated by means of the correlation coefficient (r) obtained from the calibration curve. Extraction efficiency was evaluated by passing 500 mL of synthetic samples (80 ng/mL) through a C_{18} cartridge. Repeatability was evaluated analyzing synthetic samples in duplicate, and the same day and reproducibility was evaluated analyzing synthetic samples in different days. Instrumental limits of detection (IDL) were determined according to Miller & Miler (2010), using the concentration that provides a signal that is equal to the signal corresponding to the blank ($Y_B = S_{y/x}$), plus three times the standard deviation of the blank ($S_B = a$) [9].

$$IDL = Y_B + 3^*S_B$$

$Y_B = S_{y/x}$: random error in the direction of "y"
$S_B = a$: intercept.

Standard sample blanks were injected to ensure absence of any impurities, which affect the selectivity. The retention times obtained were 9.7, 11.6, 15.8, and 16.9 min for 4NP, BPA, E_2, and EE_2, respectively (Figure 2). Linearity was determined using the correlation coefficient (r) calculated for each calibration graph, 4NP (0.9927), BPA (0.9904), E_2 (0.9917), and EE_2 (0.9800).

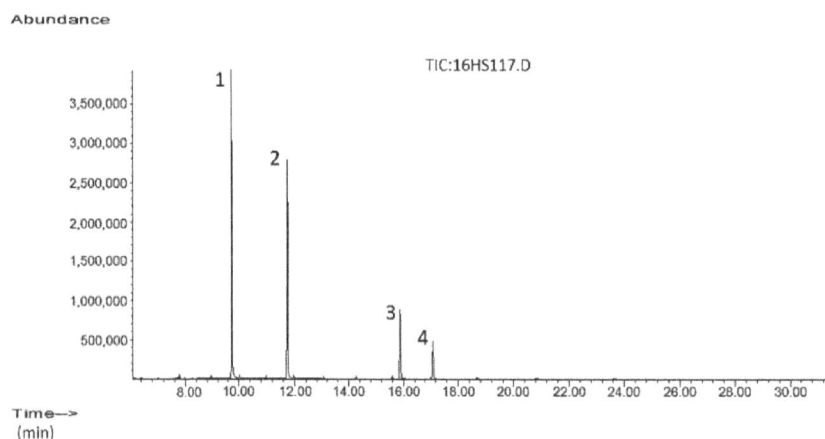

Figure 2. Retention times of endocrine disrupter chemicals (EDCs). 1: TMS-4NP, 2: TMS-BPA, 3: TMS-E_2, 4: TMS-EE_2.

Recovery percentages ranged between 78.3 (for EE_2) and 111.0% (for BPA). Precision, in terms of repeatability and reproducibility, indicated satisfactory results considering relative standard deviation, which in all cases was below 10%. This agrees with terms established by United States Protection Agency (USEPA) for environment samples that consider a maximum variation of 30% as acceptable. On the other hand, IDL ranges were between 24.7 (for EE_2) and 37.0 (for BPA) (Table 2).

Table 2. Results for evaluation of analytical method.

Compound	T_R	r	Repeatability [3] (% RSD)	Reproducibility [4] (% RSD)	% Recovery [2]	IDL
4NP	9.7	0.9927	8.2	5.2	107.6 ± 22.5	26.6
BPA	11.6	0.9904	0.9	7.2	111.0 ± 6.2	37.0
E_2	15.8	0.9917	3.1	10.5	71.8 ± 17.3	29.0
EE_2	16.9	0.9800	6.3	5.1	78.3 ± 20.5	24.7

T_R: retention time. r: correlation coefficient. IDL: instrumental detection limit (ng/mL). [2], [3] and [4]: number of dates for calculation.

3.2. Application on Natural Water Samples

Table 3 shows the concentration found in natural water samples taken in the Apatlaco river basin, as well as comparisons with respect to concentrations reported in State of Hidalgo and Xochimilco in Mexico City. In five sites, no EDCs were detected (Santa María, Callejón del diablo, Lauro Ortega, Tulipanes, and Camino Viejo a San Gaspar). In Santa María and Lauro Ortega sites, the absence of EDCs is probably because it is an uninhabited place, meanwhile, in Callejón del Diablo, Tulipanes, and Camino Viejo a San Gaspar, probably the high volume of rainfall that occurred prior to sampling could have affected the dilution of these compounds (rainfall average 37.6 mm/m^2) [10]. Only in one site studied were EDCs detected (a la Presa), probably due to house room density near the river.

Table 3. EDC concentrations detected in surface water samples and comparisons with respect to other studies carried out in Hidalgo State, México, and México City (ng mL^{-1}).

Site		4NP	BPA	E$_2$	EE$_2$	Activity
	Morelos State					
1	Santa María	ND	ND	ND	ND	Urban area
2	Callejón del Diablo	ND	ND	ND	ND	Urban area
3	10 de abril	ND	ND	ND	624.3 ± 19.9	Urban area
4	Guacamayas	ND	ND	70.1 ± 10.7	31.2 ± 9.4	Urban area
5	A la presa	85.5 ± 11.6	88.8 ± 6.2	103.6 ± 11.1	91.5 ± 9.2	Urban area
6	Josefa Ortiz I	ND	39.1 ± 6.2	39.1 ± 6.2	181.9 ± 9.7	Urban area
7	Josefa Ortiz II	ND	ND	ND	231.7 ± 10.4	Urban area
8	Lauro Ortega	ND	ND	ND	ND	Natural water
9	Paseos del río	ND	43.3 ± 6.2	37.3 ± 10.6	126.3 ± 9.3	Commercial area
10	Tulipanes	ND	ND	ND	ND	Commercial area
11	Bomberos	ND	ND	ND	159.0 ± 9.5	Urban area
12	Jiutepec centro (canal)	ND	ND	ND	138.5 ± 9.2	Urban area
13	Jiutepec	ND	ND	ND	147.9 ± 9.4	Urban area
14	Las Moras	ND	ND	ND	91.4 ± 9.2	Urban area
15	Camino viejo a San Gaspar	ND	ND	ND	ND	Urban area
16	Calera chica	ND	174.6 ± 6.2	ND	ND	Urban area
	Hidalgo State [11]					
	Residual water	16.7 ± 2.2	2.50 ± 0.4	0.022 ± 0.0	ND	Farming
	Spring water	ND	ND	ND	ND	Farming
	Xochimilco channel [12]	ND	ND	ND	ND	
4	Tlicuilli	ND	140,000	ND	ND	Livestock
10	Candelaria	ND	8420–29,350	ND	ND	Urban area
11	Santa Cruz	ND	ND	ND	ND	Urban area
12	Nuevo León	ND	ND	ND	ND	Urban area
13	Caltongo	ND	ND	ND	ND	Urban area
18	La Draga	ND	ND	ND	ND	Effluent
19	San Diego	ND	ND	ND	ND	Effluent
7 and 9	Puente Urrutia y Tlapechicalli	ND	4370–18,032	ND	ND	Farming
1, 3 and 8	Tlilac, el Bordo	ND	15,200–22,370	980–1680	ND	Farming and livestock
	el Humedal	ND	ND	ND	ND	

[11] Gibson 2007; [12] Díaz-Torres et al. 2013; ND: Not detected.

In Mexico, there is a little information about maximum concentrations of EDCs in natural water, and neither a norm that regulate them, and for this reason, in this study, we took into consideration, the USEPA regulation rules for 4NP and BPA, and the European Union regulations for E$_2$ and EE$_2$, with the purpose of making comparisons and to have reference concentrations.

The compound that presented the highest concentration was EE$_2$ (624.3 ng mL^{-1}, 10 de abril site) followed by E$_2$ (103.6 ng mL^{-1}, A la Presa site). The site that presented all compounds was "A la Presa", meanwhile, BPA was detected only in six sites (10 de abril, A la Presa, Josefa Ortiz I, Josefa Ortiz II, Paseos del río and Calera chica) in a range of concentrations between 39.1 and 174.6 ng mL^{-1}. With regard to this, the USEPA has suggested 1.78 ng mL^{-1} as the maximum concentration for BPA in surface water from United States of America [13]. For its part, 4NP was detected less frequently (only in one site, A la Presa site) at 85.6 ng mL^{-1}, and this concentration exceeds the concentration established by USEPA in surface water (6.6 ng mL^{-1}) [14].

E$_2$ it was detected only in three of the 16 analyzed samples (Guacamayas, A la Presa and Paseos del río). The concentration levels were in a range from 37.3 to 103.6 ng mL^{-1}. These concentrations were at least three magnitudes of order greater than the concentration 0.002 ng mL^{-1} suggested by EU for surface water [15].

The compound that presented the highest concentrations and frequency was EE$_2$, which was detected in 10 of 16 analyzed samples. The interval of concentrations was between 31.2 and 624.3 ng m $^{-1}$, and these values exceed, extremely, the concentration recommended by European Union (0.0001 ng mL^{-1}) [16].

The obtained results reveal that levels of EDCs in the studied sites in Morelos State, México, are higher than levels of EDCs found in other places in the republic, such as Hidalgo State, México, where the presence of 4NP, E$_2$, and EE$_2$ were reported in surface water used specifically for crop irrigation [11]. Meanwhile, the levels of BPA and E$_2$ reported in surface water from Xochimilco Lake in México City [12], are higher than the levels of EDCs observed in this study (Table 3).

3.3. Evaluation of Health Risk by EDC Exposure

To predict the healthy adverse effects of exposition to 4NP, BPA, E$_2$, and EE$_2$ present in analyzed water, it a risk evaluation was carried out in accordance with USEPA method [17]. Considering that exposure sensibility to EDC in adult and young dwellers is different, for this reason, two different scenarios were considered, one for children and the other one for adults. In both cases, the exposition frequency was 365 days. Reference body weight for adult was 70 kg, and average EDC contact was 2 L per day, meanwhile, in children, the body weight considered was 10 kg and average EDC contact was 1 L per day.

Table 4 shows the health risk values calculated for exposition to EDC in the studied sites. The site where children and adult dwellers are at major risk for contact with EDC, is "10 de abril site". It is important to highlight that this exposition is only due to synthetic hormone EE$_2$. Also, it is important to consider the shortage of information related with health risk exposure in Mexico, thus, it is relevant to create an historical record of EDC contained in surface waters, as well as to try to infer the probable health risk exposure. Although the site 10 de abril shows higher levels of health risk for exposition than all the other places, equally, it represents a risk for the exposed population.

Table 4. Health risk rate for exposition to EDCs.

Site	Compound	Concentration (ng/mL)	Exposition Rate in Adults (mg/kg*day)	Exposition Rate in Children (mg/kg*day)
10 de abril	EE$_2$	624	6.5	22.8
Guacamayas	E$_2$	70.1	0.7	2.6
	EE$_2$	31.2	0.3	1.1
A la presa	NP	85.5	0.9	3.1
	BPA	88.8	0.9	3.2
	E$_2$	104	1.1	3.8
	EE$_2$	91.5	1.0	3.3
Josefa Ortiz I	BPA	39.1	0.4	1.4
	EE$_2$	182	1.9	6.6
Josefa Ortiz II	EE$_2$	232	2.4	8.5
Paseos del río	BPA	43.3	0.5	1.6
	E$_2$	37.7	0.4	1.4
	EE$_2$	126	1.3	4.6
Bomberos	EE$_2$	159	1.7	5.8
Jiutepec centro (canal)	EE$_2$	138	1.4	5.1
Jiutepec centro	BPA	8.72	0.9	3.2
	EE$_2$	148	1.5	5.4
Las Moras	BPA	40.3	0.4	1.5
	EE$_2$	91.4	1.0	3.3
Calera Chica	BPA	175	1.8	6.4

Exposition frequency = 365 days/year; Average body weight for adults = 70 kg and 10 kg for children; Contact rate for adults = 2 L/day and 1 L/day for children.

4. Conclusions

The optimized methodology allowed analysis of NP, BPA, E_2 and EE_2 simultaneously by GC-MS. The selectivity of the method was verified by the injection of the mixture of the four compounds of interest, and the method presented good linearity and acceptable percentages of recovery according to the criteria of the EPA.

The analysis of the real samples indicates that the levels of EDCs in the bodies of natural water studied in the state of Morelos exceed the levels proposed by the USEPA and the EU, which constitutes a risk to the health of the exposed population. It is worth mentioning that this is one of the first studies carried out in this part of the republic, so it is recommended to carry out a wider diagnosis in various sources of natural water in the state.

Although, at present, there is no standard in Mexico that establishes maximum permissible levels of exposure to EDCs, it is important to calculate the exposure rates, so that the authorities responsible for creating environmental and health protection policies can use this type of information for their implementation.

Author Contributions: All authors participated in the same proportion.

Funding: This research received no external funding.

Acknowledgments: The authors would like to express their appreciation to Tec. María Gregoria Medina Pintor and M. Sc. Mònica Ivonne Arias Montoya for their support in the chromatographic analysis.

Conflicts of Interest: The authors declare no conflict of interest.

References

1. Campbell, C.G.; Borglin, S.E.; Green, F.B.; Grayson, A.; Wozei, E.; Stringfellow, W.T. Biologically directed environmental monitoring, fate, and transport of estrogenic endocrine disrupting compounds in water: A review. *Chemosphere* **2006**, *65*, 1265–1280. [CrossRef] [PubMed]

2. Azpilcueta, M.E.; Pedroza, A.; Sanchez, I.; Salced, M.; Del, R.; Trejo, R. Calidad química de agua en un área agrícola de maíz forrajero (*Zea mays* L.) en la Comarca Lagunera, México. *Rev. Int. Contam. Ambient.* **2017**, *33*, 75–83. [CrossRef]

3. Chacón, K.O.; Pinedo, C.; Rentería, M. Evaluación de elementos traza en agua de río y manantial del área minera de Ocampo, Chihuahua, México. *Rev. Int. Contam. Ambient.* **2016**, *32*, 375–384. [CrossRef]

4. Guzmán-Colis, G.; Thalasso, F.; Ramírez-López, E.M.; Rodríguez-Narciso, S.; Guerrero-Barrera, A.L.; Avelar-González, F.J. Evaluación espacio-temporal de la calidad del agua del río San Pedro en el Estado de Aguascalientes, México. *Rev. Int. Contam. Ambient.* **2011**, *27*, 115–127.

5. Mancilla-Villa, Ó.R.; Ortega-Escobar, H.M.; Ramírez-Ayala, C.; Ramos-Bello, R.; Reyes-Ortigoza, A.L. Metale pesados totales y arsénico en el agua para riego de Puebla y Veracruz, México. *Rev. Int. Contam. Ambient.* **2012**, *28*, 39–48.

6. CEAGUA. Programa Estatal Hídrico de Morelos 2014–2018. Available online: http://marcojuridico.morelos.gob.mx/archivos/reglamentos_estatales/pdf/VPHIDRICOMO.pdf (accessed on 16 January 2018).

7. CONAGUA. La Cuenca del río Apatlaco. Recuperemos el Patrimonio Ambiental de los Morelenses 2008. Available online: http://centro.paot.org.mx/documentos/semarnat/cuenca_rio_apatlaco.pdf (accessed on 11 January 2017).

8. EPA. *Method 1698: Steroids and Hormones in Water, soil, Sediment, and Biosolids by HRGC/HRMS*; EPA Method; EPA: Washington, DC, USA, 2007; pp. 1–69.

9. Miller, J.N.; Miller, J.C. *Statistics and Chemometrics for Analytical Chemistry*, 6th ed.; Pearson Education: Harlow, UK, 2010; pp. 124–126. ISBN 978-0-273-73042-2.

10. INEGI. Anuario Estadístico y Geográfico de Morelos 2016. Available online: http://internet.contenidos.inegi.org.mx/contenidos/Productos/prod_serv/contenidos/espanol/bvinegi/productos/nueva_estruc/anuarios_2016/702825084349.pdf (accessed on 25 January 2018).

11. Gibson, R. Determination of acidic pharmaceuticals and potential endocrine disrupting compounds in wastewaters and spring waters by selective elution and analysis by gas chromatography-mass spectrometry. *J. Chromatogr. A* **2007**, *1169*, 31–39. [CrossRef] [PubMed]

12. Díaz Torres, E.; Gibson, R.; González Farías, F.; Zarco Arista, A.E.; Mazari Hiriart, M. Endocrine disruptors in the Xochimilco Wetland, Mexico City. *Water Air Soil Pollut.* **2013**, *224*, 1–11. [CrossRef]

13. EPA. EPA-HQ-OPPT-2010-0812-0001 (Testing of Bisphenol A). Revised 20 April 2017. Available online: https://www.regulations.gov/document?D=EPA-HQ-OPPT-2010-0812-0001 (accessed on 2 February 2018).

14. EPA. EPA-822-R-05-005: Aquatic Life Ambient Water Quality Criteria Nonylphenol. Available online: www.epa.gov/waterscience/criteria/aqlife.html (accessed on 2 February 2018).

15. Scientific Committee on Health and Environmental Risks (SCHER). *Opinion on Draft Environmental Quality Standards under the Water Framework Directive—17β-estradiol*; European Commission, DG Health & Consumers, Directorate C: Public Health and Risk Assessment, Unit C7—Risk Assessment: Bruselas, Belgium, 2011. Available online: https://ec.europa.eu/health/scientific_committees/environmental_risks/docs/scher_o_131.pdf (accessed on 2 February 2018). [CrossRef]

16. Scientific Committee on Health and Environmental Risks (SCHER). In *Opinion on Draft Environmental Quality Standards under the Water Framework Directive—Ethinylestradiol*; European Commission, DG Health & Consumers, Directorate C: Public Health and Risk Assessment, Unit C7—Risk Assessment: Bruselas, Belgium, 2011. Available online: https://ec.europa.eu/health/scientific_committees/environmental_risks/docs/scher_o_146.pdf (accessed on 2 February 2018). [CrossRef]

17. EPA. Ecological Risk Assessment Guidance for Superfund: Process for Designing and Conducting Ecological Risk Assessments, 239. Available online: https://www.epa.gov/risk/ecological-risk-assessment-guidance-superfund-process-designing-and-conducting-ecological-risk (accessed on 2 February 2018).

## 8

NSAIDs Determination in Human Serum by GC-MS

Adamantios Krokos [1,2] (iD), **Elisavet Tsakelidou** [2], **Eleni Michopoulou** [1], **Nikolaos Raikos** [2], **Georgios Theodoridis** [1] (iD) **and Helen Gika** [2,*] (iD)

[1]   Laboratory of Analytical Chemistry, Department of Chemistry, Aristotle University of Thessaloniki, Thessaloniki 54124, Greece; akrokosa@chem.auth.gr (A.K.); elena.michopoulou@gmail.com (E.M.); gtheodor@chem.auth.gr (G.T.)

[2]   Laboratory of Forensic Medicine & Toxicology, Department of Medicine, Aristotle University of Thessaloniki, Thessaloniki 54124, Greece; tsakeliz@hotmail.com (E.T.); raikos@med.auth.gr (N.R.)

*   Correspondence: gkikae@auth.gr

Abstract: Non-steroidal anti-inflammatory drugs (NSAIDs) are being widely consumed without medical prescription and are often the cause of intoxication, usually in young children. For this, there is a special need in their determination in routine toxicology analysis. As screening methods mainly focus on drugs of abuse (DOA) that are alkaline compounds in their majority, they are not optimized for acidic drugs, such as NSAIDs. Thus, more specific methods are needed for the detection and quantification of this class of drugs. In this study, the efficient extraction of NSAIDs from blood serum and their accurate determination is studied. Optimum pH extraction conditions were studied and thereafter different derivatization procedures for their detection. From the derivatization reagents used, N,O-Bis(trimethylsilyl)trifluoroacetamide (BSTFA) with 1% Trimethylchlorosilane (TMCS) was found to be the optimum choice for the majority of the examined NSAIDs; pH of 3.7 was selected as the most efficient for the extraction step. Herein the formation of the lactam of diclofenac was also thoroughly investigated. The developed Gas Chromatography-Mass Spectrometry (GC-MS) method had a run time of 15 min with the mass spectrometer operating in Electron Impact (EI) within the mass range of 40 to 500 amu. The method was linear with R^2 above 0.991 and limits of quantitation (LOQ) ranging from 6 to 414 ng/mL. The intra-day accuracy and precision were found between 1.03%–9.79% and 88%–110%, respectively, and the inter-day accuracy and precision were between 1.87%–10.79% and 91%–113%. The optimum protocol was successfully applied to real clinical samples, where intoxication of NSAIDs was suspected.

Keywords: NSAIDs; derivatization; GC-MS; serum

---

1. Introduction

Non-steroidal anti-inflammatory drugs (NSAIDs) are acidic compounds with anti-inflammatory properties at high concentrations and several other properties at low concentrations (i.e., salicylic acid is used as anticoagulant drug) [1]. These drugs exhibit toxicity in the upper concentration levels or after a long time of intake [2,3]. Some of the toxic side effects are related with gastrointestinal disorders, intestinal ulceration, aplastic anemia, myocardial infarction, cerebrovascular events, inhibition of platelet aggregation, and renal dysfunction [4]. Especially, acetaminophen and nimesulide exhibit significant hepatotoxicity [2,5–7]. In addition, there are several cases of suicide attempts or crime commissions which are related with NSAIDs [8].

There is a number of analytical methods which report NSAIDs' determination in biological fluids, most of them related to pharmacokinetic studies or to the support of animal studies for the estimation of exposure and the investigation of potential risk after consumption [4,9].

There are reports of determinations of these analytes by immunoassays, Gas Chromatography-Flame Ionization Detection (GC-FID), High Performance Liquid Chromatography-UltraViolet/Diode Array Detection (HPLC-UV/DAD)/fluorimetric detector, spectrofluorimetry, thin layer chromatography-UV/fluorimetric detector, GC-MS, GC-MS/MS, capillary electrophoresis, LC-MS, and LC-MS/MS [10–14]. Immunoassays suffer from low selectivity due to cross-reactions, while all other methods except LC-MS/MS include time-consuming sample pretreatment. There is a good number of published multi-analyte LC-MS/MS methods for measuring NSAIDs and their metabolites that provide sensitive and reliable concentration data from biological matrices [5]. In particular, those methods are the most suitable for low concentrated salicylates originating from nutrition. However, GC-MS still represents an integral tool of a clinical and/or toxicological laboratory due to the fact that mass spectra databases are available aiding in the identification of unknowns, At the same time, latest instruments and materials hold the promise for better chromatographic separations and sensitivity, critically important for complex biological samples. Thus, GC-MS analysis provides advantages clearly important in such applications.

For thedetection of NSAID by GC-MS, various sample pretreatment protocols have been applied. Acidic liquid-liquid extraction is often used, however some researchers have also developed simple SPE protocols instead [3,11]. Furthermore, it has been shown that the selection of a suitable derivatization agent is a key factor for the sensitivity of their detection. Apart from this, there are several issues which are related with the stability of these compounds in the GC-MS conditions.

In this paper, a multi-analyte GC-MS method was developed for the simultaneous quantification of eight NSAIDs, based on a specific pro-analysis procedure. The optimum conditions were selected based on experiments focusing on the sample treatment, to obtain a method able to address the needs of a toxicological analysis for these challenging acidic analytes. The method provides a valuable tool for the determination of eight different NSAIDS by GC-MS, with the pros that the GC-MS technique offers, as stressed above, the additional potential of identification of metabolites, degradation products, or others.

2. Materials and Methods

2.1. Chemicals and Reagents

All solvents were of analytical or LC-MS grade. Acetonitrile (ACN) LC-MS was purchased from Chem Lab NV, (Zedelgem, Belgium). Nordiazepam-d_5 solution 1 mg/mL in methanol reference, material reference standards (>99%) of acetaminophen (APAP), acetyl salicylic acid (ASA), salicylic acid (SA), ibuprofen (IBP), diclofenac (DCF), nimesulide (NI), niflumic acid (NFA), mefenamic acid (MFA), and naproxen (NAP) were purchased from Sigma-Aldrich (Saint Louis, MO, USA). Ethyl acetate (99%) was supplied from Penta (Livingston, NJ, USA). All derivatization reagents; N,O-Bis(trimethylsilyl)trifluoroacetamide with 1% trimethylchlorosilane (BSTFA & 1% TMCS), N-tert-Butyldimethylsilyl-N-methyltrifluoroacetamide (MTBSTFA), Pentafluoropropionic anhydride (PFPA), 2,2,3,3,3-Pentafluoro-1-propanol (PFPOH), Trifluoroacetic anhydride (TFAA), Heptafluorobutyric anhydride (HFBA), were for GC derivatization ≥ 99% grade and were purchased from Sigma-Aldrich (Saint Louis, MO, USA). Serum samples were obtained from 4 cases that arrived in AHEPA University General Hospital and were suspected for intoxication. Drug-free human serum was obtained from healthy donors and before its use it was screened by GC/MS for the presence of the NSAIDs.

2.2. Preparation of the Standard Solutions

Stock solutions of all compounds were prepared in ACN at 10,000 μg/mL. From these, a mix solution containing all nine drugs was prepared and diluted to the following concentrations: 1000 μg/mL, 200 μg/mL, 100 μg/mL, 40 μg/mL, 20 μg/mL, 4 μg/mL, and 2 μg/mL.

2.3. Sample Preparation

For the selection of the optimum pH conditions, 200 µL of spiked human serum at 200 µg/mL was adjusted at pH of 3.7, 4.7, and 5.7 with the addition of 100 µL of HCOOH/HCOONa buffer solutions prepared at these pH values (by mixing appropriate values of 0.2 M solutions of HCOOH and HCOONa). Then 1 mL of ethyl acetate was added and the mixture was shaken for 10 min. After centrifugation for 5 min at $6300 \times g$ the organic phase was collected and dried under gentle nitrogen stream at room temperature. The obtained residue was redissolved either in 50 µL of ethyl acetate or in 50 µL of derivatization reagents, as described in Section 2.5. In any case, 1 µL of the extract was injected in the GC-MS system.

2.4. Derivatization Procedure

For the selection of the optimum derivatization procedure, five different reagents were used: (a) BSTFA with 1% TMCS, (b) MTBSTFA, (c) HFBA, (d) PFPA with PFPOH, and (e)TFAA. The procedure followed for silylation was as follows; addition of 50 µL BSTFA, 1% TMCS, or 50 µL MTBSTFA in the dry residue and after vortexing the mixture was heated for 20 min at 70 °C, finally 1 µL was injected in the system. Acetylation was performed with the addition of either 50 µL of HFBA, or of 30 µL PFPA with 20 µL PFPOH, or of 50 µL of TFAA. The mixture was heated for 20 min at 70 °C and then after cooling down was evaporated and the residue was dissolved in 50 µL of ethyl acetate. From this extract, 1 µL was injected to GC-MS.

2.5. GC/MS Analysis

GC-MS analysis was performed on an Agilent Technologies 7890A GC, equipped with a CTC autosampler and combined with a 5975C inert XL EI/CI MSD with Triple-Axis Detector (Agilent Technologies, Santa Clara, CA, USA). GC separations were performed on a 30 m Agilent J&W HP-5ms UI capillary column, with a film thickness of 0.25 µm and an i.d. of 0.25 µm. Back-flash was performed with a 1.5 m deactivated Agilent column with a film thickness of 0.18 mm. The method had a duration of 15 min with the following temperature program: Initial oven temperature at 120 °C, hold for 1 min, and then increase to 300 °C with a 15 °C/min rate. A back-flash step followed, at 300 °C for 10 min. Injection of 1 µL of sample was made on a PTV injector operating from 200 °C to 320 °C. The mass spectrometer (MS) was operated at electron impact ionization mode (EI, 70 eV) and the mass scan range was from 40 to 500 amu.

3. Results

The majority of the studied drugs are categorized in the acidic class of compounds as they contain carboxyl moieties, except from APAP and NI. Their chemical structures, together with their pKa values, can be seen in Table 1. These are expected to be extracted more efficiently by the organic solvent under acidic conditions, where the ionization of their carboxyl group is suppressed. However, as their physicochemical properties are varying due to their structure and the presence of different substitution groups, the optimum pH for their extraction is needed to be examined.

Table 1. Chemical structures and pKa values of the studied non-steroidal anti-inflammatory drugs (NSAIDs).

Compound	Chemical Structure	pKa
APAP		9.38

Table 1. *Cont.*

Compound	Chemical Structure	pKa
DCF		4.15
IBP		4.91
NI		6.86
NFA		1.88
MFA		4.2
NAP		4.15
ASA		3.49
SA		2.97

3.1. Extraction pH

Under three different pH conditions, extraction of the eight NSAIDs from a serum sample spiked at 200 µg/mL was performed, and the extracts were thereafter analyzed without any derivatization. As expected, detection sensitivity of the underivatized drugs is low, however it was acceptable for comparative purposes. Apart from the eight drugs, two more peaks were considered: (1) The peak which corresponds to salicylic acid, a product occurring from ASA in the sample, and (2) the diclofenac-lactam peak, which is obtained by DCF partial conversion during GC analysis. Table 2 provided the peak areas of the studied drugs at three different pHs tested for their extraction. As can be seen, more acidic pH conditions favor the extraction of the majority of the drugs resulting in higher detected peak areas. NI, however, provides higher signal under pH 4.7. For the case of diclofenac, its lactam is detected in higher proportion under all three pH conditions. Apart from that, the methylester of dichlofenac (DCF-ME) is also detected with relatively high signal. Salicylic acid is detected with higher amount under the more acidic pHs compared to acetyl salicylic acid.

Table 2. Absolute area of each NSAID at different pH extraction conditions.

Compound	Absolute Area (10^6)		
	pH = 3.7	pH = 4.7	pH = 5.7
APAP	282.2	261.5	139.0
DCF-lactam/DCF-ME/DCF	78.3/52/24	56.1/27/35	10.8/3.9/-
NAP	28.2	6.2	0.8
NFA	12.0	3.4	0.5
ASA	4.1	1	0.2
SA	10.1	1.1	0.2
IBP	267.6	158.6	20.4
MFA	77.7	74.6	4.4
NI	123.7	185.6	23.6

3.2. Analysis Without Derivatization

When the extracted NSAIDs were analyzed directly without prior derivatization, the detection sensitivity was not satisfactory for the majority of the eight drugs. The chromatographic peaks were broad with excessive tailing, which was attributed to intramolecular interactions and the interaction of polar carboxylic groups of drugs with the column's supporting material. Detection of such compounds can be challenging as several transformation products of the drugs were also detected. During GC/MS analysis, NFA, MFA, and NAP were not found stable without derivatization, as their decarboxylated compounds ($-CO_2$) were also detected. This has also been reported in previous studies [15]. As an example, together with NAP detected at 9.86 min, its NAP-CO_2 degradation product is also detected at 7.04 min with characteristic ions at m/z 184 and 141. Another issue is that ASA is converted to SA over time in both aqueous and organic solutions, thus it can also be detected due to that reason. This can mislead its determination, due to the fact that SA is always detected in serum after administration as it is the active metabolite of ASA. Because of that, freshly prepared solutions of ASA should always be used, and SA should also be determined when ASA is present in serum.

In addition, for the cases of ASA, SA, MFA, NFA, IBP, NAP, and DCF, their methyl ester products were also detected in the spiked extract.

As it concerns DCF, it was observed that when the analysis was performed directly in the serum extract without prior derivatization, only a small amount of DCF could be detected, whereas the highest amount was detected as diclofenac lactam. This was firstly reported by El Haj et al. in 1999, who studied the methanolic solutions of DCF [15]. The authors attributed the formation of lactam to the high temperatures applied in the inlet during the GC-MS analysis. Here, in order to investigate this phenomenon, various analyses were conducted and the findings are discussed below.

3.3. Derivatization Study

In order to enhance the chromatographic peak characteristics of the compounds and the detection sensitivity, the derivatization procedure was performed with different reagents. The aim was to select the optimum procedure and facilitate the simultaneous determination of the eight NSAIDs. In spiked serum samples at 200 μg/mL the derivatization procedures were conducted as described in Section 2.4, and the obtained chromatographic traces with the five reagents are summarized in Table 3. The chromatographic peaks were assigned based on the GC-MS libraries used [NIST v11 and Mass Spectral Library of Drugs, Poisons, Pesticides, Pollutants, and Their Metabolites, 3th Edition (PMW_Tox3.l)]. All the obtained peaks could be identified, apart from the acetyl derivatives, the spectra of which didn't exist in any of the used MS libraries, web-based libraries, or in the literature. In Table 3, the retention time and the characteristic ions of each drug or drug derivative are juxtaposed for the five derivatization procedures and that without any derivatization. As can be seen, AS and ASA, could not be detected at all with HFBA, TFA, or PFPA-PFPOH derivatization. In some cases, more than one derivative was detected, for example for APAP with MTBSTFA, while for others the underivatized drug was also detected, such as for NI with silylation reagents. In the case of PFPA-PFPOH, derivatization of NFA gives two peaks which cannot be assigned based on their spectra, as acetyl-derivatives are not registered in the MS spectral libraries used. The major peak is that at 6.9 min, which was attributed to NFA-pfp, while the other peak at 8 min can be another derivative of NFA. Ideally, the optimum procedure should lead to a sole derivative, whereas when the underivatized compound or more than one derivative is detected, the complexity of the detection is increased and the reproducibility of the obtained results is hindered. Based on the obtained results, HFBA, TFA, and PFPA-PFPOH are not considered to be the most appropriate for the simultaneous detection and determination of the eight NSAIDs for the above mentioned reasons.

As it concerns the detection sensitivity of the tested derivatization protocols, silylation provided better results in comparison to the other reagents for all NSAIDs except from IBP. The latter formed a derivative with PFPA/PFPOH which exhibited higher peak area when compared to the BSTFA and MTBSTFA derivatives. Between the two silylation reagents, BSTFA was selected as the best one. In Figure 1, a characteristic total ion chromatogram of serum sample spiked with the studied NSAIDs is presented. The selection was based on the fact that the majority of the peaks had higher peak areas and that MTBSTFA, in the case of APAP, forms two derivatives at the same peak height. Only NI derivatization with BSTFA seems to have low yield, as it gives a small peak of the derivative and the largest peak as NI. However, this is observed for the other silylation reagent as well, whereas NI does not derivatized at all with HFBA, PFPA/PFPOH, and TFA. This means that NI will finally be determined by considering the peak of the underivatized drug. In Table 4, the obtained peak areas are given for the eight drugs, where the higher peak areas can be seen for the silyl-derivatives. In the cases where the peak of the underivatized drug is detected this is noted by an asterisk.

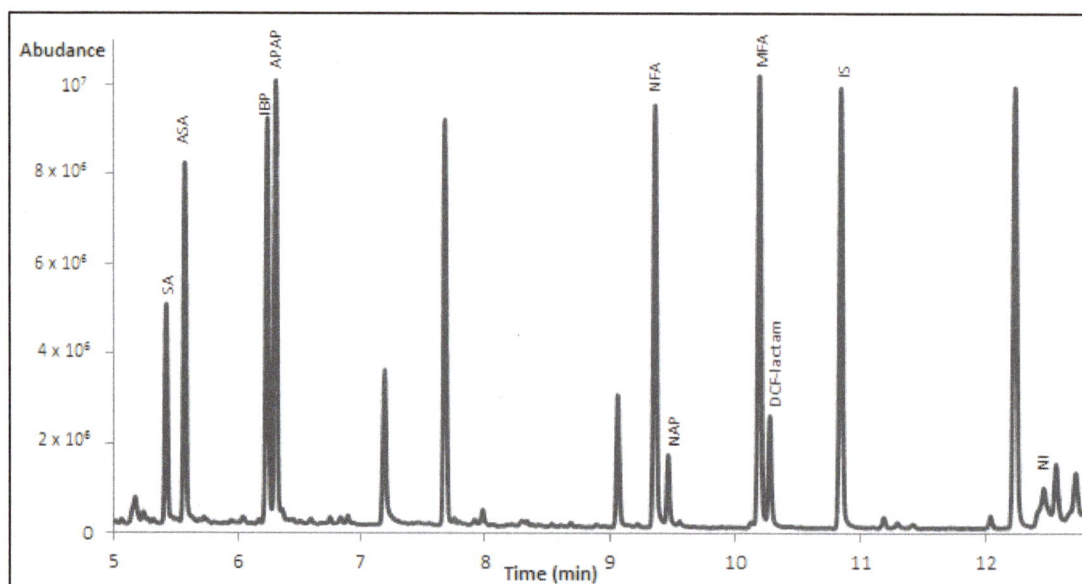

Figure 1. TIC of a serum sample spiked with the studied non-steroidal anti-inflammatory drugs (NSAIDs) at 5 μg/mL, derivatized with BSTFA & 1% TMCS.

Table 3. Retention time, quantifier ion, and two qualifier ions monitored for each NSAID and their derivatives.

Compound	No Derivatization	BSTFA, 1% TMCS	MTBSTFA	HFBA	TFAA	PFPA-PFPOH
	R.T./**Qf**, QI ions	R.T./**Qf**, QI ions/der	R.T./**Qf**, QI ions/der	R.T./**Qf**, QI ions/der	R.T./**Qf**, QI ions/der	R.T./**Qf**, QI ions/der
ASA	4.09/**120**,138,92	5.55/**195**,120,210/ASA-tms	7.20/**195**,237/135/ASA-tbdms	-	-	-
SA	5.24/**120**,92,138	5.48/**267**,135,91/SA-2tms	8.61/**309**,195,209/SA-2tbdms	-	-	-
IBP	6.02/**161**,163,206	6.22/**160**,263,353/IBP-tms	7.85/**263**,75,117/IBP-tbdms	6.02/**161**,163,206/IBP	7.87/**263**,161,117/IBP-tfa 6.02/**161**,163,206/IBP	4.99/**161**,295,239/IBP-pfp
APAP	6.89/**109**,151,80	6.29/**206**,280,295/APAP-2tms	8.91/**208**,265/166/APAP-tbdms 9.49/**322**,308,248/APAP-2tbdms	6.16/**108**,305,347/APAP-hfb 4.31/**304**,109,322/no ID*	5.81/**108**,205,247/APAP-tfa	5.89/**108**,255,297/APAP-pfp
NFA	9.10/**237**,263,282	9.34/**236**,263,353/NFA-tms	10.79/**245**,339,265/NFA-tbdms	7.00/**433**,265,168/NFA-hfb	6.90/**333**,265,168/NFA-tfa	6.92/**236**,264,441/NFA-pfp 8.00/**413**,263,236/no ID*
MFA	10.32/**223**,208,241	10.17/**223**,313,208/MFA-tms	11.60/**224**,209,298/MFA-tbdms	8.10/**224**,393,209/MFA-hfb	9.83/**222**,319,250/MFA-tfa 8.03/**294**,224,209/no ID*	8.60/**398**,250,222/MFA-pfp
DCF	10.36/**214**,242,277/DCFlactam 13.01/**214**,242,295/DCF	10.30/**349**,190,314/DCF-lactam-tms 10.82/**214**,242,367/DCF-tms	11.62/**300**,302,391/DCF-lactam-tbdms 12.24/**352**,214,409/DCF-tbdms	10.36/**214**,242,277/DCFlactam	10.36/**214**,242,277/DCF-lactam	10.36/**214**,242,277/DCF-lactam
NAP	9.86/**185**,230,170	9.46/**185**,243,302/NAP-tms	10.84/**287**,185,141/NAP-tbdms	9.86/**185**,230,170/NAP 7.04/**184**,141,169/NAP-CO$_2$	9.86/**185**,230,170/NAP 7.04/**184**,141,169/NAP-CO$_2$	8.24/**185**,170,141/NAP-pfp
NI	12.46/**154**,229,308	12.52/**380**,365,228/NI-tms 12.46/**154**,229,308/NI	12.52/**287**,241,344/NI-tbdms 12.46/**154**,229,308/NI	12.46/**154**,229,308/NI	12.46/**154**,229,308/NI	12.46/**154**,229,308/NI

Qf: Quantifier, QI: Qualifier, der: Derivative. * Unidentified peak which corresponds to a derivative.

Table 4. Peak areas of the obtained derivatives with the five derivatization protocols applied in the eight NSAIDs.

Compound	Peak Area 10^6				
	BSTFA	MTBSTFA	HFBA	PFPA-PFPOH	TFAA
APAP	330	194/190	190	171	90
IBP	362	266	137 *	455	20.7/219 *
MFA	443	350	188	244	153
NFA	381	350	97.5	310/12 *	22.4
NAP	196	190	8 */4	185	8 */3
ASA	425	420	-	-	-
SA	350	345	-	-	-
NI	115 */2.5	107 */3	130 *	140 *	100 *
DCF-lactam	440	402	306 *	280 *	322 *

* Underivatized compound.

3.4. Application of the Method

With the aim to apply the method to real clinical human samples, recovery of the optimum procedure was examined and linearity and limits of detection and quantification of the method was assessed. Recovery (R%) was experimentally calculated and expressed as the percentage ratio of the peak areas of the serum spiked before extraction at 10 µg/mL to the peak areas of the serum extract spiked after extraction. The R% ranged from 51.50% for ASA to 85.81% for NAP. The R% for all the studied drugs are presented in Table 5.

Table 5. Recovery % of the studied NSAIDs.

Compound	Recovery % (before LLE/after LLE)
APAP	61.35
ASA	51.50
DCF	70.30
IBP	63.17
MFA	64.27
NAP	85.81
NFA	65.15
NI	52.90
SA	59.28

Linearity of the method was evaluated by analyzing human serum samples spiked with a mixture of NSAIDs at concentrations of 0.5, 1, 5, 10, 25, and 50 µg/mL using as internal standard 20 µL nordiazepm-D_5 (C = 5 µg/mL). BSTFA derivatization was applied with 1% TMCS as described in Sections 2.3 and 2.4, and the peak areas ratios were considered for quantitation. For the case of NI, the peak of underivatized drug was considered, as in low concentrations the derivative is not detected at all. The limits of quantitation (LOQ) were experimentally calculated as signal to noise ratio 10:1 and the limits of detection, and limits of detection (LOD) as signal to noise 3:1. The equations of calibration curves based on linear regression, LOQs, and LODs for all drugs are given in Table 6.

Selectivity was determined on the drug-free blood serum samples obtained from six different healthy volunteers from the laboratory staff. No traces of the studied NSAIDs or other interferences could be detected.

Here it should be noted that freshly prepared ASA solutions in ACN were used for the determination of ASA, as it was observed that ASA transforms to SA. This has been observed to proceed faster in methanol than in ACN [16]. As SA is also present in real samples, its co-determination is required for both reasons.

Table 6. Linearity features of the method such as R^2, limits of detection and quantitation, and the linear equation for all drugs.

Compounds	Linear Equation	R^2	LOD (ng/mL)	LOQ (ng/mL)
ASA-tms	y = 0.578x + 0.407	0.994	10	30
SA-2tms	y = 0.456x + 0.965	0.998	5	15
APAP-2tms	y = 0.8222x + 0.8157	0.997	2	6
DCF-lactam-tms	y = 0.3448x + 0.7265	0.991	3	10
MFA-tms	y = 0.4625x + 0.9759	0.993	12	40
NFA-tms	y = 0.516x + 1.0058	0.992	13	43
IBP-tms	y = 0.567x + 1.7578	0.993	2	6
NAP-tms	y = 0.3187x + 1.1305	0.996	10	34
NI *	y = 0.7793x − 1.4552	0.999	124	414

* Nimesulide was determined underivatized.

The accuracy and precision of the method was evaluated within a day ($n = 4$) and over a period of a week ($n = 3$) at low, medium, and high concentration levels (1, 10, 50 µg/mL). The intra-day accuracy and precision were found to be between 1.03%–9.79% and 88%–110%, respectively, while the inter-day accuracy and precision were between 1.87%–10.79% and 91%–113%. In Table 7, the data from accuracy and precision assays are presented.

Table 7. Intra- and inter-assay data of the method.

Compound	Added (µg/mL)	Mean Found (µg/mL)	SD	CV %	Accuracy %	Overall Mean Found (µg/mL)	SD	CV %	Accuracy %
		Intra-Assay ($n = 4$)				Inter-Assay ($n = 3$)			
ASA	1	1.09	0.02	1.83	109.00	1.07	0.02	1.87	107.00
	10	11.00	0.60	5.45	110.00	11.30	0.61	5.40	113.00
	50	48.00	1.81	3.77	96.00	49.10	2.10	4.28	98.20
SA	1	0.98	0.05	5.10	98.00	0.99	0.06	6.06	99.00
	10	9.85	0.51	5.18	98.50	9.87	0.54	5.47	98.70
	50	51.10	1.20	2.35	102.20	50.80	1.30	2.56	101.60
APAP	1	0.97	0.01	1.03	97.00	1.00	0.03	3.00	100.00
	10	10.70	0.30	2.80	107.00	10.40	0.32	3.08	104.00
	50	49.80	0.90	1.81	99.60	50.20	1.20	2.39	100.40
DCF	1	0.88	0.04	4.55	88.00	0.91	0.06	6.59	91.00
	10	10.62	0.24	2.26	106.20	10.42	0.35	3.36	104.20
	50	49.20	1.50	3.05	98.40	49.30	1.90	3.85	98.60
MFA	1	1.12	0.05	4.46	112.00	1.05	0.06	5.71	105.00
	10	9.30	0.60	6.45	93.00	9.70	0.63	6.49	97.00
	50	51.30	2.40	4.68	102.60	50.90	3.10	6.09	101.80
NFA	1	0.98	0.03	3.06	98.00	1.01	0.05	4.95	101.00
	10	10.20	0.46	4.51	102.00	10.30	0.64	6.21	103.00
	50	51.60	3.20	6.20	103.20	51.60	3.50	6.78	103.20
IBP	1	0.94	0.07	7.45	94.00	0.96	0.08	8.33	96.00
	10	9.76	0.75	7.68	97.60	9.83	0.89	9.05	98.30
	50	49.80	2.60	5.22	99.60	50.30	3.20	6.36	100.60
NAP	1	1.02	0.05	4.90	102.00	1.06	0.08	7.55	106.00
	10	9.79	0.46	4.70	97.90	9.86	0.53	5.38	98.60
	50	50.50	4.60	9.11	101.00	50.81	2.26	4.45	101.62
NI	1	0.95	0.06	6.32	95.00	0.98	0.07	7.14	98.00
	10	9.86	0.42	4.26	98.60	9.88	0.56	5.67	98.80
	50	52.10	5.10	9.79	104.20	51.90	5.60	10.79	103.80

The method was then applied to four clinical cases where exposure to NSAIDs was suspected. The findings are presented in Table 8. S1 and S2 correspond to serum samples which were taken 12 h after APAP ingestion, and they were found below toxic levels (below 50 µg/mL, Rumack-Matthew nomogram). S3 corresponds to a case of a 13 year old girl claiming a suicide attempt with DCF, however the concentration after 3 h at 2.31 μg/mL was at the therapeutic levels. S4 corresponds to a patient treated with ASA (the sample was taken 40 min after ingestion) and the concentration was found to be below the therapeutic levels. The extracted ion chromatograms of S2 and S3 are presented in Figure 2a,b.

Table 8. Concentrations found with the applied method in the serum cases positive to NSAIDs.

Sample/Compound Found	Concentration Found (μg/mL) ± SD
S1/APAP	18.2 ± 0.3
S2/APAP	7.50 ± 0.02
S3/DCF	2.31 ± 0.05
S4/ASA	2.0 ± 0.2
S4/SA	0.21 ± 0.02

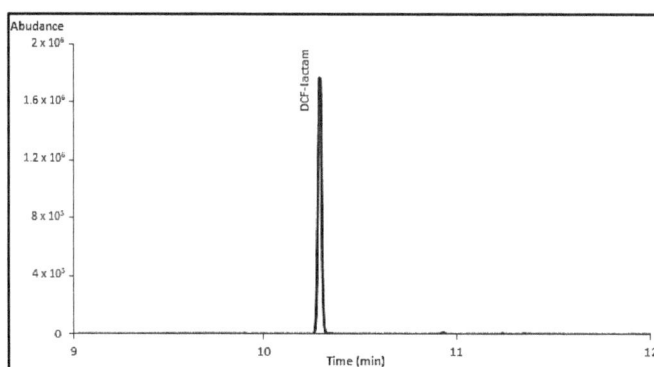

(a)

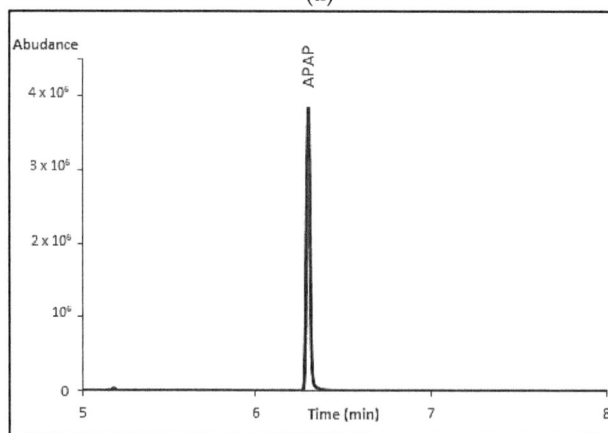

(b)

Figure 2. Extracted ion chromatograms of real serum sample (**a**) S3 at 349 m/z found positive for diclofenac determined at 2.31 μg/mL and (**b**) S2 at 206 m/z found positive for APAP determined at 7.5 μg/mL.

4. Discussion

According to our findings, more efficient extraction of the studied drugs was performed under an acidic pH of 3.7. NI showed also high extraction recovery in higher pH as well.

For most of the studied drugs, derivatization was required for their sensitive detection, however NI provided high, sharp peak in the initial extract and can be best determined without any derivatization.

Some of our findings lead us to the conclusion that determination of these drugs needs cautious sample treatment and data interpretation. We have observed, like previous findings [16], that ASA is not stable in solution and it is hydrolysed into SA, and finally over time both compounds are in equilibrium in the solution. Degradation of ASA has been studied thoroughly [16–18], however there are recent studies where the authors overlook this parameter [1,19]. In order to overcome this transformation, freshly prepared solution should be used for ASA.

Without derivatization, the majority of the NSAIDs seem to be unstable under the high temperature conditions in GC-MS, and a variety of different derivatives are also detected.

The decarboxylated products of NAP, MFA, and NFA are detected to elute some minutes earlier than the parent compound peaks. The formation of these compounds have been reported by others [15] and most probably is due to the fact that the underivatized compounds are labile compared to their derivatives under the same high temperature conditions in GC-MS.

In addition, for almost all the studied drugs, their methylesters were also detected. The esterification process seems to take place at the high temperature conditions in GC-MS, especially in methanolic solutions, whereas this doesn't seem to happen for the derivatives. For this, ACN was used as a solvent.

What is most interesting is that DCF transformed to its dehydrated product, the lactam. To investigate this, a series of experiments was conducted. First, a solution of DCF was analyzed directly with GC-MS, and it was observed that almost all of the diclofenac was transformed to its lactam form. When the same solution was analyzed after derivatization with BSTFA + 1% TMCS only diclofenac-tms was determined, indicating stability of the molecule with derivatization. Then a real human serum sample positive to DCF and a human serum sample spiked with DCF were analyzed after derivatization with BSTFA + 1% TMCS. In both cases, and in contrast to the previous finding, DCF-lactam-tms was mainly detected. Contrastingly, in an aqueous solution of DCF, which was analyzed after extraction and derivatization, similarly to a real sample, only DCF-tms could be detected. This could mean that diclofenac-tms is more stable than the underivatized drug, which is dehydrated under high temperatures in the GC-MS. However, it seems that the serum matrix under acidic conditions enhances the full conversion of DCF to lactam, as DCF-lactam-tms is the sole peak detected in this case, whereas in the absence of serum matrix DCF-tms is detected. The transformation of DCF to DCF-lactam in water samples, pharmaceutical dosage forms, and urine was reported previously but it hasn't been thoroughly studied [11,16]. In a study, DCF-lactam was falsely reported as DCF [20], and in another study DCF-lactam-tms wasn't detected at all [21].

5. Conclusions

Based on our findings, the most efficient sample preparation protocol for the accurate determination of the studied commonly prescribed NSAIDs is derivatization, more specifically with BSTFA, except from NI where no derivatization step is needed, after acidic extraction.

For some of these drugs, cautious handling is needed, as ASA hydrolyses quickly in solution and DCF is converted to its lactam form in the serum matrix under acidic pH. This conversion is enhanced at high temperatures when its carboxyl group is not protected.

The method needs only 200 μL of blood serum and can determine even trace amounts of the studied compounds.

Author Contributions: Conceptualization, A.K. and H.G.; Methodology, A.K. and N.R.; Validation, A.K. and E.T.; Formal Analysis, A.K. and E.M.; Resources, N.R.; Data Curation, A.K. and E.M.; Writing-Original Draft Preparation, A.K. and H.G.; Writing-Review & Editing, G.T.; Supervision, G.T. and H.G.

Conflicts of Interest: The authors declare no conflict of interest.

Abbreviations

ASA   Acetyl salicylic acid
SA   Salicylic acid
NAP   Naproxen
IBP   Ibuprofen
APAP   Acetaminophen
NFA   Niflumic acid
MFA   Mefenamic acid
NI   Nimesulide
DCF   Diclofenac

References

1. Sirok, D.; Pátfalusi, M.; Szeleczky, G.; Somorjai, G.; Greskovits, D.; Monostory, K. Robust and sensitive LC/MS-MS method for simultaneous detection of acetylsalicylic acid and salicylic acid in human plasma. *Microchem. J.* **2018**, *136*, 200–208. [CrossRef]

2. Bylda, C.; Thiele, R.; Kobold, U.; Volmer, D.A. Simultaneous quantification of acetaminophen and structurally related compounds in human serum and plasma. *Drug Test. Anal.* **2014**, *6*, 451–460. [CrossRef] [PubMed]

3. Yilmaz, B.; Sahin, H.; Erdem, A.F. Determination of naproxen in human plasma by GC-MS. *J. Sep. Sci.* **2014**, *37*, 997–1003. [CrossRef] [PubMed]

4. Vinci, F.; Fabbrocino, S.; Fiori, M.; Serpe, L.; Gallo, P. Determination of fourteen non-steroidal anti-inflammatory drugs in animal serum and plasma by liquid chromatography/mass spectrometry. *Rapid Commun. Mass Spectrom.* **2006**, *20*, 3412–3420. [CrossRef] [PubMed]

5. Sun, X.; Xue, K.L.; Jiao, X.Y.; Chen, Q.; Xu, L.; Zheng, H.; Ding, Y.F. Simultaneous determination of nimesulide and its four possible metabolites in human plasma by LC-MS/MS and its application in a study of pharmacokinetics. *J. Chromatogr. B* **2016**, *1027*, 139–148. [CrossRef] [PubMed]

6. Taylor, R.R.; Hoffman, K.L.; Schniedewind, B.; Clavijo, C.; Galinkin, J.L.; Christians, U. Comparison of the quantification of acetaminophen in plasma, cerebrospinal fluid and dried blood spots using high-performance liquid chromatography-tandem mass spectrometry. *J. Pharm. Biomed. Anal.* **2013**, *83*, 1–9. [CrossRef] [PubMed]

7. Chhonker, Y.S.; Pandey, C.P.; Chandasana, H.; Laxman, T.S.; Prasad, Y.D.; Narain, V.S.; Dikshit, M.; Bhatta, R.S. Simultaneous quantitation of acetylsalicylic acid and clopidogrel along with their metabolites in human plasma using liquid chromatography tandem mass spectrometry. *Biomed. Chromatogr.* **2016**, *30*, 466–473. [CrossRef] [PubMed]

8. Hložek, T.; Bursová, M.; Čabala, R. Fast ibuprofen, ketoprofen and naproxen simultaneous determination in human serum for clinical toxicology by GC–FID. *Clin. Biochem.* **2014**, *47*, 109–111. [CrossRef] [PubMed]

9. Chang, K.C.; Lin, J.S.; Cheng, C. Online eluent-switching technique coupled anion-exchange liquid chromatography–ion trap tandem mass spectrometry for analysis of non-steroidal anti-inflammatory drugs in pig serum. *J. Chromatogr. A* **2015**, *1422*, 222–229. [CrossRef] [PubMed]

10. Elsinghorst, P.W.; Kinzig, M.; Rodamer, M.; Holzgrabe, U.; Sörgel, F. An LC-MS/MS procedure for the quantification of naproxen in human plasma: Development, validation, comparison with other methods, and application to a pharmacokinetic study. *J. Chromatogr. B* **2011**, *879*, 1686–1696. [CrossRef] [PubMed]

11. Ding, Y.; Garcia, C.D. Determination of Nonsteroidal Anti-inflammatory Drugs in Serum by Microchip Capillary Electrophoresis with Electrochemical Detection. *Electroanalysis* **2006**, *18*, 2202–2209. [CrossRef]

12. Payán, M.R.; López, M.Á.B.; Fernández-Torres, R.; Bernal, J.L.P.; Mochón, M.C. HPLC determination of ibuprofen, diclofenac and salicylic acid using hollow fiber-based liquid phase microextraction (HF-LPME). *Anal. Chim. Acta* **2009**, *653*, 184–190.

13. Way, B.A.; Wilhite, T.R.; Smith, C.H.; Landt, M. Measurement of plasma ibuprofen by gas chromatography-mass spectrometry. *J. Clin. Lab. Anal.* **1997**, *11*, 336–339. [CrossRef]

14. Bhushan, R.; Joshi, S.; Arora, M.; Gupta, M. Study of the liquid chromatographic separation and determination of NSAID. *JPC J. Planar Chromatogr. Mod. TLC* **2005**, *18*, 164–166. [CrossRef]

15. El Haj, B.M.; Al Ainri, A.M.; Hassan, M.H.; Bin Khadem, R.K.; Marzouq, M.S. The GC/MS analysis of some commonly used non-steriodal anti-inflammatory drugs (NSAIDs) in pharmaceutical dosage forms and in urine. *Forensic Sci. Int.* **1999**, *105*, 141–153. [CrossRef]
16. Skibinski, R.; Komsta, L. The stability and degradation kinetics of acetylsalicylic acid in different organic solutions revisited—An UHPLC-ESI-QTOF spectrometry study. *Curr. Issues Pharm. Med. Sci.* **2016**, *29*, 39–41. [CrossRef]
17. Marra, M.C.; Cunha, R.R.; Vidal, D.T.; Munoz, R.A.; do Lago, C.L.; Richter, E.M. Ultra-fast determination of caffeine, dipyrone, and acetylsalicylic acid by capillary electrophoresis with capacitively coupled contactless conductivity detection and identification of degradation products. *J. Chromatogr. A* **2014**, *1327*, 149–154. [CrossRef] [PubMed]
18. Abuirjeie, M.A.; Abdel-hamid, M.E.; Ibrahim, E.-S.A. Simultaneous High-Performance Liquid Chromatographic Assay of Acetaminophen, Acetylsalicylic Acid, Caffeine, and D-Propoxyphene Hydrochloride. *Anal. Lett.* **1989**, *22*, 365–375. [CrossRef]
19. Kees, F.; Jehnich, D.; Grobecker, H. Simultaneous determination of acetylsalicylic acid and salicylic acid in human plasma by high-performance liquid chromatography. *J. Chromatogr. B* **1996**, *677*, 172–177. [CrossRef]
20. Yilmaz, B.; Ciltas, U. Determination of diclofenac in pharmaceutical preparations by voltammetry and gas chromatography methods. *J. Pharm. Anal.* **2015**, *5*, 153–160. [CrossRef] [PubMed]
21. Yilmaz, B. GC-MS Determination of Diclofenac in Human Plasma. *Chromatographia* **2010**, *71*, 549–551. [CrossRef]

## 9

Application of Headspace Gas Chromatography-Ion Mobility Spectrometry for the Determination of Ignitable Liquids from Fire Debris

María José Aliaño-González, Marta Ferreiro-González * [ID], Gerardo F. Barbero [ID], Miguel Palma and Carmelo G. Barroso

Department of Analytical Chemistry, Faculty of Sciences, ceiA3, IVAGRO, University of Cadiz, 11510 Puerto Real, Cadiz, Spain; mariajose.alianogonzalez@alum.uca.es (M.J.A.-G.); gerardo.fernandez@uca.es (G.F.B.); miguel.palma@uca.es (M.P.); carmelo.garcia@uca.es (C.G.B.)
* Correspondence: marta.ferreiro@uca.es

Abstract: A fast and correct identification of ignitable liquid residues in fire debris investigation is of high importance in forensic research. Advanced fast analytical methods combined with chemometric tools are usually applied for these purposes. In the present study, the Headspace Gas Chromatography-Ion Mobility Spectrometry (HS-GC-IMS) combined with chemometrics is proposed as a promising technique for the identification of ignitable liquid residues in fire debris samples. Fire debris samples were created in the laboratory, according to the Destructive Distillation Method for Burning that is provided by the Bureau of Forensic Fire and Explosives. Four different substrates (pine wood, cork, paper, and cotton sheet) and four ignitable liquids of dissimilar composition (gasoline, diesel, ethanol, and paraffin) were used to create the fire debris. The Total Ion Current (TIC) Chromatogram combined with different chemometric tools (hierarchical cluster analysis and linear discriminant analysis) allowed for a full discrimination between samples that were burned with and without ignitable liquids. Additionally, a good identification (95% correct discrimination) for the specific ignitable liquid residues in the samples was achieved. Based on these results, the chromatographic data from HS-GC-IMS have been demonstrated to be very useful for the identification and discrimination of ignitable liquids residues. The main advantages of this approach vs. traditional methodology are that no sample manipulation or solvent is required; it is also faster, cheaper, and easy to use for routine analyses.

Keywords: headspace gas chromatography-ion mobility spectrometry; ignitable liquid residues; fire debris; multivariate analyses; total ion current chromatogram

1. Introduction

The identification of ignitable liquids (ILs) commonly used as accelerant by arsonists is a hard task for forensic analysts. Most of the ILs that are used as accelerants are petroleum-based products. So, the fast and continues development of petroleum industry and materials results in an increase of new liquids and new interference matrix that challenge even more the proper identification of the IL in fire debris analyses.

In addition, fire debris analysis is complicated due to many other reasons: the destructive nature of the fire, the high temperatures reached at the scene that usually evaporate almost all of the liquids, the use of foams and water by the firefighters [1,2], or the degradation phenomena, such as weathering or biodegradation that the samples can suffer after the fire [3–5].

For those reasons, it is necessary to keep investigating new reliable and rapid analytical methods that allow for the detection and identification of the IL [6,7]. A fast and proper identification of the IL allows for making decisions about the legal responsibilities of the incident [8,9]. Most of the researches about the identification and discrimination of both neat IL and ignitable liquid residues (ILRs) in fire debris that are described in the bibliography are based on gas chromatography-mass spectroscopy (GC-MS) [10,11]. Indeed, the American Society for Testing and Materials (ASTM) standard provides guidelines for the analysis of ILs from fire debris based on GC-MS [12–14]. The ASTM also provides a classification scheme in which the ILRs are classified into 8 major classes of IL: gasoline, petroleum distillates, isoparaffinic products, aromatic products, naphthenic-paraffinic products, normal alkane products, oxygenated solvents, and miscellaneous category [15]. Some databases, in particular, the Ignitable Liquids Database and Reference Collection (ILRC) created and maintained by the National Center for Forensic Science (NCFS) in collaboration with the Scientific Working Group for Fire and Explosions (SWGFEX), are available to assist fire debris analysts with the identification of ILRs form fire debris by following the ASTM E1618 [15] (https://ilrc.ucf.edu/). The classification of each IL is based on visual pattern recognition of the total ion chromatograms (TICs), extracted ion profiling (EIPs), and target compound analysis [12,16,17].

So far, because ILRs are located fire debris solid samples, a first extraction method is usually required for a preconcentration of the samples before GC analysis. Different methods have been used with the purpose of separating the ILR from fire debris by passive headspace with different adsorbents, such as solid phase microextraction (SPME) fiber [14], Tenax [18], or activated carbon strips (ACS) [13]. Although these procedures work well, they still present some disadvantages, like the long adsorption times and the use of a solvent for desorbing the ACS, or the high price, the lowest robustness of the fibers as well as the lifetime of them in the case of SPME.

Regardless the preconcetration method, the analysis is done by GC-MS, and then, the interpretation of the results is based on the ASTM 1618 standard with the aim of classifying the ILR into one of the ASTM classes. The interpretation of the results is still a challenge for the analyst, since the approach is not only time consuming and complicated due to the interference matrix, moreover, it is highly related to the experience of the analyst running the analysis [19]. For this reason, in the last decade new alternatives or complementary approaches have been studied. The use of chemometric tools, in particular, pattern recognition techniques in order to obtain useful information of the total data matrix has been increased in the last years. The most commonly chemometric tools that are used are principal component analysis (PCA), cluster analysis (CA), discriminant analysis (DA), qualitative data analysis (QDA), or soft independent modelling by class analogy (SIMCA) [4,8,20,21].

The electronic nose (eNose) based on headspace concentration with mass spectrometry (HS-MS eNose) in combination with pattern recognition techniques has been successfully applied for direct analyses of ILR in fire debris samples in order to provide fast and almost fully automated results [6,7]. Headspace Gas Chromatography-Ion Mobility Spectrometry (HS-GC-IMS) uses the same sample preparation step than the eNose, i.e., headspace generation, and then, a gas sample is directly injected into the system. HS-GC-IMS is a novel technique with numerous advantages [22], as a high sensitivity (ppb range), fast analysis (a few minutes), option for portable devices, and no residues are generated. Due to these advantages, HS-GC-IMS has been applied in different fields as the oil industry [23,24], agrifood research [25–27], biomedical investigations [28,29], and in other forensic fields [30–32]. Nevertheless, it has not been previously applied for fire debris analyses.

The aim of the present work is to study the capacity of the HS-GC-IMS in combination with chemometrics tools (cluster analysis and discriminant analysis) for the determination and discrimination of different ILR in fire debris.

2. Materials and Methods

2.1. Samples

Four different substrates were chosen for this study (pinewood, cork, paper, and cotton sheet). All of the substrates were burned alone (without using any IL as accelerant) and then with each of the four ignitable liquids chosen for this study (gasoline, diesel, paraffin, and ethanol). Substrates and ILs were all purchased in local Spanish stores.

Fire debris samples were generated in the laboratory following Destructive Distillation Method for Burning provided by the Bureau of Forensic Fire and Explosives [17], inside of a laboratory hood with strict control of temperature (25 °C). For that, 5 cm × 5 cm pieces of substrates were placed on the bottom of one-quart paint cans and 500 μL of the corresponding ignitable liquid were added to the support. Cans were covered with a lid that contained nine 1 mm diameter holes.

Heat was applied to the bottom of the can with a flame from a propane torch held at a distance of 4 cm from the bottom. When smoke appeared, it was considered that the fire "starts" and samples were allowed to burn for two additional minutes. After this time, cans were allowed to cool down to room temperature and the perforated lids were replaced with an intact lid in order to avoid losing volatile compounds from the headspace.

Once the cans were cool, fire debris residues were extracted and placed into 10 mL vials (Agilent Crosslab) to their posterior analysis by HS-GC-IMS.

All of the experiments were carried out by duplicate and from each experiment; two pieces were obtained from different places of the can in order to guarantee no influence of the position of the support respect to the flame. A total of four burned samples were generated under the same conditions. So, a total of 80 fire debris samples were created.

2.2. HS-GC-IMS Analysis Acquisition

The 80 fire debris samples were analyzed by Headspace-Gas Chromatography-Ion Mobility Spectrometry FlavourSpec (G.A.S., Dortmund, Germany). No pretreatment was carried out for fire debris samples. Vials with fire debris were directly placed in the autosampler oven to be heated and agitated in order to generate headspace. Conditions for the creation of headspace were 75 °C of incubation temperature for 20 min with agitation. A total of 200 μL of headspace were injected with a splitless method of 500 μL/s while using a gas syringe at 80 °C, heated up to 5 °C more than sampling temperature in order to avoid condensation phenomena. Between each sample injection, gas syringe was flushed with carried gas for five minutes to avoid cross-contamination. Conditions of analysis inside of HS-GC-IMS were as follows: column temperature of 40 °C and IMS temperature of 45 °C. The drift gas flow was maintained at 150 mL/min during the whole analyses. The initial carrier gas program started with a flow of 2 mL/min for a minute, followed by a ramp carrier gas program of 5 mL/min held for 4 min and then a flow of 150 mL/min, which was held for the final 10 min of the analysis. Gas chromatography column was multicapillary MCC OV-5 (G.A.S., Dortmund, Germany). Drift gas and carrier gas selected were nitrogen with a purity of 99.999%, with a nitrogen generator (G.A.S., Dortmund, Germany). The ionization method used was by ^{3}H Tritium beta radiation.

2.3. Data Analysis

HS-GC-IMS implies a 2-dimensional separation of chemical compounds (retention time vs. drift time) as shown in Figure S1. In this case, two-dimensional chromatograms were obtained for each sample (intensity vs. retention time). They included information from 4286 retention time points from 0 to 899.85 s, recording a data from each 0.21 s. Total signal from the IMS system was used, meaning that no spectrometry information was used but the sum of signals as any drift time in the ion mobility spectrometer. Each chromatogram was normalized by assigning one unit to the maximum intensity. Information from the total ion current (TIC) chromatographic were arranged in data matrixes named $D_{m \times n}$ where m is the number of fire debris samples and n is the number of retention times.

TIC chromatographic data were analyzed by chemometric tools. In particular, hierarchical cluster analysis (HCA) and linear discriminant analysis (LDA), while using the statistical computer package SPSS 22.0 (SPSS Inc., Armonk, NY, USA) were performed.

3. Results and Discussion

Data from the total ion current (TIC) chromatograms were treated while using chemometric tools. No clear chromatographic peaks were obtained since very fast chromatographic analyses were run and the total signal from the IMS was used for this study. Therefore, chemometric tools were mandatory to extract the useful information from the TIC chromatograms. A hierarchical cluster analysis (HCA) using the whole data matrix ($D_{80 \times 4286}$) was carried out. HCA is a non-supervised chemometric tool, and the purpose pursued was to search whether the differences detected in the TIC chromatograms were enough to detect and discriminate ILRs from fire debris samples. Ward's method and Squared Euclidian distance were used during the HCA. Results from these analyses were represented by a dendrogram displayed in Figure 1.

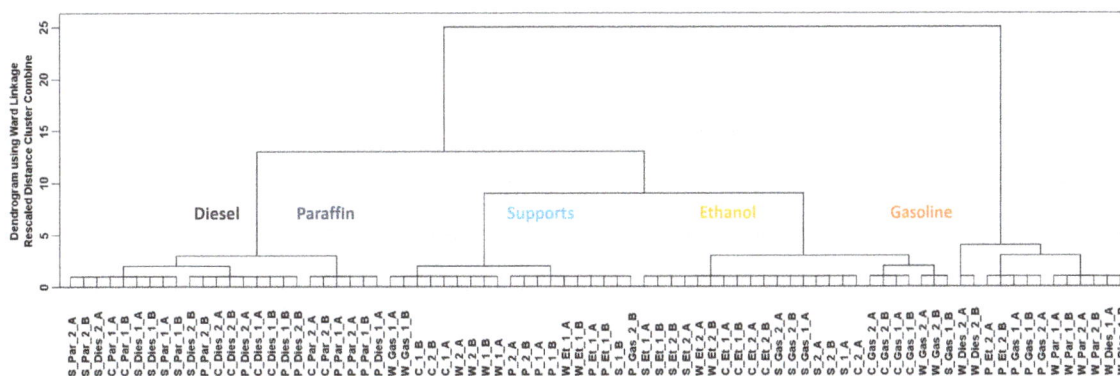

Figure 1. Dendrogram obtained from hierarchical cluster analysis (HCA) on the fire debris samples ($D_{80 \times 4286}$) using TIC chromatograms. (Samples are denoted as the IL code followed by the substrate: Gasoline (Gas), Diesel (Dies), Paraffin (Par) and Ethanol (Et), cotton sheet (S), paper (P) for pinewood (W) and for cork (C). A y B is used for the replicates.

Most of the supports burned without any IL (12 of 16 in total) are clustered together, as well as samples that were burned with ethanol (14 from 16). However, samples that were burned with diesel and paraffin are grouped in the same clusters. Diesel and paraffin are both heavy petroleum distillates so they have similar chemical composition, therefore they should provide similar related compounds to the ILR. Samples burned with gasoline are divided in two different clusters. These two groups of samples burned with gasoline as well as some of the misclassified samples seem to be related to the type of substrate used for burning and not to the IL used since all the cork and wood samples are clustered together regardless the IL. Although a full separation of the samples was not obtained, the results from this non-supervised technique suggest that the TIC chromatograms could be used to detect ILRs in fire debris or even to identify the ILRs.

Next, supervised pattern recognition tools to determine whether data from TIC chromatograms allow for the discrimination between samples was performed. For this purpose, a linear discriminant analysis (LDA) was carried out. A smoothing treatment for data was carried out and average of each ten columns was obtained by reducing the number of retention times to 428. Two classes were established a priori, i.e., samples burned using ILs and samples burned without ILs.

During the LDA, 60% of the samples were randomly chosen as a calibration set and the remaining 40% of the samples were used to validate the model. Stepwise method for the LDA was applied. A full classification between the two groups of samples was achieved (100%), first during the model development process, and later during the validation step. The score plot with all of the samples is represented in Figure 2. Samples that were used to calibrate the model were represented in blue in the case of the substrates burned alone and in yellow in the case of the ILRs. Samples that were used to validate the model were represented in a circle blue in the case of the supports burned alone and in a circle yellow in ILRs. A full discrimination between the two classes was obtained in the case of the calibration samples, as well as in the validation samples.

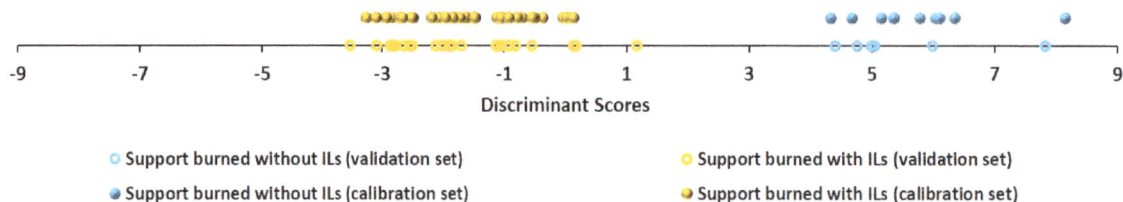

○ Support burned without ILs (validation set)   ○ Support burned with ILs (validation set)
● Support burned without ILs (calibration set)   ● Support burned with ILs (calibration set)

Figure 2. Discriminant scores obtained from linear discriminant analysis (LDA) for all of the fire debris samples ($D_{80 \times 428}$).

Fisher's linear discriminant functions were extracted from the LDA. A total of eight retention times showed a high contribution to those discriminant functions. Based on the selected retention times, average intensities for each retention time in the two groups were calculated and normalized to the maximum intensity inside of each group. In this way, two different fingerprints were generated (Figure 3), allowing for the discrimination between samples that were burned with IL and samples burned without IL. Therefore, the HS-GC-IMS based method can be used for the detection of ILR in fire debris.

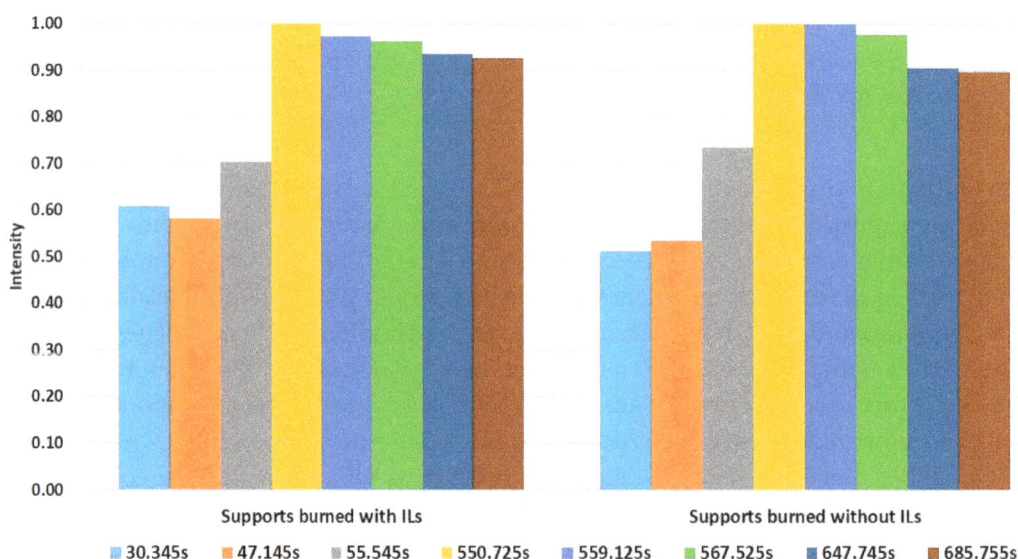

Figure 3. Fingerprints obtained from Fisher's linear discriminant function for samples burned with or without ILs.

Next, based on the results from the HCA, additional chemometric work was developed trying to discriminate between the different types of ILRs in the samples. A new LDA was carried out, but only using the samples that were burned with IL using again average retention times. Four groups were stablished (one for each kind of ILRs): gasoline (Gas_ILR), diesel (Die_ILR), ethanol (Et_ILR), and paraffin (Par_ILR). Cross-validation method (leave one out) and stepwise method was selected in the LDA. As shown in Table 1, a 98.4% of correct classification in the calibration set and a 98.4% for the cross-validated set was obtained.

Table 1. Classification results from the LDA using the 64 samples burned with ignitable liquids (IL).

			Classification Results [a,c]				
		GR	Predicted Group Membership				Total
			Gas_ILR	Dies_ILR	Et_ILR	Par_ILR	
Original	Count	Gas_ILR	16	0	0	0	16
		Dies_ILR	0	15	0	1	16
		Et_ILR	0	0	16	0	16
		Par_ILR	0	0	0	16	16
	%	Gas_ILR	100.0	0	0	0	100.0
		Dies_ILR	0	93.8	0	6.3	100.0
		Et_ILR	0	0	100.0	0	100.0
		Par_ILR	0	0	0	100.0	100.0
Cross-validated [b]	Count	Gas_ILR	16	0	0	0	16
		Dies_ILR	0	15	0	1	16
		Et_ILR	0	0	16	0	16
		Par_ILR	0	0	0	16	16
	%	Gas_ILR	100.0	0	0	0	100.0
		Dies_ILR	0	93.8	0	6.3	100.0
		Et_ILR	0	0	100.0	0	100.0
		Par_ILR	0	0	0	100.0	100.0

[a] 98.4% of original grouped cases correctly classified. [b] Cross validation is done only for those cases in the analysis. In cross validation. each case is classified by the functions derived from all cases other than that case. [c] 98.4% of cross-validated grouped cases correctly classified.

It can be observed that all misclassification occurs between groups of diesel and paraffin. During the model development, a sample burned with diesel was classified as sample burned with paraffin. Later, during the cross validation step, one sample burned with diesel was classified as being burned with paraffin again.

The score plot for all of the fire samples when using the three discriminant functions (FC1, FC2, and FC3) is displayed in Figure 4. It can be seen that, although they are not overlapped samples with diesel ILRs and paraffin, ILRs fell relatively close. As it was explained before, diesel and paraffin are chemically more similar in comparison to the other IL used in this study. FC3 is the function that shows higher influence for the discrimination between these two groups, while FC1 allows for the discrimination among samples that were burned with gasoline, with ethanol and the rest. Therefore, using the chromatographic data from the HS-GC-IMS, identifying the ILR in the fire debris is possible. However, a larger data set of fire debris samples would be needed to keep testing this novel technique, including several additional IL to follow the classification by the ASTM E1618.

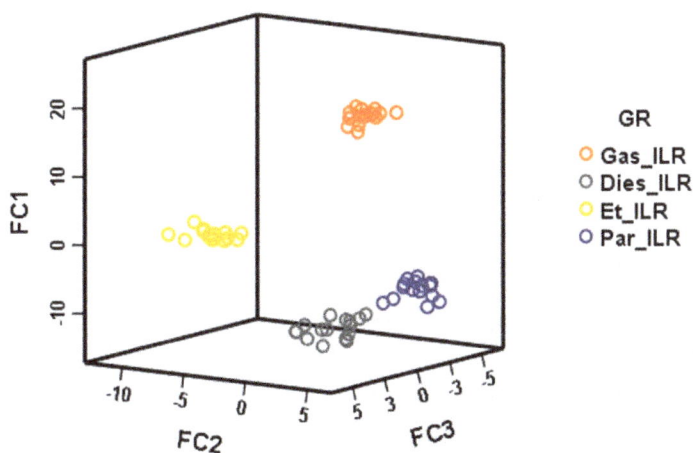

Figure 4. Discriminant scores obtained from LDA for burned samples ($D_{64 \times 428}$).

4. Conclusions

It has been demonstrated that the results from the fire debris analysis by HS-GC-IMS allows for the detection of ILR in the samples. The four ILs that were used in this study can be detected in the fire debris samples using chemometric tools without sample preparation and in 15 min.

Chemometric analysis of the HS-GC-IMS results also allows for the identification of the IL used among gasoline, diesel, paraffin, and ethanol with high reliability (98% good classification results).

Based on these results, HS-GC-IMS can be considered as a new promising tool in fire debris analyses. Besides, this technique has several advantages in comparison to traditional methods since HS-GC-IMS is a green technique (solvents are not required), fast, easy to use, and there are portable devices so the analyses could be applied at the fire scene.

Author Contributions: Data curation, M.J.A.-G. and M.P.; Formal analysis, M.F.-G. and G.F.B.; Investigation, M.J.A.-G. and M.F.-G.; Methodology, G.F.B. and M.P.; Resources, C.G.B.; Supervision, M.F.-G., G.F.B., M.P. and C.G.B.; Writing—original draft, M.J.A.-G.; Writing—review & editing, M.F.-G. and M.P.

Funding: This research was funded by the University of Cadiz grant number PR2016-17.

Conflicts of Interest: The authors declare no conflict of interest.

References

1.   Dennis, D.-M.K.; Williams, M.R.; Sigman, M.E. Investigative probabilistic inferences of smokeless powder manufacturers utilizing a bayesian network. *Forensic Chem.* **2017**, *3*, 41–51. [CrossRef]

2.   Falatova, B.; Ferreiro-González, M.; Martín-Alberca, C.; Kačíková, D.; Galla, Š.; Palma, M.G.; Barroso, C. Effects of fire suppression agents and weathering in the analysis of fire debris by HS-MS eNose. *Sensors* **2018**, *18*, 1933. [CrossRef] [PubMed]

3.   Martín-Alberca, C.; Carrascosa, H.; San Román, I.; Bartolomé, L.; García-Ruiz, C. Acid alteration of several ignitable liquids of potential use in arsons. *Sci. Justice* **2018**, *58*, 7–16. [CrossRef] [PubMed]

4.   Martín-Alberca, C.; García-Ruiz, C.; Delémont, O. Study of chemical modifications in acidified ignitable liquids analysed by GC-MS. *Sci. Justice* **2015**, *55*, 446–455. [CrossRef] [PubMed]

5.   Kindell, J.H.; Williams, M.R.; Sigman, M.E. Biodegradation of representative ignitable liquid components on soil. *Forensic Chem.* **2017**, *6*, 19–27. [CrossRef]

6.   Ferreiro-Gonzalez, M.; Ayuso, J.; Alvarez, J.A.; Palma, M.; Barroso, C.G. Application of an HS-MS for the detection of ignitable liquids from fire debris. *Talanta* **2015**, *142*, 150–156. [CrossRef] [PubMed]

7.   Ferreiro-Gonzalez, M.; Barbero, G.F.; Palma, M.; Ayuso, J.; Alvarez, J.A.; Barroso, C.G. Determination of ignitable liquids in fire debris: Direct analysis by electronic nose. *Sensors* **2016**, *16*, 695. [CrossRef] [PubMed]

8.   Green, M.K.; Kuk, R.J.; Wagner, J.R. Collection and analysis of fire debris evidence to detect methamphetamine, pseudoephedrine, and ignitable liquids in fire scenes at suspected clandestine laboratories. *Forensic Chem.* **2017**, *4*, 82–88. [CrossRef]

9.   Choi, S.; Yoh, J.J. Fire debris analysis for forensic fire investigation using laser induced breakdown spectroscopy. *Spectrochim. Acta Part B Atomic Spectrosc.* **2017**, *134*, 75–80. [CrossRef]

10.   Martín-Alberca, C.; Ortega-Ojeda, F.E.; García-Ruiz, C. Analytical tools for the analysis of fire debris. A review: 2008–2015. *Anal. Chim. Acta* **2016**, *928*, 1–19. [CrossRef] [PubMed]

11.   Sturaro, A.; Vianello, A.; Denti, P.; Rella, R. Fire debris analysis and scene reconstruction. *Sci. Justice* **2013**, *53*, 201–205. [CrossRef] [PubMed]

12.   Stauffer, É.; Lentini, J.J. ASTM standards for fire debris analysis: A review. *Forensic Sci. Int.* **2003**, *132*, 63–67. [CrossRef]

13.   American Society for Testing and Materials (ASTM). *Standard Practice for Separation of Ignitable Liquid Residues from Fire Debris Samples by Passive Headspace Concentration with Activated Charcoal*; ASTM E1412 (2012); ASTM International: West Conshohocken, PA, USA, 2012.

14.   American Society for Testing and Materials (ASTM). *Standard Practice for Separation and Concentration of Ignitable Liquid Residues from Fire Debris Samples by Passive Headspace Concentration with Solid Phase Microextraction (SPME)*; ASTM E2154 (2008); ASTM International: West Conshohocken, PA, USA, 2001.

15. American Society for Testing and Materials (ASTM). *Standard Test Method for Ignitable Liquid Residues in Extracts from Fire Debris Samples by Gas Chromatography-Mass Spectrometry*; ASTM E1618 (2014); ASTM International: West Conshohocken, PA, USA, 2014.

16. Hupp, A.M.; Marshall, L.J.; Campbell, D.I.; Smith, R.W.; McGuffin, V.L. Chemometric analysis of diesel fuel for forensic and environmental applications. *Anal. Chim. Acta* **2008**, *606*, 159–171. [CrossRef] [PubMed]

17. Williams, M.R.; Sigman, M.E.; Lewis, J.; Pitan, K.M. Combined target factor analysis and bayesian soft-classification of interference-contaminated samples: Forensic fire debris analysis. *Forensic Sci. Int.* **2012**, *222*, 373–386. [CrossRef] [PubMed]

18. Borusiewicz, R.; Zadora, G.; Zieba-Palus, J. Application of headspace analysis with passive adsorption for forensic purposes in the automated thermal desorption-gas chromatography-mass spectrometry system. *Chromatographia* **2004**, *60*, 133–142. [CrossRef]

19. Lopatka, M.; Sigman, M.E.; Sjerps, M.J.; Williams, M.R.; Vivó-Truyols, G. Class-conditional feature modeling for ignitable liquid classification with substantial substrate contribution in fire debris analysis. *Forensic Sci. Int.* **2015**, *252*, 177–186. [CrossRef] [PubMed]

20. Sigman, M.E.; Williams, M.R. Assessing evidentiary value in fire debris analysis by chemometric and likelihood ratio approaches. *Forensic Sci. Int.* **2016**, *264*, 113–121. [CrossRef] [PubMed]

21. González-Rodríguez, J.; Sissons, N.; Robinson, S. Fire debris analysis by raman spectroscopy and chemometrics. *J. Anal. Appl. Pyrolysis* **2011**, *91*, 210–218. [CrossRef]

22. Li, F.; Xie, Z.; Schmidt, H.; Sielemann, S.; Baumbach, J.I. Ion mobility spectrometer for online monitoring of trace compounds. *Spectrochim. Acta Part B Atomic Spectrosc.* **2002**, *57*, 1563–1574. [CrossRef]

23. Garrido-Delgado, R.; del MarDobao-Prieto, M.; Arce, L.; Valcárcel, M. Determination of volatile compounds by GC-IMS to assign the quality of virgin olive oil. *Food Chem.* **2015**, *187*, 572–579. [CrossRef] [PubMed]

24. Garrido-Delgado, R.; Mercader-Trejo, F.; Sielemann, S.; de Bruyn, W.; Arce, L.; Valcárcel, M. Direct classification of olive oils by using two types of ion mobility spectrometers. *Anal. Chim. Acta* **2011**, *696*, 108–115. [CrossRef] [PubMed]

25. Vautz, W.; Baumbach, J.I.; Jung, J. Continuous monitoring of the fermentation of beer by ion mobility spectrometry. *Int. J. Ion Mobil. Spectrom.* **2004**, *7*, 3–5.

26. Camara, M.; Gharbi, N.; Cocco, E.; Guignard, C.; Behr, M.; Evers, D.; Orlewski, P. Fast screening for presence of muddy/earthy odorants in wine and in wine must using a hyphenated gas chromatography-differential ion mobility spectrometry (GC-DMS). *Int. J. Ion Mobil. Spectrom.* **2011**, *14*, 39–47. [CrossRef]

27. Márquez-Sillero, I.; Cárdenas, S.; Valcárcel, M. Direct determination of 2,4,6-tricholoroanisole in wines by single-drop ionic liquid microextraction coupled with multicapillary column separation and ion mobility spectrometry detection. *J. Chromatogr. A* **2011**, *1218*, 7574–7580. [CrossRef] [PubMed]

28. Mochalski, P.; Wiesenhofer, H.; Allers, M.; Zimmermann, S.; Güntner, A.T.; Pineau, N.J.; Lederer, W.; Agapiou, A.; Mayhew, C.A.; Ruzsanyi, V. Monitoring of selected skin- and breath-borne volatile organic compounds emitted from the human body using gas chromatography ion mobility spectrometry (GC-IMS). *J. Chromatogr. B* **2018**, *1076*, 29–34. [CrossRef] [PubMed]

29. Thompson, R.; Perry, J.D.; Stanforth, S.P.; Dean, J.R. Rapid detection of hydrogen sulfide produced by pathogenic bacteria in focused growth media using SHS-MCC-GC-IMS. *Microchem. J.* **2018**, *140*, 232–240. [CrossRef]

30. Armenta, S.; de la Guardia, M.; Alcalà, M.; Blanco, M.; Perez-Alfonso, C.; Galipienso, N. Ion mobility spectrometry evaluation of cocaine occupational exposure in forensic laboratories. *Talanta* **2014**, *130*, 251–258. [CrossRef] [PubMed]

31. Keller, T.; Keller, A.; Tutsch-Bauer, E.; Monticelli, F. Application of ion mobility spectrometry in cases of forensic interest. *Forensic Sci. Int.* **2006**, *161*, 130–140. [CrossRef] [PubMed]

32. Moran, J.; McCall, H.; Yeager, B.; Bell, S. Characterization and validation of ion mobility spectrometry in methamphetamine clandestine laboratory remediation. *Talanta* **2012**, *100*, 196–206. [CrossRef] [PubMed]

## 10

Sampling Dynamics for Volatile Organic Compounds Using Headspace Solid-Phase Microextraction Arrow for Microbiological Samples

Kevin E. Eckert [1], **David O. Carter** [2] **and Katelynn A. Perrault** [1,*]🆔

[1] Laboratory of Forensic and Bioanalytical Chemistry, Forensic Sciences Unit, Division of Natural Sciences and Mathematics, Chaminade University of Honolulu, 3140 Waialae Ave., Honolulu, HI 96816, USA; kevin.eckert@student.chaminade.edu

[2] Laboratory of Forensic Taphonomy, Forensic Sciences Unit, Division of Natural Sciences and Mathematics, Chaminade University of Honolulu, 3140 Waialae Avenue, Honolulu, HI 96816, USA; david.carter@chaminade.edu

* Correspondence: Katelynn.perrault@chaminade.edu

Abstract: Volatile organic compounds (VOCs) are monitored in numerous fields using several commercially-available sampling options. Sorbent-based sampling techniques, such as solid-phase microextraction (SPME), provide pre-concentration and focusing of VOCs prior to gas chromatography–mass spectrometry (GC–MS) analysis. This study investigated the dynamics of SPME Arrow, which exhibits an increased sorbent phase volume and improved durability compared to traditional SPME fibers. A volatile reference mixture (VRM) and saturated alkanes mix (SAM) were used to investigate optimal parameters for microbiological VOC profiling in combination with GC–MS analysis. Fiber type, extraction time, desorption time, carryover, and reproducibility were characterized, in addition to a comparison with traditional SPME fibers. The developed method was then applied to longitudinal monitoring of *Bacillus subtilis* cultures, which represents a ubiquitous microbe in medical, forensic, and agricultural applications. The carbon wide range/polydimethylsiloxane (CWR/PDMS) fiber was found to be optimal for the range of expected VOCs in microbiological profiling, and a statistically significant increase in the majority of VOCs monitored was observed. *B. subtilis* cultures released a total of 25 VOCs of interest, across three different temporal trend categories (produced, consumed, and equilibrated). This work will assist in providing foundational data for the use of SPME Arrow in future microbiological applications.

Keywords: microbial VOCs; SPME Arrow; gas chromatography–mass spectrometry (GC–MS); sampling optimization; *Bacillus subtilis*

1. Introduction

The sampling and analysis of volatile organic compounds (VOCs) is crucial to the success of a number of fields and industries, including the food and fragrance industries [1,2], environmental monitoring [3–7], and plant ecology [8,9]. The value of VOCs is also more recently emerging in numerous other fields such as forensic science [10,11], biomedical disease detection [12,13], and archaeology [14–17], where methodologies are adopted from other industries to provide significant steps forward in understanding and aiding human science. Headspace sampling is often used in these fields because analytes can easily be separated from their matrices in an effective manner with minimal contact and/or contamination. After collection, VOCs are analyzed using gas chromatography–mass spectrometry (GC–MS), which is well suited for volatile analysis because separations are conducted in the gas phase. Minimal sample preparation is required for VOC analyses, as volatiles can be trapped and introduced into the instrument without the introduction of solvents or lengthy purification steps. To date, numerous sampling techniques exist to sample VOCs. Direct analysis techniques can be used, such as collecting the sample into an air canister, polymer bag, or container; subsequently, injection can be achieved by injection using a gastight syringe, air server, or dynamic headspace attachment. In these methods, an aliquot of gas or air containing VOCs is collected and/or introduced directly into the GC–MS. This allows for very volatile compounds to be targeted, but does not provide significant benefits for sample enrichment and focusing [18]. This can sometimes lead to poor chromatographic characteristics, such as peak overloading or reduced repeatability.

Sorbent-based sampling techniques are often preferential due to their ability to pre-concentrate or enrich the sample prior to injection [18], as well as providing reduced volume and improved stability for sample transport and/or storage [19]. Solid-phase microextraction (SPME) was first developed by Pawliszyn et al. in 1990 [20–22] and has gained high popularity for VOC sampling in nearly all applicable fields. While the sorbent capacity of solid-phase microextraction is less than that of techniques such as sorbent tube collection, the micro-solid phase nature of the technique has many practical and analytical advantages, including (1) sensitivity on small samples; (2) reduced sampling time; and (3) minimal equipment required [19]. Criticisms of SPME for VOC analysis include the poor mechanical robustness of the apparatus for long-term sampling (i.e., prone to breakage, difficult to perform in-field sampling, septum coring, etc.) [23], as well as the relative minimal capacity offered by the fibers in comparison to higher capacity sorbent techniques such as stir bar sorptive extraction (SBSE) or sorbent tubes [24–26].

A recent commercial development, known as SPME Arrow, offers the same static enrichment of sample headspace as the original SPME design; however, the arrow exhibits an increase in the amount of sorbent coating (i.e., phase volume) and altered tip design for improved mechanical robustness and reliability [27]. The altered sorptive phase dimensions provide a higher VOC loading capacity for trace samples, allowing for either increased sensitivity or reduced sampling time in comparison to the original SPME design [28]. The tip of the fiber is pointed, which decreases damage to vial and inlet septa, further reducing coring and thereby maintenance costs [28]. This is considered to be of particular value in scenarios where (1) the concentration of analytes of interest is too low upon the headspace equilibration with the sorptive phase of traditional SPME fibers, (2) throughput is a major consideration for sampling equilibration time, and/or (3) reduced down time and costs of inlet maintenance are desired.

Though no literature currently exists on the use of SPME Arrow to collect VOCs from biological matrices, it could provide significant benefits. Many forensic and medical studies that analyze biological samples are currently interested in profiling volatiles from different species of bacteria. In forensic science, this information can assist in the improvement of search and recovery procedures [29–31], since many microbes have now been closely associated with deceased individuals [32]. In medicine, the volatiles given off by microbes and human cell lines can be key indicators of health or disease status [33–36]. When investigating sources of this nature, a number of challenges are present including the high diversity in the VOC profile, the dynamic range of concentrations of VOCs, the large number of samples that must be investigated based on project design, as well as the sensitive and critical nature of each individual sample [35,37]. When dealing with volatile sampling, repeated sample injection may not be possible, and so the reliability and robustness of the sampling technique is extremely important. An increase in sensitivity based on higher VOC loading during the collection process will be a major advantage in dealing with the dynamic range experienced with such samples. As such, it is likely that SPME Arrow will become a more common sampling technique used in these fields over traditional SPME fibers, particularly for biological matrices.

B. subtilis (Bacillales: Bacillaceae) is a species of bacteria that is facultatively aerobic, Gram-positive, motile, and rod shaped (typically 0.7–0.8 μm × 2.0–3.0 μm) [38]. *B. subtilis* was selected for the current study because it is widely distributed throughout nature [39], associated with decomposing remains [40], and serves as the type species for the genus *Bacillus* [38]. This species can colonize diverse habitats because it has the ability to decompose a wide range of organic and inorganic resources including protein, amino acids, ammonium, and nitrate; all of these compounds are forensically-relevant as they are associated with decomposing remains and represent key steps in the nitrogen cycle [41]. Additionally, in medical contexts, *B. subtilis* has been shown to be associated with cases of food poisoning due to spoiled meat, dermatitis, respiratory problems, tumor infection, bacteremia, and septicemia [42]. Improving our ability to profile temporal trends in volatile byproducts during the metabolic activity of *B. subtilis* using a more sensitive and robust SPME Arrow technology has implications across all these areas.

There currently exists minimal published data on the SPME Arrow technique to demonstrate optimal conditions for VOC analysis from biological matrices, as well as the dynamics of different VOCs that may be encountered during the sampling process. In particular, the limited studies that highlight SPME Arrow [18,27,28,43] do not target compounds encountered in life sciences applications such as forensic science and biomedical disease detection. The objective of this work was to provide preliminary data on the headspace SPME collection of a range of common VOCs encountered in life sciences applications using this high capacity SPME Arrow technique. A volatile reference mixture (VRM) was created to be used in the assessment of the SPME Arrow collection parameters. Fiber type, exposure time, desorption time, carryover, and method reproducibility were investigated to better understand fiber dynamics. A comparison to traditional SPME fiber design was also performed. Finally, the optimized SPME Arrow collection parameters were applied to the VOC profiling of a microbiological source, the headspace of cultured *B. subtilis*, which is a common microbe associated with forensic, medical, and agricultural applications.

2. Materials and Methods

2.1. Volatiles Reference Mix (VRM)

The VRM was created from a combination of custom mixtures and individual compounds. Mix 1 contained 2-ethyl-1-hexanol, 1-propanol, 2-propanol, 2-butanone, cyclohexane, and 2-methylfuran in P&T methanol/water (90:10), each at a concentration of 1000 µg/mL (GC grade, Restek Corporation, Bellefonte, PA, USA). Mix 2 contained styrene, 2-methylpentane, 3-methylpentane, 2,4-dimethylheptane, 2-methylhexane, naphthalene, and 1,2,3-trimethylbenzene in P&T methanol, each at a concentration of 1000 µg/mL (GC Grade Quality, Restek Corporation). Mix 3 was a commercially available standard containing benzene, ethylbenzene, toluene, m-xylene, o-xylene, and p-xylene in P&T methanol, each at a concentration of 2000 µg/mL (certified reference mixture grade, Restek Corporation). Individual compounds used were heptanal, dimethyl trisulfide, hexanal, and isoprene (analytical standard grade, Sigma-Aldrich, St. Louis, MO, USA). The solvent used for dilution was HPLC-grade methanol (J.T. Baker, Center Valley, PA, USA).

From the commercial mixes, custom mixes, and individual standards, a series of mixed standards were made for SPME Arrow evaluation. VRM 1 contained mix 1, mix 2, heptanal, dimethyl trisulfide, hexanal, and isoprene each at a concentration of 100 µg/mL, and mix 3 at a concentration of 200 µg/mL. VRM 2 contained mix 1, mix 2, heptanal, dimethyl trisulfide, hexanal, and isoprene each at a concentration of 10 µg/mL, and mix 3 at a concentration of 20 µg/mL. VRM 3 contained mix 1, mix 2, heptanal, dimethyl trisulfide, hexanal, and isoprene each at a concentration of 1.1 µg/mL, and mix 3 at a concentration of 2.2 µg/mL.

In addition to the VRM, a commercial saturated alkanes mix (SAM$_C$) containing heptane, octane, nonane, decane, undecane, dodecane, tridecane, tetradecane, pentadecane, hexadecane, heptadecane, octadecane, nonadecane, eicosane, heneicosane, docosane, tricosane, tetracosane, pentacosane, hexacosane, heptacosane, octacosane, nonacosane, and triacontane was used, each at a concentration of 1000 µg/mL in hexane (certified reference material grade, Supelco, Bellefonte, PA, USA). The solvent used for dilution was hexane (CHROMASOLV® Plus, Sigma Aldrich, St. Louis, MO, USA).

From the commercial mix, a series of diluted standards were made for SPME Arrow evaluation. SAM 1 contained SAM_C at a concentration of 100 µg/mL. SAM 2 contained SAM_C at a concentration of 10 µg/mL. SAM 3 contained SAM_C at a concentration of 1.1 µg/mL.

All mixtures were prepared fresh in glass vials (within one month of use) and stored in a freezer between analyses.

2.2. Gas Chromatography–Mass Spectrometry (GC–MS) Analysis and Data Processing

The inlet temperature was 250 °C for all trials. The Thermo TRACE Ultra inlet conversion kit (Restek Corporation) and a splitless liner (Restek Corporation, $2.0 \times 8.0 \times 105$ mm) were used for all SPME injections. A Rxi-624Sil MS capillary column was used (Restek Corporation, 30 m $\times$ 0.25 mm ID $\times$ 1.4 µm film thickness) inside of a Focus GC coupled to a DSQ II (dual-stage quadrupole) mass selective detector (MSD) (Thermo Scientific, Bellefonte, PA, USA). A helium carrier gas flow rate of 1.0 mL/min was used (Airgas, Radnor, PA, USA). The oven temperature was held at 35 °C for 5 min and then increased by 5 °C/min to 240 °C, where it was held for 5 min. The MS transfer line and source temperatures were held at 250 °C and 200 °C, respectively, and the MSD was operated in full electron ionization (EI) scan mode from 40 to 450 m/z at a scan rate of 5 scans/s.

Data acquisition was performed using Thermo XCalibur version 3.0.63. Raw files were imported into Thermo Chromeleon version 7.25 SR5 for further processing. MS detection was performed using the Cobra detection algorithm. Automatic baseline correction was used with peak dependent correlation set at n = 3 spectra for each of the left region bunch, right region bunch, and peak spectrum bunch. Tentative peak identifications were made using the National Institute of Standards and Technology (NIST) 2014 Mass Spectral Library with a forward match factor threshold of 700 and reverse match factor threshold of 700. Data were exported as *.csv files and statistics were generated as described in Section 2.5.

2.3. Optimization of SPME Arrow Method

The SPME Arrows under study were coated with one of four different stationary phases: 120 µm carbon wide range (CWR)/polydimethylsiloxane (PDMS), 100 µm PDMS, 100 µm polyacrylate (PA), and 120 µm divinylbenzene (DVB)/PDMS (Restek Corporation). Selection of SPME Arrow stationary phase was performed by static exposure of CWR/PDMS, PDMS, PA, and DVB/PDMS Arrows for 5 min inside sealed headspace vials (Restek Corporation, 20 mL, 22 $\times$ 75 mm) containing 10.0 µL of VRM 1 and 1.0 µL of SAM_C added with a micropipettor prior to sampling at room temperature.

Following fiber selection, the optimum exposure time was determined for the selected fiber. A range of exposure times were tested including 30 s and 1 min to 15 min (in 1 min intervals). For each trial, the SPME Arrow was exposed inside a sealed headspace vial containing 1.0 µL of VRM 2 and 1.0 µL of SAM 2 added with a micropipettor prior to sampling at room temperature. To determine the optimum time the SPME Arrow needed for complete VOC desorption to occur in the GC inlet, a range of inlet desorption times were tested: 1, 5, 15, 20, 25, 30, 45, 60, and 90 s. The SPME Arrow was first conditioned at 270 °C for 8 min in the GC inlet, and then exposed for 5 min inside a sealed headspace vial containing 1.0 µL of VRM 1 and 1.0 µL of SAM 1 added with a micropipettor prior to sampling at room temperature. Carryover was also tested by running a series of fiber blanks after saturating the SPME Arrow with VRM and SAM and performing the optimized desorption time to confirm that optimal desorption removed all analytes prior to further analysis. Following exposure, the SPME Arrows were analyzed using the GC–MS method in Section 2.2.

Intraday and interday repeatability using the final sampling parameters were tested using n = 3 replicate vials spiked with 9.0 µL of VRM 3 and 9.0 µL SAM 3 via injection through the septum using a 10 µL syringe (Trajan Scientific and Medical, Ringwood, Victoria, Australia). Finally, a 23 ga. CWR/PDMS PAL SPME Fiber (Restek Corporation) was also compared to the corresponding CWR/PDMS PAL SPME Arrow (Restek Corporation) to assess the magnitude of increased loading using the SPME Arrows. This test was carried out on a single day using n = 3 replicate vials spiked with 9.0 μL of VRM 3 and 9.0 μL of SAM 3 via injection through the septum using a 10 μL syringe (Trajan Scientific and Medical).

2.4. Application to Bacillus subtilis Headspace

B. subtilis was isolated in previous work from previous pig (*Sus scrofa domesticus*) decomposition studies at the university taphonomic research facility. The microbial growth medium used in the current study was standard nutrient agar (HiMedia Laboratories Pvt. Ltd., Mumbai, India). 4.0 mL of growth medium was added to sterilized 20 mL headspace vials (Restek Corporation) as a slant using a sterile serological pipette (VWR International, Radnor, PA, USA). After cooling and solidifying, a VWR Symphony incubator was used to incubate the vials at 24 °C for 72 h to verify they were free of biological contamination. The headspace vials were then inoculated with *B. subtilis* uniformly over the growth medium using a metal loop. The vials were sealed and sampled using the SPME Arrow using the optimized parameters from the above tests. These parameters were: CWR/PDMS stationary phase, pre-conditioning time of 30 min at 270 °C, exposure time of 5 min, and GC inlet desorption time of 3 min. After sampling, the headspace vial was opened inside the sterilized laminar flow cabinet to purge the headspace of VOCs for 5 min. The headspace vial was resealed and incubated at 24 °C for 24 h. Vials were prepared in triplicate and a single injection of each vial was resampled every 24 h over a five-day period. Uninoculated vials were analyzed in triplicate in a manner identical to the *B. subtilis* vials.

2.5. Statistical Analysis

Statistical analysis was performed on data by importing *.csv files into Prism 7 (Graphpad Software, La Jolla, CA, USA) for univariate analysis and graphing, and the Unscrambler X version 10.5 (CAMO Software, Oslo, Norway) for multivariate analysis. Univariate statistical analysis was performed on raw peak areas. For multivariate analysis, data were normalized using unit vector normalization, scaled by standard deviation and mean centered. The pre-treated data was then subjected to principal component analysis (PCA) for further visualization and characterization. One PCA was performed including all control and bacteria data to demonstrate the differences between vials containing blank agar and those containing bacterial cultures. A second PCA was performed including only bacteria data to demonstrate the differences between samples on each of the days of analysis, in order to highlight the transition in VOCs across the growth phase of the bacteria. Outliers were verified to not exist in both data sets using Hotelling's T^2 ellipse within the Unscrambler X software.

3. Results and Discussion

3.1. Arrow Method Optimization

A variety of sorbent materials are now commercially available as part of the PAL SPME Arrow product line. As such, it was necessary to verify suitability for biological applications, specifically for the range of volatiles expected from microbial sources as described for this study. The results of fiber comparison for the SPME 1.1 mm Arrow range are shown in Figure 1a. The CWR/PDMS fiber recovered the highest overall abundance of combined compounds. This was largely attributed to an above average recovery of analytes across all compound classes, whereas other fibers exhibited low recoveries within specific compound groups. For example, the PA fiber had high abundance for $>C_{12}$ n-alkanes; however, short chain alkanes ($<C_{12}$) as well as branched and cyclic alkanes—such as 2-methylpentane, 3-methyl-pentane, 3-methylhexane, and cyclohexane—could not be detected from the PA fiber. The high total peak area of the PA fiber is largely attributed to high recovery of $>C_{13}$ alkanes. The DVB/PDMS and PDMS fibers exhibited lower recoveries than CWR/PDMS and polyacrylate for aromatics, aldehydes, sulfides, and alcohols. Overall, the CWR/PDMS fiber was found to exhibit the highest efficacy for compounds expected within biological matrices such as the headspace of bacterial cultures. The longest n-alkane recovered by the CWR/PDMS fiber was n-heptadecane at a retention time of approximately 41.5 min; therefore, it is unlikely that compounds with a retention index above 1700 would be recovered if the GC–MS conditions employed herein are applied in future work. The GC–MS method extended beyond this time to 52 min, however, no larger alkanes were detected.

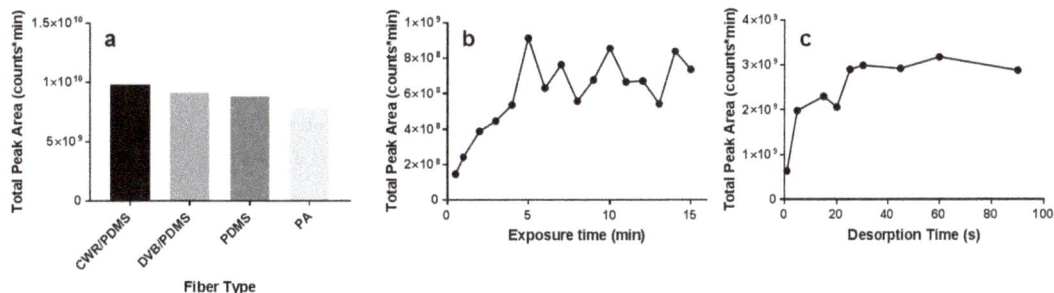

Figure 1. (a) Arrow fiber performance based on total summed peak area of all compounds in the volatiles reference mix (VRM) 1 and the commercial saturated alkanes mix (SAM_C). (b) Total peak area measured for carbon wide range (CWR)/polydimethylsiloxane (PDMS) Arrow at different exposure times to VRM 2 and SAM 2. (c) Total peak area measured for CWR/PDMS Arrow at 5 min exposure time and varied desorption times after saturation in VRM 1 and SAM 1.

Further testing of the CWR/PDMS Arrow demonstrated that exposure time peaked in performance after 5 min (Figure 1b). Though fluctuations in response occurred with longer exposure times, extending the exposure time beyond 5 min did not result in any substantial increase in VOC recovery. Additionally, a saturated CWR/PDMS Arrow was found not to require a lengthy desorption time in the inlet for maximum desorption (Figure 1c). Sufficient desorption occurred between 40–60 s; however, as the manufacturer recommended a 3 min desorption to ensure conditioning of the Arrow prior to next use, a 3 min desorption was chosen. Carryover testing also confirmed effective conditioning of the fiber prior to next usage. A chromatogram displaying the results of the final SPME Arrow parameters is displayed in Figure 2.

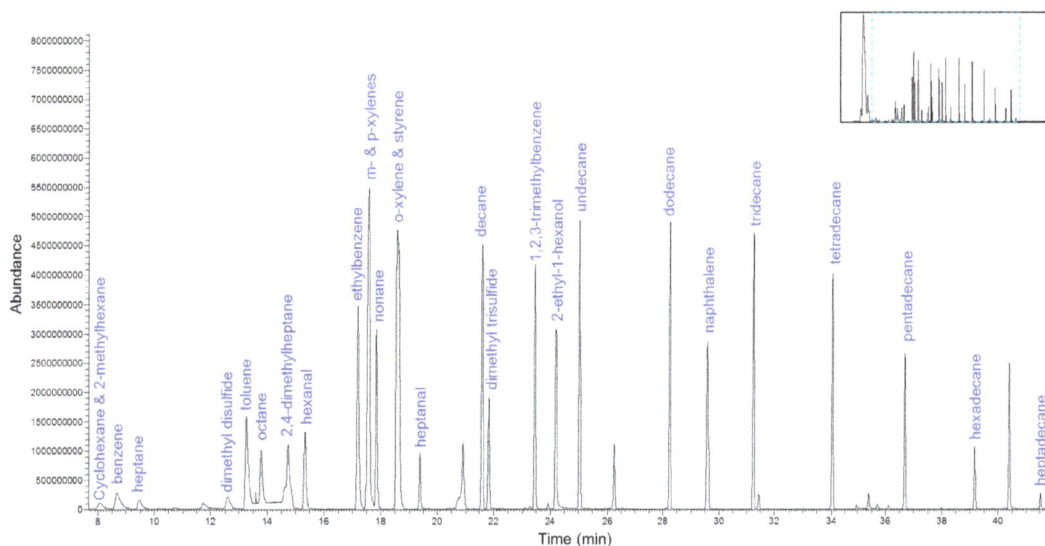

Figure 2. Total ion current chromatogram of the volatiles reference mix (VRM) 1 and the saturated alkanes mix (SAM) 1 using carbon wide range (CWR)/polydimethylsiloxane (PDMS) solid-phase microextraction (SPME) Arrow.

The comparison of CWR/PDMS Arrows to the CWR/PDMS traditional SPME fibers is displayed in Figure 3. In this case, the same method (5 min exposure time, 3 min desorption time) was used. For all compounds except for benzene, naphthalene, and $>C_{13}$ alkanes, the SPME Arrow provided significant increase in recovery ($p < 0.05$ for Student's t-test). Often, the SPME Arrow recovered more than double the total peak area detected with the traditional SPME fiber. This demonstrates the higher capacity of sorbent present on the SPME Arrow fibers, specifically for the range of compounds expected in biological matrices.

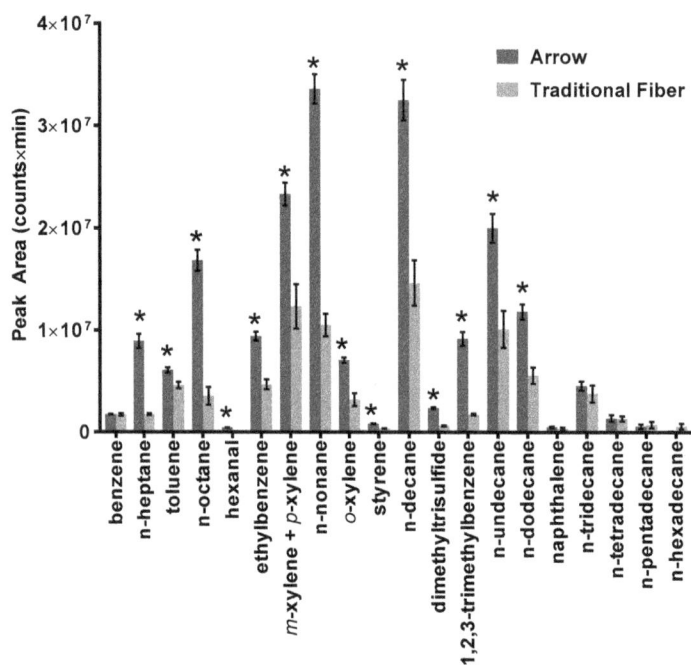

Figure 3. Comparison data for the carbon wide range (CWR)/polydimethylsiloxane (PDMS) fiber using the volatiles reference mix (VRM) 3 and the saturated alkanes mix (SAM) 3. Error bars represent standard deviation based on n = 3 samples. Fibers in both categories were exposed at room temperature for 5 min and analyzed using the same GC–MS method. Statistical significance (*) was determined using the multiple t-test option, correcting for multiple comparisons using the Holm–Sidak method with $\alpha = 0.05$. Each compound was analyzed individually, without assuming a consistent standard deviation.

In addition, it was originally hypothesized that larger recovered peak areas may result in a difference in the reproducibility of the SPME Arrows compared to the traditional SPME fibers. However, the standard deviation of peak areas appeared to be comparable for the two fiber types (Figure 3). The % RSD on the peak area of each compound was found to be, on average, less than 5.8% for intraday variation (n = 3 replicates measured on three individual days). Naphthalene, 1,2-3-trimethylbenzene, and the n-alkanes larger than C_{12} were the only compounds that exhibited intraday variation greater than 10%. Naphthalene exhibited the highest intraday variation with a % RSD of 24.5% on the first of three days of analysis. The lowest intraday variation was for nonane with a% RSD of 0.6% on the first of three days of analysis. The interday variation was slightly higher for each compound than intraday variation. The average of the interday % RSD for all compounds was 9.2%. With the exception of naphthalene and the n-alkanes higher than C_{12}, the interday variation was always less than 15% RSD. The naphthalene and higher n-alkane data being higher in variation could possibly be attributed to the fact that these compounds had inherently low recoveries on the CWR/PDMS fiber; therefore, variation in the peak area represented a larger proportion of error compared to the average. The tabulated % RSD for intraday and interday variation by each compound can be found in Supplementary Material, Table S1.

In this study, some compounds on the wide end of the volatility range were not recovered in significantly higher quantities when comparing SPME Arrow with traditional SPME. Benzene and naphthalene were two compounds that were expected to be recovered in higher quantity by SPME Arrow that were not. A previous study comparing some similar volatiles demonstrated a 2.5 fold increase for benzene and 32 fold increase for naphthalene when comparing SPME Arrow to traditional SPME fibers [18]. However, this previous study compared the two types of fibers for the PDMS sorbent [18] and not for the CWR/PDMS fiber used in the study herein. The PDMS Arrow used in the previous study also had a fiber phase volume of 10.2 µL [18], whereas the current commercial version of the CWR/PDMS Arrow has a phase volume of 3.8 µL. This is in contrast to the traditional SPME fiber phase volume of 0.6 µL used in both studies. The differences in phase coating and phase volume likely contributing to the inconsistency in results compared to the previous report. However, it is important to stress that the responses recovered for these two compounds were still sufficient for detection and quantification, and therefore, this does not represent a significant challenge.

3.2. Bacillus subtilis Profiling

The application of this method to the profiling of B. subtilis can provide valuable information about the volatiles produced throughout its growth cycle. Table 1 depicts the compounds that were detected using the above optimized method. Originally, it was hypothesized that the blank agar samples would produce low level VOCs that would remain stable over time. It was also expected that the bacteria would produce additional VOCs that would surpass levels in the uninoculated samples or not be detected in the uninoculated samples. However, this only represented a single category of compound production and compounds were classified into one of three categories as depicted in Figure 4. The initial category was referred to as Category 1 ("produced VOCs") and represents the situation where a VOC was not produced, or was produced at lower levels by comparison, in the uninoculated controls and at higher abundance by the bacteria. In Category 2 VOCs ("consumed VOCs"), the level of the compound was higher in the uninoculated controls and found at a lower concentration in the bacterial samples. Finally, Category 3 VOCs ("equilibrated VOCs") appeared to be those which were either higher or lower in the bacterial samples initially, but eventually equilibrated with the level of the VOC in the uninoculated controls with time. Note that for Category 3 compounds, no significant difference between the two data series was found for any of these VOCs. For all Category 1 ('produced') and Category 2 ('consumed') compounds, a significant difference between the time series for controls and bacterial samples was detected (Table 1). The only exception to this was for 3-methyl-2-pentanone (Category 1), likely due to the minimal difference observed between the controls on days 0 and 1 impacting the overall statistical difference in the time series. However, the difference in 3-methyl-2-pentanone production by the bacteria in comparison to the controls did appear to be substantial on the later days monitored (days 2–4). The peak for 3-methyl-2-pentanone also eluted at a very similar retention time to the toluene peak. Though different quant ions were used in integration for these two compounds, an improved chromatographic separation in this region may also reveal a more significant trend in 3-methyl-2-pentanone.

Table 1. Compounds of interest identified from solid-phase microextraction (SPME) Arrow sampling of *Bacillus subtilis* over a five-day period. Compound labels refer to labels in Figures 4 and 5. The number of samples used to calculate retention time and retention index for each compound is listed for each value. The longitudinal data for each compound of interest was tested for significance (*) when comparing the level of the compound in the uninoculated control to the bacterial sample over five days. Significance was tested using an unpaired two-tailed Student's *t*-test with $\alpha = 0.05$.

Label	Compound	CAS	Retention Time (min)	Retention Index	Trend Category
1	1,3-pentadiene	504-60-9	3.282 ± 0.015 (n = 6)	591.2 ± 0.2 (n = 6)	1 *
2	2-methylfuran	534-22-5	6.049 ± 0.017 (n = 9)	639.5 ± 0.3 (n = 9)	3
3	butanal	123-72-8	6.452 ± 0.016 (n = 12)	647.0 ± 0.3 (n = 12)	2 *
4	2-butanone	78-93-3	6.774 ± 0.027 (n = 14)	653.0 ± 0.5 (n = 14)	2 *
5	trichloromethane	67-66-3	7.495 ± 0.008 (n = 27)	666.7 ± 0.2 (n = 27)	3
6	3-methylbutanal	590-86-3	8.986 ± 0.005 (n = 18)	696.3 ± 0.1 (n = 18)	2 *
7	2-methylbutanal	96-17-3	9.298 ± 0.008 (n = 18)	702.6 ± 0.2 (n = 18)	2 *
8	1-butanol	71-36-3	10.052 ± 0.008 (n = 12)	718.2 ± 0.2 (n = 12)	1 *
9	bromodichloromethane	75-27-4	11.294 ± 0.009 (n = 27)	744.8 ± 0.2 (n = 27)	3
10	1,3-diazine	289-95-2	12.157 ± 0.007 (n = 27)	763.8 ± 0.2 (n = 27)	2 *
11	dimethyl disulfide	624-92-0	12.387 ± 0.005 (n = 30)	769.0 ± 0.1 (n = 30)	3
12	1-methylcyclohexene	591-49-1	12.632 ± 0.002 (n = 7)	774.5 ± 0.1 (n = 7)	1 *
13	2-methylbutanenitrile	18936-17-9	12.794 ± 0.004 (n = 10)	778.2 ± 0.1 (n = 10)	2 *
14	toluene	108-88-3	13.105 ± 0.006 (n = 29)	785.3 ± 0.1 (n = 29)	3
15	3-methyl-2-pentanone	209.282-1	13.277 ± 0.005 (n = 8)	789.2 ± 0.1 (n = 8)	1
16	furfural	98-01-1	17.408 ± 0.005 (n = 17)	890.1 ± 0.1 (n = 17)	2 *
17	2-heptanone	110-43-0	18.913 (n = 1)	929.5 (n = 1)	1 *
18	2,5-dimethylpyrazine	123-32-0	19.281 ± 0.003 (n = 30)	939.4 ± 0.1 (n = 30)	2 *
19	methoxy-phenyl-oxime		21.691 ± 0.006 (n = 6)	1006.3 ± 0.2 (n = 6)	2 *
20	benzaldehyde	100-52-7	22.087 ± 0.003 (n = 18)	1017.6 ± 0.1 (n = 18)	2 *
21	2-ethyl-5-methylpyrazine	13360-64-0	22.535 ± 0.004 (n = 25)	1030.6 ± 0.1 (n = 25)	2 *
22	limonene		22.969 ± 0.004 (n = 15)	1043.3 ± 0.1 (n = 15)	2 *
23	dodecane	112-40-3	28.078 ± 0.003 (n = 29)	1201.7 ± 0.1 (n = 29)	3
24	decanal	112-31-2	29.572 ± 0.003 (n = 15)	1251.2 ± 0.1 (n = 15)	2 *
25	butylated hydroxytoluene	128-37-0	37.804 ± 0.004 (n = 12)	1550.2 ± 0.2 (n = 12)	2 *

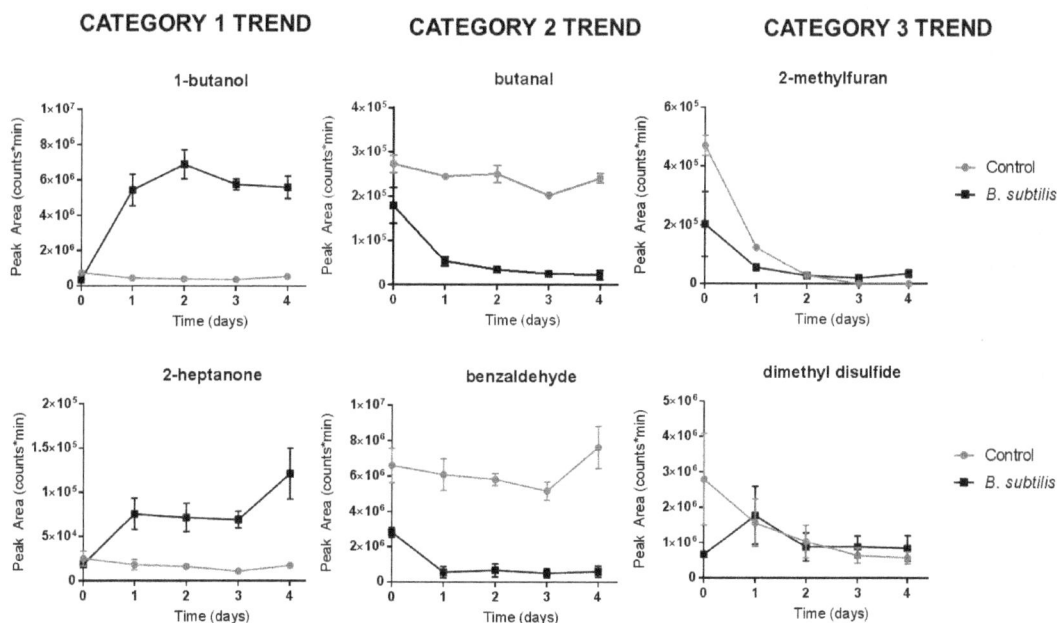

Figure 4. Example volatile organic compounds (VOCs) detected from *Bacillus subtilis* in each of the three categories of compound production. Category 1 compounds ('produced VOCs') were higher in bacterial samples than in control samples. Category 2 compounds ('consumed VOCs') were higher in control samples than in bacterial samples. Category 3 compounds ('equilibrated VOCs') in bacterial samples eventually equilibrated with control samples early in the longitudinal data and did not exhibit a significant difference to the controls over time. Compound trends for all 25 compounds of interest are shown in the Supplementary Material, Figures S1–S3.

When monitoring trends in VOC profile, several analytical priorities exist. First, it is important to use a sampling method that allows a wide range of compounds to be recovered with high efficiency. In this study, this was achieved by implementing the SPME Arrow technique and choosing a fiber type that targeted a wide range of volatiles that were specifically of interest and known to be produced by bacterial metabolism. Second, it is important that the reproducibility of a higher capacity sampling technique remains high priority. In this study, the magnitude of difference in certain VOCs over time and in comparison to blank agar was often very minimal. However, these minimal differences could be relevant quantities to life science applications and therefore, without high precision in the sampling technique, it may not be possible to detect subtle yet important differences in VOC amount. In this study, the reproducibility of SPME Arrow was verified to be adequate for usage in microbiological sampling. Finally, optimal and precise recovery is needed for a quantitative approach to VOC production in order to compare absolute concentrations of VOCs across studies. While absolute quantification was not performed in this study, the analytical parameters established for the SPME Arrow and GC–MS approach indicate that absolute quantification of VOCs is possible and a priority for future research studies. This has recently been highlighted as being critical for advancement of life sciences applications such as monitoring of breath VOCs for medical diagnostics [44]. However, the sentiments in this call for action can also be extended to other areas of importance such as the monitoring of VOCs from decomposing remains for forensic search and recovery [45]. In these application areas, microbial communities play a significant role in VOC production and, therefore, linkage of VOC trends through robust quantification is an important consideration for future advancement in these fields.

This preliminary study on *B. subtilis* identified 25 compounds in the course of the trial. Some of these compounds exhibited significant trends over time, while some did not differ significantly from levels in the uninoculated controls. It is important to note that this was a preliminary investigation using a new sampling technique, and that changing the conditions used in this study will likely impact the identified VOCs. For example, if more advanced separation technology was used, such as comprehensive two-dimensional gas chromatography–mass spectrometry (GC×GC–MS), it is possible that further deconvolution of low-level co-eluting peaks may uncover additional compounds or allow more robust compound identifications. One particular example of this was the lack of a significant difference over time being detected for the compound 3-methyl-2-pentanone, despite the substantial increase in amount of this compound on days 2–4 of *B. subtilis* monitoring. Since this particular compound has a very close retention time to toluene, but a very different chemical structure, a two-dimensional separation would benefit the chromatographic separation of these peaks and potentially improve detection of a significant trend. In addition, other experimental parameters may also contribute to the specificity of detected compounds. In this study, bacteria were cultured on standard nutrient agar; however, the use of different growth media (e.g., tryptic soy agar, blood agar), would have provided the bacteria with different sources of nutrients for their metabolism, likely impacting detected VOCs. Future studies will be aimed at identifying the variation in VOC profile with different growth media. Despite these parameters that could change the resulting VOC profile detected, this study provides preliminary data on SPME Arrow sampling dynamics from microbiological sources which will provide a foundation for future studies.

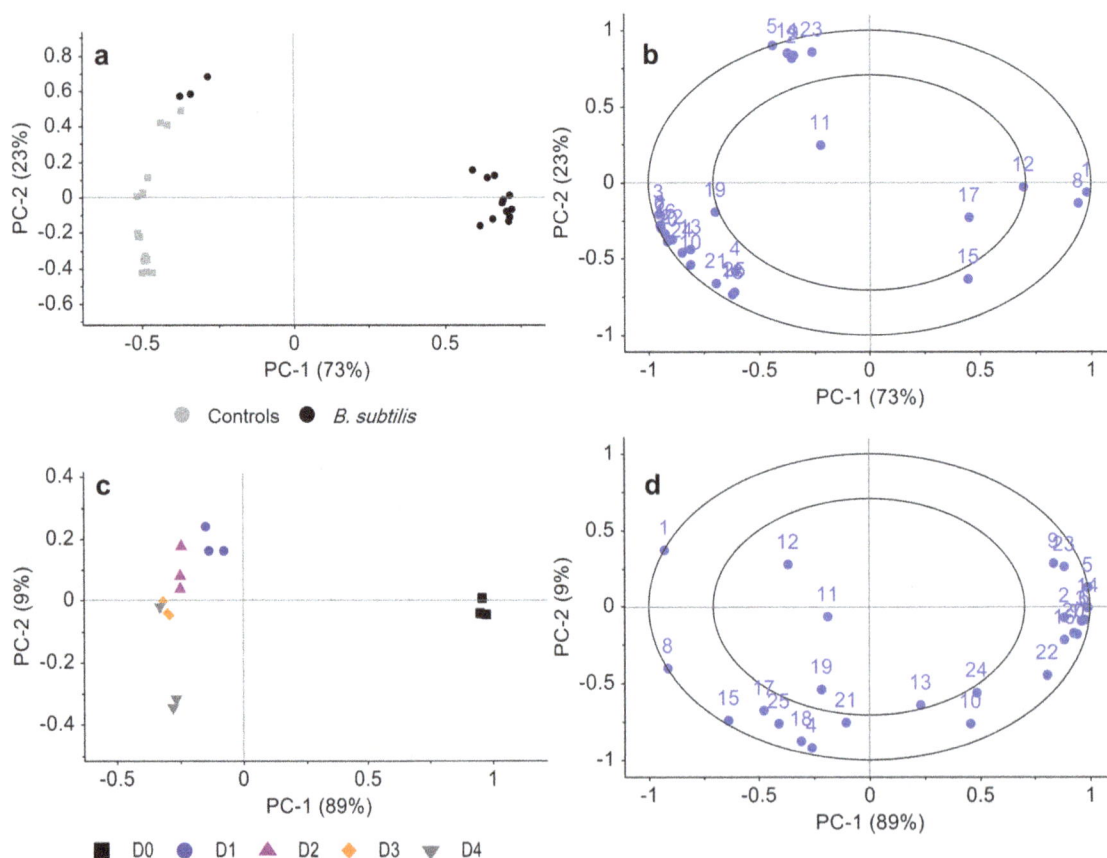

Figure 5. Principal component analysis (PCA) scores plots (left) and loadings plots (right) for data obtained from SPME Arrow profiling of *Bacillus subtilis* samples. For all control and bacterial samples, the PCA scores (**a**) and loadings (**b**) are shown on the top. For the bacterial samples with controls removed, the PCA scores (**c**) and loadings (**d**) are shown on the bottom. Number labels on both loadings plots correspond to compound identities listed in Table 1.

Considering the multivariate nature of the data obtained from these 25 compounds of interest, visualizing the changes in the overall VOC profile of *B. subtilis* over time was desired. PCA was conducted in order to observe these multivariate trends (Figure 5). Overall, the bacterial samples on Day 0 (immediately after culturing) appeared to produce a VOC profile very similar to the control samples on Day 0. The blank agar samples shifted in VOC profile over time; however, the trend for bacterial samples was distinctly different from the shift in VOC profile of controls (Figure 5a). The compounds largely responsible for differentiating the *B. subtilis* samples from the controls in the direction towards the right were the Category 1 compounds (i.e., those VOCs produced in high abundance in the bacteria samples and low abundance in the controls). After removing the controls from the analysis and investigating the longitudinal trends of *B. subtilis*, these Category 1 compounds were responsible for distinguishing the difference between Day 0 and the following days of monitoring; however, the remaining Category 2 and Category 3 compounds appeared to also be important in distinguishing the temporal differences between days 1–4. A number of Category 2 compounds appeared on the far right side of the loadings plot (Figure 5d), as these were the compounds which were consumed in the bacteria samples and present at lower abundance in comparison to the blanks, resulting in the appearance of the PCA scores for later time points on the left side of the scores plot (Figure 5c). Some Category 2 compounds also appeared in the lower left quadrant of the loadings plot (Figure 5d) indicating that they were important markers of the longitudinal progression in VOC profile of *B. subtilis*. Though not performed in this study, future monitoring of bacterial growth phase, microscopically or through spectrophotometric analysis, may also assist in better understanding these longitudinal trends.

4. Conclusions

SPME is a sorbent-based VOC collection technique that has gained widespread popularity across many life science applications due to its ability to focus and enrich the sample using minimal specialized equipment. SPME Arrow offers these same advantages, yet with an increase in sorbent phase volume and alterations to the SPME fiber design that have the ability to increase its mechanical robustness. As such, many life scientists may desire more information about the resulting effects of advancing to an SPME Arrow approach in the future.

This study provides foundational data on the sampling dynamics of commercially-available SPME Arrow in biological applications. The CWR/PDMS fiber was found to be a suitable fiber type for the expected range of volatiles. A sampling extraction time was optimized at 5 min (room temperature) with a 3 min inlet desorption time. Carryover of VOCs was not observed using the optimized method. A significant increase in recovery was observed for most analytes using the CWR/PDMS SPME Arrow. For some compounds where a significant increase was not observed (e.g., benzene, naphthalene), recovery was found to be comparable to the traditional SPME fiber and remained above sufficient levels for analysis. The optimized method produced intraday repeatability values of peak area less than 10% and interday repeatability values of peak area less than 15%, indicating sufficient precision for monitoring longitudinal trends in biological applications.

Finally, the optimized SPME Arrow method was applied to *B. subtilis* cultures and monitored over a five-day period. A total of 25 VOCs were identified across three longitudinal trend categories. In the 'produced VOCs' category, compounds were detected higher in abundance in the bacterial samples than in the control samples over time. In the 'consumed VOCs' category, compounds were detected in lower abundance in the bacterial samples than in the control samples over time. In the 'equilibrated VOCs' category, compounds were either higher or lower in the bacterial samples initially, and eventually equilibrated with levels in the control samples. Compounds in the 'produced VOCs' and 'consumed VOCs' categories generally represented compounds that exhibited a significant difference between the two types of samples, while 'equilibrated VOCs' did not. The SPME Arrow method allowed for a sensitive, selective, and repeatable VOC sampling methodology to be used on bacterial samples, which is imperative to future quantitative approaches using SPME Arrow for microbiological VOC monitoring.

Supplementary Materials: The following are available online at http://www.mdpi.com/2297-8739/5/3/45/s1, Figure S1: Category 1 compounds ('produced VOCs') detected from the headspace of *B. subtilis* samples incubated at 24 °C and monitored over a five day period using the optimized SPME Arrow method, Figure S2: Category

2 compounds ('consumed VOCs') detected from the headspace of *B. subtilis* samples incubated at 24 °C and monitored over a five day period using the optimized SPME Arrow method, Figure S3: Category 3 compounds ('equilibrated VOCs') detected from the headspace of *B. subtilis* samples incubated at 24 °C and monitored over a five day period using the optimized SPME Arrow method, Table S1: Tabulated average peak area, standard deviation (s), and % relative standard deviation (% RSD) for intraday and interday repeatability for reference volatile organic compounds using the optimized SPME Arrow method.

Author Contributions: Conceptualization, K.E.E., D.O.C., and K.A.P.; Data curation, K.E.E., D.O.C., and K.A.P.; Formal analysis, K.E.E., D.O.C., and K.A.P.; Funding acquisition, K.A.P.; Methodology, D.O.C.; Project administration, K.A.P.; Resources, D.O.C.; Supervision, K.A.P.; Writing—original draft, K.E.E. and K.A.P.; Writing—review & editing, K.E.E., D.O.C., and K.A.P.

Funding: This research was funded by the Air Force Research Laboratory grant number CHAM 13-S7700-01-C2, and the Restek Academic Support Program.

Acknowledgments: The authors would like to acknowledge support from the Air Force Minority Leader's Research Competitiveness Program and Restek Corporation's Academic Support Program. The authors would like to recognize the late Carl Sung for his dedication to this research program during its early conception. Hilary Corcoran is acknowledged for her laboratory support. Additional acknowledgements to Hyo Park for his support in assisting with the volatiles reference mix list. Merlin Instrument Company also donated 1.1 mm SPME Arrow microseals for sample injection in this study.

Conflicts of Interest: Restek Corporation donated the chemical standards used in this study through the Restek Academic Support Program. Restek Corporation had no role in the design of the study; in the collection, analysis, or interpretation of data; in the writing of the manuscript; or in the decision to publish the results.

References

1. Tranchida, P.Q.; Dugo, P.; Dugo, G.; Mondello, L. Comprehensive two-dimensional chromatography in food analysis. *J. Chromatogr. A* **2004**, *1054*, 3–16. [CrossRef]

2. Tranchida, P.Q.; Donato, P.; Cacciola, F.; Beccaria, M.; Dugo, P.; Mondello, L. Potential of comprehensive chromatography in food analysis. *Trends Anal. Chem.* **2013**, *52*, 186–205. [CrossRef]

3. Serrano, A.; Gallego, M. Sorption study of 25 volatile organic compounds in several Mediterranean soils using headspace-gas chromatography–mass spectrometry. *J. Chromatogr. A* **2006**, *1118*, 261–270. [CrossRef] [PubMed]

4. Hewitt, A.D. Comparison of sample preparation methods for the analysis of volatile organic compounds in soil samples: Solvent extraction vs. vapor partitioning. *Environ. Sci. Technol.* **1998**, *32*, 143–149. [CrossRef]

5. Kurán, P.; Sojak, L. Environmental analysis of volatile organic compounds in water and sediment by gas chromatography. *J. Chromatogr. A* **1996**, *733*, 119–141. [CrossRef]

6. Voice, T.C.; Kolb, B. Static and dynamic headspace analysis of volatile organic compounds in soils. *Environ. Sci. Technol.* **1993**, *27*, 709–713. [CrossRef]

7. Kleeberg, K.K.; Liu, Y.; Jans, M.; Schlegelmilch, M.; Streese, J.; Stegmann, R. Development of a simple and sensitive method for the characterization of odorous waste gas emissions by means of solid-phase microextraction (SPME) and GC–MS/olfactometry. *Waste Manag.* **2005**, *25*, 872–879. [CrossRef] [PubMed]

8. Tholl, D.; Boland, W.; Hansel, A.; Loreto, F.; Röse, U.S.R.; Schnitzler, J.-P. Practical approaches to plant volatile analysis. *Plant J.* **2006**, *45*, 540–560. [CrossRef] [PubMed]

9. Heath, R.R.; Manukian, A. An automated system for use in collecting volatile chemicals released from plants. *J. Chem. Ecol.* **1994**, *20*, 593–608. [CrossRef] [PubMed]

10. Iqbal, M.A.; Nizio, K.D.; Ueland, M.; Forbes, S.L. Forensic decomposition odour profiling: A review of experimental designs and analytical techniques. *Trends Anal. Chem.* **2017**, *91*, 112–124. [CrossRef]

11. Verheggen, F.; Perrault, K.A.; Caparros Megido, R.; Dubois, L.M.; Francis, F.; Haubruge, E.; Forbes, S.L.; Focant, J.-F.; Stefanuto, P.-H. The odour of death: An overview of current knowledge on characterization and applications. *Bioscience* **2017**, *67*, 600–613. [CrossRef]

12. Cao, W.; Duan, Y. Breath analysis: Potential for clinical diagnosis and exposure assessment. *Clin. Chem.* **2006**, *52*, 800–811. [CrossRef] [PubMed]

13. Bijland, L.R.; Bomers, M.K.; Smulders, Y.M. Smelling the diagnosis A review on the use of scent in diagnosing disease. *Neth. J. Med.* **2013**' *71*, 300–307. [PubMed]

14. Hamm, S.; Bleton, J.; Connan, J.; Tchapla, A. A chemical investigation by headspace SPME and GC–MS of volatile and semi-volatile terpenes in various olibanum samples. *Phytochemistry* **2005**, *66*, 1499–1514. [CrossRef] [PubMed]

15. Cnuts, D.; Perrault, K.A.; Stefanuto, P.-H.; Dubois, L.M.; Focant, J.-F.; Rots, V. Fingerprinting glues using HS-SPME GC×GC-HRTOFMS: A new powerful method allows tracking glues back in time. *Archaeometry* **2018**, 1–16. [CrossRef]

16. Perrault, K.A.; Stefanuto, P.; Dubois, L.; Cnuts, D.; Rots, V.; Focant, J.-F. A new approach for the characterization of organic residues from stone tools using GC×GC-TOFMS. *Separations* **2016**, *3*, 1–16. [CrossRef]

17. Jerković, I.; Marijanović, Z.; Gugić, M.; Roje, M. Chemical profile of the organic residue from ancient amphora found in the Adriatic Sea determined by direct GC and GC–MS analysis. *Molecules* **2011**, *16*, 7936–7948. [CrossRef] [PubMed]

18. Kremser, A.; Jochmann, M.A.; Schmidt, T.C. Systematic comparison of static and dynamic headspace sampling techniques for gas chromatography. *Anal. Bioanal. Chem.* **2016**, *408*, 6567–6579. [CrossRef] [PubMed]

19. Harper, M. Sorbent trapping of volatile organic compounds from air. *J. Chromatogr. A* **2000**, *885*, 129–151. [CrossRef]

20. Arthur, C.L.; Pawliszyn, J. Solid phase microextraction with thermal desorption using fused silica optical fibers. *Anal. Chem.* **1990**, *62*, 2145–2148. [CrossRef]

21. Zhang, Z.; Pawliszyn, J. Headspace solid-phase microextraction. *Anal. Chem.* **1993**, *65*, 1843–1852. [CrossRef]

22. Pawliszyn, J. Theory of solid-phase microextraction. *J. Chromatogr. Sci.* **2012**, *38*, 13–59. [CrossRef]

23. Bagheri, H.; Piri-Moghadam, H.; Naderi, M. Towards greater mechanical, thermal and chemical stability in solid-phase microextraction. *Trends Anal. Chem.* **2012**, *34*, 126–139. [CrossRef]

24. Spietelun, A.; Kloskowski, A.; Chrzanowski, W.; Namieśnik, J. Understanding solid-phase microextraction: Key factors influencing the extraction process and trends in improving the technique. *Chem. Rev.* **2013**, *113*, 1667–1685. [CrossRef] [PubMed]

25. Baltussen, E.; Sandra, P.; David, F.; Cramers, C. Stir bar sorptive extraction (SBSE), a novel extraction technique for aqueous samples: Theory and principles. *J. Microcolumn Sep.* **1999**, *11*, 737–747. [CrossRef]

26. Tienpont, B.; David, F.; Bicchi, C.; Sandra, P. High capacity headspace sorptive extraction. *J. Microcolumn Sep.* **2000**, *12*, 577–584. [CrossRef]

27. Helin, A.; Rönkkö, T.; Parshintsev, J.; Hartonen, K.; Schilling, B.; Läubli, T.; Riekkola, M.L. Solid phase microextraction arrow for the sampling of volatile amines in wastewater and atmosphere. *J. Chromatogr. A* **2015**, *1426*, 56–63. [CrossRef] [PubMed]

28. Kremser, A.; Jochmann, M.A.; Schmidt, T.C. PAL SPME Arrow—Evaluation of a novel solid-phase microextraction device for freely dissolved PAHs in water. *Anal. Bioanal. Chem.* **2016**, *408*, 943–952. [CrossRef] [PubMed]

29. Forbes, S.L.; Troobnikoff, A.N.; Ueland, M.; Nizio, K.D.; Perrault, K.A. Profiling the decomposition odour at the grave surface before and after probing. *Forensic Sci. Int.* **2016**, *259*, 193–199. [CrossRef] [PubMed]

30. Armstrong, P.; Nizio, K.D.; Perrault, K.A.; Forbes, S.L. Establishing the volatile profile of pig carcasses as analogues for human decomposition during the early postmortem period. *Heliyon* **2016**, *2*, e00070. [CrossRef] [PubMed]

31. Perrault, K.A.; Stefanuto, P.-H.; Stuart, B.H.; Rai, T.; Focant, J.-F.; Forbes, S.L. Detection of decomposition volatile organic compounds in soil following removal of remains from a surface deposition site. *Forensic Sci. Med. Pathol.* **2015**, *11*, 376–387. [CrossRef] [PubMed]

32. Carter, D.O.; Tomberlin, J.K.; Benbow, M.E.; Metcalf, J.L. (Eds.) *Forensic Microbiology*, 1st ed.; John Wiley & Sons Ltd.: Chichester, UK, 2017; ISBN 978-1-119-06255-4.

33. Bean, H.D.; Dimandja, J.-M.D.; Hill, J.E. Bacterial volatile discovery using solid phase microextraction and comprehensive two-dimensional gas chromatography–time-of-flight mass spectrometry. *J. Chromatogr. B* **2012**, *901*, 41–46. [CrossRef] [PubMed]

34. Bean, H.D.; Rees, C.A.; Hill, J.E. Comparative analysis of the volatile metabolomes of Pseudomonas aeruginosa clinical isolates. *J. Breath Res.* **2016**, *10*, 47102. [CrossRef] [PubMed]

35. Nizio, K.D.; Perrault, K.A.; Troobnikoff, A.N.; Ueland, M.; Shoma, S.; Iredell, J.R.; Middleton, P.G.; Forbes, S.L. In vitro volatile organic compound profiling using GC×GC-TOFMS to differentiate bacteria associated with lung infections: A proof-of-concept study. *J. Breath Res.* **2016**, *10*, 26008. [CrossRef] [PubMed]

36. Filipiak, W.; Sponring, A.; Filipiak, A.; Ager, C.; Schubert, J.; Miekisch, W.; Amann, A.; Troppmair, J. TD-GC–MS analysis of volatile metabolites of human lung cancer and normal cells in vitro. *Cancer Epidemiol. Biomark. Prev.* **2010**, *19*, 182–195. [CrossRef] [PubMed]

37. Perrault, K.A.; Nizio, K.D.; Forbes, S.L. A comparison of one-dimensional and comprehensive two-dimensional gas chromatography for decomposition odour profiling using inter-year replicate field trials. *Chromatographia* **2015**, *78*, 1057–1070. [CrossRef]

38. Logan, N.A.; De Vos, P. Genus I: Bacillus. In *Bergey's Manual of Systematic Bacteriology*; De Vos, P., Garrity, G.M., Jones, D., Krieg, N.R., Ludwig, W., Rainey, F.A., Schleifer, K.-H., Whitman, W.B., Eds.; Springer: New York, NY, USA, 2019; pp. 21–128.

39. Dyer, B.D. *A Field Guide to Bacteria*; Cornell University Press: Ithaca, NY, USA, 2003.

40. Chun, L.P.; Miguel, M.J.; Junkins, E.N.; Forbes, S.L.; Carter, D.O. An initial investigation into the ecology of culturable aerobic postmortem bacteria. *Sci. Justice* **2015**, *55*, 394–401. [CrossRef] [PubMed]

41. Carter, D.O.; Yellowlees, D.; Tibbett, M. Cadaver decomposition in terrestrial ecosystems. *Naturwissenschaften* **2007**, *94*, 12–24. [CrossRef] [PubMed]

42. Logan, N.A. Bacillus species of medical and veterinary importance. *J. Med. Microbiol.* **1988**, *25*, 157–165. [CrossRef] [PubMed]

43. Barreira, L.M.F.; Parshintsev, J.; Kärkkäinen, N.; Hartonen, K.; Jussila, M.; Kajos, M.; Kulmala, M.; Riekkola, M.L. Field measurements of biogenic volatile organic compounds in the atmosphere by dynamic solid-phase microextraction and portable gas chromatography-mass spectrometry. *Atmos. Environ.* **2015**, *115*, 214–222. [CrossRef]

44. Smith, D.; Španěl, P. On the importance of accurate quantification of individual volatile metabolites in exhaled breath. *J. Breath Res.* **2017**, *11*, 4. [CrossRef] [PubMed]

45. Stefanuto, P.-H.; Perrault, K.A.; Lloyd, R.M.; Stuart, B.H.; Rai, T.; Forbes, S.L.; Focant, J.-F. Exploring new dimensions in cadaveric decomposition odour analysis. *Anal. Methods* **2015**, *7*, 2287–2294. [CrossRef]

## 11

Developing a Method for the Collection and Analysis of Burnt Remains for the Detection and Identification of Ignitable Liquid Residues Using Body Bags, Dynamic Headspace Sampling, and TD-GC×GC-TOFMS

Katie D. Nizio *[iD] and **Shari L. Forbes** [iD]

Centre for Forensic Science, University of Technology Sydney, P.O. Box 123, Broadway, NSW 2007, Australia; Shari.Forbes@uts.edu.au
* Correspondence: KatieDNizio@gmail.com

Abstract: In cases of suspected arson, a body may be intentionally burnt to cause loss of life, dispose of remains, or conceal identification. A primary focus of a fire investigation, particularly involving human remains, is to establish the cause of the fire; this often includes the forensic analysis of fire debris for the detection of ignitable liquid residues (ILRs). Commercial containers for the collection of fire debris evidence include metal cans, glass jars, and polymer/nylon bags of limited size. This presents a complication in cases where the fire debris consists of an intact, or partially intact, human cadaver. This study proposed the use of a body bag as an alternative sampling container. A method was developed and tested for the collection and analysis of ILRs from burnt porcine remains contained within a body bag using dynamic headspace sampling (using an Easy-VOC™ hand-held manually operated grab-sampler and stainless steel sorbent tubes containing Tenax TA) followed by thermal desorption comprehensive two-dimensional gas chromatography–time-of-flight mass spectrometry (TD-GC×GC-TOFMS). The results demonstrated that a body bag containing remains burnt with gasoline tested positive for the presence of gasoline, while blank body bag controls and a body bag containing remains burnt without gasoline tested negative. The proposed method permits the collection of headspace samples from burnt remains before the remains are removed from the crime scene, limiting the potential for contamination and the loss of volatiles during transit and storage.

Keywords: forensic chemistry; fire debris analysis; fire debris packaging; burnt remains; ignitable liquid residues (ILRs); volatile organic compounds (VOCs); dynamic headspace sampling; thermal desorption; comprehensive two-dimensional gas chromatography (GC×GC); time-of-flight mass spectrometry (TOFMS)

---

1. Introduction

The detection and identification of ignitable liquid residues (ILRs) from fire debris can be a significant challenge for fire investigators and forensic practitioners. Detection of ILRs is important in cases of arson whereby a structure, vehicle, or other property has been intentionally destroyed, often through the application of accelerants [1]. Arson can also lead to the loss of life, whether accidentally or intentionally, to dispose of the remains, destroy evidence, or conceal the victim's identity. When a fire investigation involves a fatality, the body or remains must be recovered to determine the cause and manner of death [2]. Additionally, detection of ILRs from the body or tissues may assist in understanding the circumstances surrounding death.

The identification of ignitable liquids requires the extraction of highly volatile organic compounds (VOCs) from fire debris collected at a scene [3,4]. A range of containers may be used to collect fire debris depending on the volume of debris and state of recovery [5–7]. These typically include metal cans [8], glass jars (e.g., Mason jars) [2,9], and polymer/nylon bags [6,10–13] of varying size, though, generally, not large enough for the collection of burnt human remains. While detection of ILRs from human remains can be conducted through the analysis of postmortem blood samples [14–16] or other postmortem tissues excised during the autopsy [2], contamination or a loss of volatiles may occur during these processes. Nevertheless, the recovery of victim remains from a fire scene usually involves the use of a body bag to transport the remains to the morgue so that an autopsy can be performed; therefore, this study introduces the use of a body bag as an integral addition to the fire debris analysts' packaging arsenal. The body bag not only provides a packaging container large enough for the collection of burnt victim remains, but also permits sample collection at the scene, reducing the potential for contamination and the loss of volatiles.

The extraction of VOCs from fire debris can be undertaken using a range of sample collection techniques. Common methods include direct headspace sampling [2,17], activated charcoal strips [1,3,18], dynamic headspace sampling [19–21], and solid-phase microextraction (SPME) [4,8,22,23]. For trace analysis, methods that concentrate the sample, such as dynamic headspace sampling and SPME, are preferred for the analysis of ILRs. Sample collection is followed by analysis using gas chromatography-mass spectrometry (GC-MS) as outlined in ASTM E1618-14 [24]. However, fire debris typically generates a highly complex mixture of VOCs, and GC-MS can produce unresolved chromatograms whereby trace levels of ignitable liquids may be masked by background pyrolysis products. Additionally, human remains produce a range of natural VOCs when burnt [25] that must be distinguished from the VOCs present in gasoline or other ignitable liquids, further confounding the characterization of accelerant use at a fire scene. The use of comprehensive two-dimensional gas chromatography–time-of-flight mass spectrometry (GC×GC-TOFMS) has been shown to alleviate such issues through increased peak capacity, improved resolution, enhanced sensitivity, and structured chromatograms, allowing for the rapid characterization of ignitable liquids present [26].

This study proposes a new method involving collection of VOCs directly from the body bag using dynamic headspace sampling, combined with analysis by GC×GC-TOFMS. As in previous studies [2,25], pig carcasses were used as analogues for human remains. Due to its prevalence in arson cases [7], gasoline was chosen as the accelerant of interest in this study. The benefit of this technique is the ability to sample directly from the body bag at the crime scene, and prior to transport and storage, to reduce the loss of key compounds for characterization of ILRs. Sample collection directly onto a sorbent tube allows the tube to be sealed and stored safely for several weeks prior to the analysis being conducted. This method has the potential to significantly increase the likelihood of detecting and identifying ILRs from victims of arson.

2. Materials and Methods

2.1. Experimental Design

Due to the legal and ethical regulations associated with the acquisition and burning of human cadavers, pig carcasses were selected as a substitute for human remains in this study. Two adult, domestic pig carcasses (*Sus scrofa domesticus* L.), weighing approximately 50–60 kg each, were purchased, postmortem, by way of excess stock from Hawkesbury Valley Meat Processors (licensed abattoir in Sydney, NSW, Australia). In accordance with the guidelines of the *Australian code for the care and use of animals for scientific purposes* [27], animal ethics approval was not required for this study since the pig carcasses used herein were: (1) acquired postmortem; and (2) not killed specifically for the purposes of this research.

To mimic a clothed human cadaver, each pig carcass was clothed in a white 100% cotton t-shirt and a pair of black polyester briefs (Kmart, Sydney, NSW, Australia) prior to burning. The experimental burns were performed outdoors by trained fire and rescue personnel at the Fire & Rescue NSW Fire Investigation Research Unit (Londonderry, NSW, Australia) within 24 h of death. Each pig carcass was burned independently on top of a 2 m × 2 m stainless steel tray, which was used to contain any fluid leakage from the pigs, as well as to prevent the fires from spreading out of control. The pig carcasses were placed on their sides for burning, and leaf litter and brush were placed underneath and on top of each pig carcass to assist in sustaining the fire.

The first experimental burn was conducted in the absence of any ignitable liquids, with a propane gas torch used to ignite the brush surrounding the pig carcass. The second experimental burn was conducted in the presence of ~720 mL unleaded 91-octane gasoline (purchased locally). A hand-held butane gas lighter was used to ignite the gasoline immediately after it was poured onto the pig carcass and surrounding brush. A small sample of the gasoline (~30 mL) was retained as a reference sample. The temperature of each experimental burn was measured and recorded using a DT85 data-logger and five Type N thermocouples (20 m × 3 mm; TC Measurement & Control, Melbourne, VIC, Australia) placed above the upper and lower torso regions and underneath the hind legs, torso, and head of each pig carcass. Each fire was extinguished with water after 20 min of burning.

2.2. Sample Collection

Prior to conducting the experimental burns, control headspace samples were collected in triplicate, from two brand-new polyethylene tarpaulin body bags (175 cm × 70 cm; Pro-Pac Packaging Limited, Wetherill Park, NSW, Australia) for the purpose of measuring the background VOC profile produced by the body bags and surrounding environment. Each body bag was left open to the outdoor environment (Figure 1a) for 15 min, before being zipped closed and sampled through a small opening in the body bag closure (Figure 1b). The headspace within each body bag was sampled dynamically onto a stainless steel sorbent tube containing Tenax TA (35/60 mesh; Markes International Ltd., Llantrisant, RCT, UK) using an Easy-VOC™ grab-sampler (Figure 1c) (Markes international Ltd.) in 5 × 100 mL headspace aliquots, sampled successively (total headspace volume sampled = 500 mL).

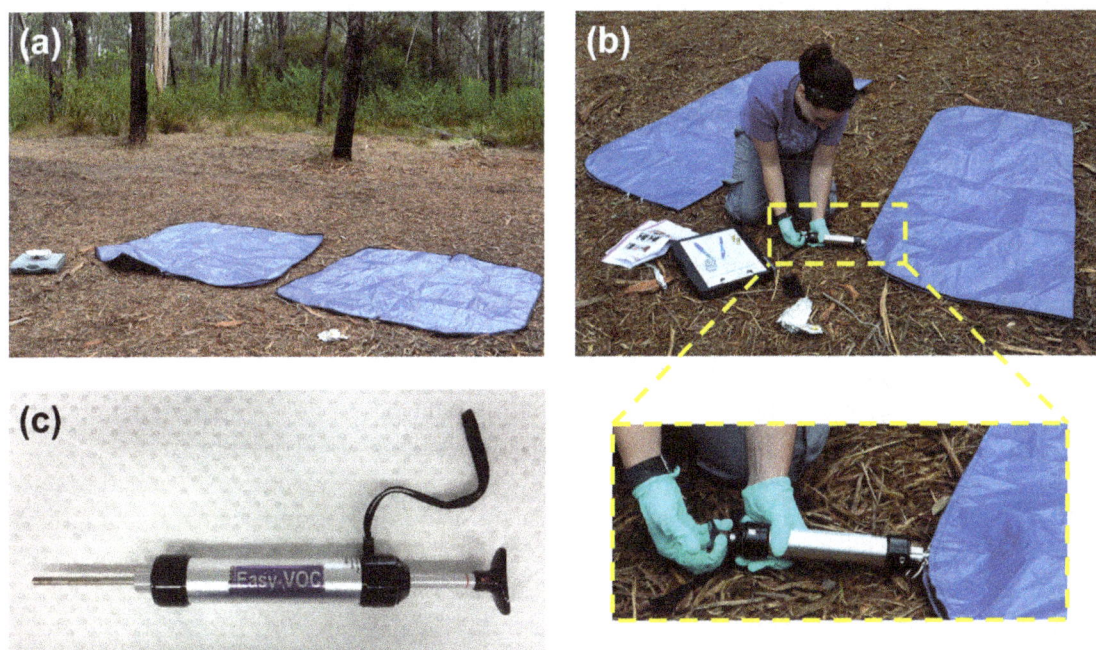

Figure 1. Collection of control headspace samples from empty body bags: (**a**) body bags open to environment prior to sample collection; (**b**) headspace sample collection through small opening in body bag closure; and (**c**) Easy-VOC™ grab-sampler with sorbent tube attached.

After burning, each pig carcass was left to cool for 30 min before being placed into one of the polyethylene tarpaulin body bags. The headspace within the body bag was permitted to accumulate for 15 min. A total of five replicate experimental headspace samples were collected from each pig carcass using the procedure previously described for the control samples. For the purposes of consistency, all experimental headspace samples were collected from the corner of the body bag nearest to the head. The headspace within each body bag was allowed to re-accumulate for 5 min between samples.

Following sample collection, each sorbent tube was sealed with brass long-term storage caps, wrapped in aluminum foil, and placed in a glass air-tight container for transportation to the laboratory in accordance with the U.S. Environmental Protection Agency Method TO-17 [28]. Sorbent tubes were stored at 4 °C in this condition in the laboratory, until GC×GC-TOFMS analysis was performed. Adhering to the guidelines of the *Australian code for the care and use of animals for scientific purposes* [27], the two pig carcasses burned in this study were utilized in a subsequent study investigating the influence of fire modification on the odor of decomposition [29], in an attempt to reduce the number of pig carcasses utilized in the authors' research studies.

2.3. TD-GC×GC-TOFMS Analysis

Sorbent tube thermal desorption (TD) was achieved via a UNITY 2 Thermal Desorber and Series 2 ULTRA multitube autosampler (Markes International Ltd.), followed by sample analysis on a Pegasus® 4D GC×GC-TOFMS system (LECO, Castle Hill, NSW, Australia). The TD unit and autosampler were coupled to the GC×GC-TOFMS system via a 1 m uncoated fused silica transfer line (Markes International Ltd.) maintained at 140 °C, and connected to the first dimension (^{1}D) GC column inside the ^{1}D GC oven using an Ultimate Union Kit (Agilent Technologies, Mulgrave, NSW, Australia). Markes Maverick control software for Unity 2 (version 4.1.29; Markes International Ltd.) was used to control the TD unit and autosampler, and ChromaTOF® software (version 4.51.6.0; LECO) was used for GC×GC-TOFMS control.

Prior to first use, all freshly packed sorbent tubes were initially conditioned under a 50 mL/min flow of high purity helium (BOC, Sydney, NSW, Australia) for 120 min at 320 °C, followed by 30 min at 335 °C, according to the manufacturer's recommendations. Thermal desorption of the sorbent tubes was operated under double split flow conditions (i.e., split flow during both tube and trap desorption steps). Three different tube desorption/trap desorption split flow conditions were tested using high purity helium (BOC): (1) 20 mL/min/20 mL/min (split ratio = 15.4:1); (2) 20 mL/min/ 50 mL/min (split ratio = 36.4:1); (3) 50 mL/min/100 mL/min (split ratio = 102.0:1). For all split flow conditions tested, tube desorption was performed at a temperature of 320 °C for 15 min with collection onto a Tenax TA/Carbograph 1TD general purpose cold trap maintained at −10 °C, followed by trap desorption at 320 °C for 3 min. Split flow conditions were evaluated using the samples collected from the pig carcass burnt in the presence of gasoline. Optimal desorption conditions were determined to be a double split flow of 50 mL/min during tube desorption, and 100 mL/min during trap desorption, in order to reduce overloading observed from the gasoline trace (see Section 3.2). Thermal desorption of the reference gasoline sample (i.e., 0.2 μL 91-octane gasoline spiked directly onto sorbent tube) was performed under a double split flow of 100 mL/min during tube desorption, and 100 mL/min during trap desorption (split ratio = 153.0:1), to further reduce overloading of the gasoline trace.

GC×GC-TOFMS analysis was performed according to a previously published method designed to achieve a near-theoretical maximum in peak capacity gain for the forensic analysis of ignitable liquids [26]. Briefly, the column configuration consisted of a 60 m × 0.25 mm inner diameter (i.d.), 0.50 μm film thickness (d_f) Rxi®-1ms column in the ^{1}D and a 1.1 m × 0.25 mm i.d., 0.5 μm d_f Stabilwax® column in the second dimension (^{2}D) (Restek Corporation, Bellefonte, PA, USA). The ^{1}D and ^{2}D columns were connected directly before the modulator using a SilTite™ μ-Union (SGE Analytical Science, Wetherill Park, NSW, Australia). A constant flow of 2.0 mL/min of high purity helium (BOC) was used as the carrier gas. The ^{1}D oven was held at 40 °C for 3 min, and then ramped at 3.5 °C/min to a final temperature of 255 °C, which was held for a further 5 min (total GC×GC-TOFMS runtime

= 69.43 min). Relative to the ^{1}D oven, the ^{2}D oven and modulator were programmed to have +5 °C and +20 °C offsets, respectively. The modulation period was 2 s with a 0.5 s hot pulse time and 0.5 s cooling time between stages. Mass spectra were acquired at a rate of 400 spectra/s for m/z 35–400. The MS transfer line and ion source were maintained at 240 °C and 225 °C, respectively. Electron ionization was achieved at 70 eV with the detector operating at a 200 V offset above the optimized detector voltage.

2.4. Data Processing

Data processing and chromatographic visualization were achieved using ChromaTOF® software (version 4.51.6.0; LECO). To assist with chromatographic visualization of trace components, a color scale of 0–10% of the normalized signal intensity was utilized. The baseline of each chromatogram was smoothed automatically, with an 80% offset. Peak searching was performed using an expected peak width of 8 s in the ^{1}D and 0.1 s in the ^{2}D, with a signal-to-noise ratio (S/N) cut-off of 150 for base peaks and 20 for subpeaks. Subpeaks were combined with each other, and their corresponding base peak when a mass spectral match of 65% or greater was detected. A forward search of the 2011 National Institute of Standards and Technology (NIST) mass spectral library database was used to tentatively identify peaks with a minimum similarity match of 80% or greater.

Chromatographic alignment was carried out using the Statistical Compare software feature within ChromaTOF® (LECO). Samples were input into Statistical Compare and separated into four classes: (1) reference gasoline (n = 3); (2) body bag control (n = 6); (3) pig burnt *without* gasoline (n = 3); and (4) pig burnt *with* gasoline (n = 3). During chromatographic alignment, maximum retention time deviations permitted between samples were restricted to 2 s (i.e., one modulation period) in the ^{1}D and 0 s in the ^{2}D. Peak re-searching was performed during chromatographic alignment using a lower S/N cut-off of 20 to search for peaks not found during the initial peak finding step. Peak alignment required a minimum similarity match of 60% or greater. Analytes that did not meet this mass spectral match threshold, and that were not detected in at least three samples across the four classes, were removed from the final compound list. Statistical Compare was used to calculate the ratio of between-class variance to within-class variance (i.e., the Fisher ratio) for each analyte using analyte peak areas (calculated using unique mass). Analytes with Fisher ratios above a critical value (F_{crit}—computed in Microsoft Excel using the F-distribution) were considered to be class-distinguishing analytes (i.e., analytes that statistically differed in abundance between the defined classes) [30–32]. Analytes with a Fisher ratio above the F_{crit} threshold (i.e., F_{crit} = 3.59) were exported as a *.csv file and imported into Microsoft Excel for the manual removal of chromatographic artefacts (e.g., column or sorbent bleed) and further processing.

Principal component analysis (PCA) was applied to the analytes of interest (i.e., only those analytes that were retained after Fisher ratio filtering) using The Unscrambler® X (version 10.3.31813.89; CAMO Software, Oslo, Norway). Prior to PCA, data pre-processing steps performed in The Unscrambler® X included mean centering, variance scaling, and unit vector normalization [33]. Following PCA, the dataset was evaluated and confirmed to contain no outlying points using the Hotelling's T^2 95% confidence limit.

3. Results and Discussion

3.1. Experimental Burn Conditions and Observations

The thermocouples placed underneath the pig carcass burnt *without* gasoline recorded fire temperatures ranging from 91.8 to 832.2 °C, while the thermocouple used to measure the temperature of the fire directly above the lower torso measured a maximum temperature of 648 °C (note that the thermocouple placed above the upper torso failed to record a reading during the fire—the reason for the malfunction is unknown). The overall level of burning observed varied (Figure 2a,b), with the topside presenting the most consistent degree of burning. On the Crow–Glassman scale (CGS) [34], the level of fire damage to the head, neck, and torso regions of the remains was categorized as level 2, while the degree of fire damage to the limbs, which remained intact, was categorized as level 1. The polyester briefs were entirely consumed during the fire, while a small portion of the 100% cotton t-shirt was found charred underneath the upper torso of the pig carcass.

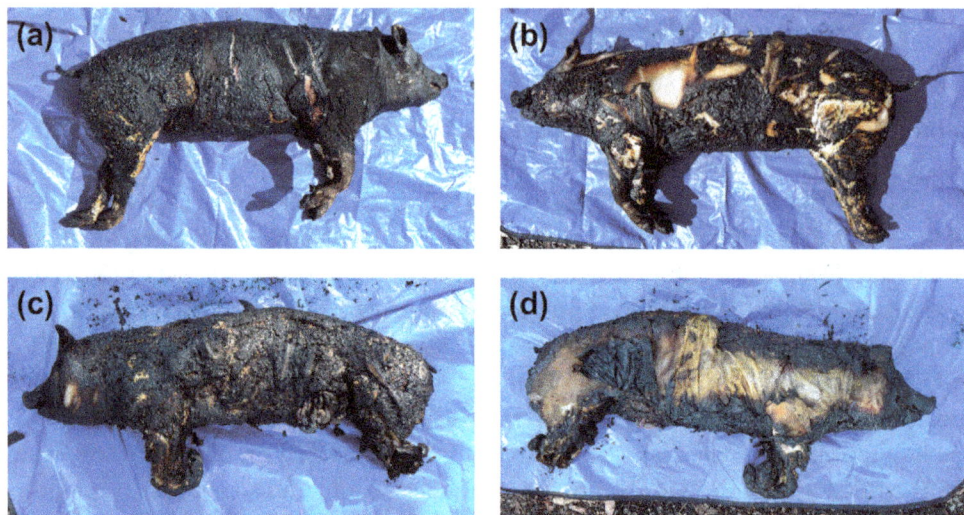

Figure 2. Photographs of the (**a**) topside and (**b**) underside of the pig carcass burnt *without* gasoline and the (**c**) topside and (**d**) underside of the pig carcass burnt *with* gasoline.

Lower overall temperatures were recorded for the fire conducted *with* gasoline, with fire temperatures ranging from 250.0 to 451.8 °C underneath the pig carcass, and 536.7–669.2 °C above the pig carcass. Fire damage was largely observed to be limited to the topside of the pig carcass (Figure 2c), with the exception of the head and limbs, which showed extensive fire damage on both sides (Figure 2c,d). After burning, a tear in the abdominal region resulting in intestinal herniation (i.e., exposure of the visceral organs—Figure 2c) was observed. Both the anterior and posterior limbs were very delicate following burning (Figure 2c,d), with the front limbs displaying cadaveric spasm, and elements of the hind legs disarticulated and unrecoverable after the fire. Overall, the level of fire damage was categorized as CGS level 2 for the head and limbs, as well as the neck, torso, and posterior regions on the topside of the remains. Fire damage to the neck, torso, and posterior regions on the underside of the remains were categorized as CGS level 1. A small portion of the polyester briefs was recovered intact and unburnt underneath the remains, along with the entire right side of the 100% cotton t-shirt, which showed extensive charring along the edges.

Depending upon the resting position of the remains during burning, and the combustibility of the substrate upon which the remains are burned (e.g., a non-combustible floor), fire modification is often limited to a single side of the remains (i.e., the side which is not in contact with the substrate) and clothing shielded from the fire (i.e., located between the remains and the substrate) may also be recoverable. In this study, the pig carcass burnt *without* gasoline was burnt on top of a pile of brush considerably larger than that used in the second burn for the pig carcass burnt *with* gasoline, where the pile of brush was concentrated more on top of the pig carcass, as opposed to underneath. As a result, the pig carcass burnt *without* gasoline exhibited burns on both sides of the remains with very little clothing recoverable, while the pig carcass burnt *with* gasoline exhibited burns predominantly limited to one side of the remains with larger pieces of unburnt clothing intact and recoverable as previously described. This information is important to consider when conducting experimental burns, with the desired burn pattern dependent upon the scenario being simulated, also bearing in mind the large variability that occurs in every fire due to the uncontrollable variables of combustion. Given this variability, it is imperative that experimental burns, such as those conducted herein, are replicated, to provide a statistically significant dataset.

This information is also important to consider when collecting fire debris for the analysis of ILRs in cases of suspected arson. Following burning, any recoverable clothing and/or unburnt substrate below the remains, where ignitable liquids may pool and be absorbed, should be considered for collection and subsequent analysis of ILRs. However, in cases where clothing and substrate are unrecoverable, or are known to produce volatiles that interfere with ILR detection and identification, (e.g., carpet and carpet padding [35]), the intact or partially intact remains may present the only source of fire debris available for analysis and, as such, is the focus of this study.

3.2. Sample Collection Method

The collection of headspace samples was achieved using an Easy-VOC™ hand-held manually operated grab-sampler from Markes International Ltd. (Figure 1c). The Easy-VOC™ device permits precise volumes of air, sampled in 50 or 100 mL aliquots, to be sampled (and thus pre-concentrated) directly onto a sorbent tube. If required, larger volumes of air can be sampled by drawing a series of aliquots onto the same sorbent tube in rapid succession, increasing sensitivity for trace-level target compounds. Since the Easy-VOC™ device does not require electrical power or batteries, it is an ideal choice for collecting samples in the field.

In this study, five 100 mL aliquots of headspace were sampled successively onto each sorbent tube (total headspace volume sampled = 500 mL), and a total of five replicate headspace samples were collected from each set of remains. The first three replicates collected from the pig carcass burnt *with* gasoline were used to evaluate the split ratio (i.e., tube and trap desorption split flow conditions for thermal desorption) for optimal detection of the gasoline trace. The final two replicates collected were analyzed with the optimal split ratio, to provide a confirmation of results and to provide replicates for multivariate data analysis (see Section 3.4).

A split ratio of 15.4:1 was tested first (i.e., tube desorption/trap desorption split flow = 20 mL/min/ 20 mL/min), resulting in overloaded and co-eluting peaks (Figure 3a). To reduce overloading, the split ratio was increased to 36.4:1 (i.e., tube desorption/trap desorption split flow = 20 mL/min/ 50 mL/min), but overloading was still observed for some of the more concentrated volatiles (Figure 3b). The split ratio was further increased to 102.0:1 (i.e., tube desorption/trap desorption split flow = 50 mL/min/100 mL/min), producing sharp, focused peaks with minimal overloading observed (Figure 3c). As the split ratio was increased to reduce overloading of the more concentrated volatiles, the number of compounds detected overall inadvertently decreased, due to the loss of trace volatiles. This can be observed in Figure 3, with the loss of peak markers particularly in the latter half of the contour plots (e.g., loss of the C_5-alkyl benzenes [26] in Figure 3b,c). The increased split ratio and subsequent loss of volatiles detected, however, did not significantly affect the overall detection of the targeted gasoline compounds (see ASTM E1618-14—Table 3 [24]) used to positively identify the presence of gasoline in this study (see Section 3.3), with the exception of the loss of 5-methylindane. The loss of 5-methylindane was accepted, and the split ratio of 102.0:1 was chosen as optimal, in favor of the sharp, focused peaks produced, which provided stronger overall mass spectral matches for tentative identification of peaks and was more amenable to subsequent multivariate data analysis (see Section 3.4).

Given the overloading observed at the lower split ratios tested, it stands to reason that a smaller volume, and thus fewer aliquots, of headspace could have been sampled. Headspace sampling in this study lasted approximately 5 min per sample/sorbent tube (i.e., ~1 min on average per aliquot of headspace sampled). Accounting for initial headspace accumulation (15 min) and re-accumulation between samples (5 min), the entire sample collection procedure (for five replicates) took ~60 min. Sampling a lower volume of headspace via the collection of fewer aliquots per sample may be desirable for a more rapid sample collection procedure; a lower overall sample volume may also reduce or eliminate the need for headspace re-accumulation between samples further reducing the timeline for sample collection. It is noted that the final three replicate samples collected in this trial were analyzed in order of collection to see if the detected signal decreased over time—i.e., diminishing odor due to removal of headspace from body bag with each sample collected. However, no diminishing signal was observed.

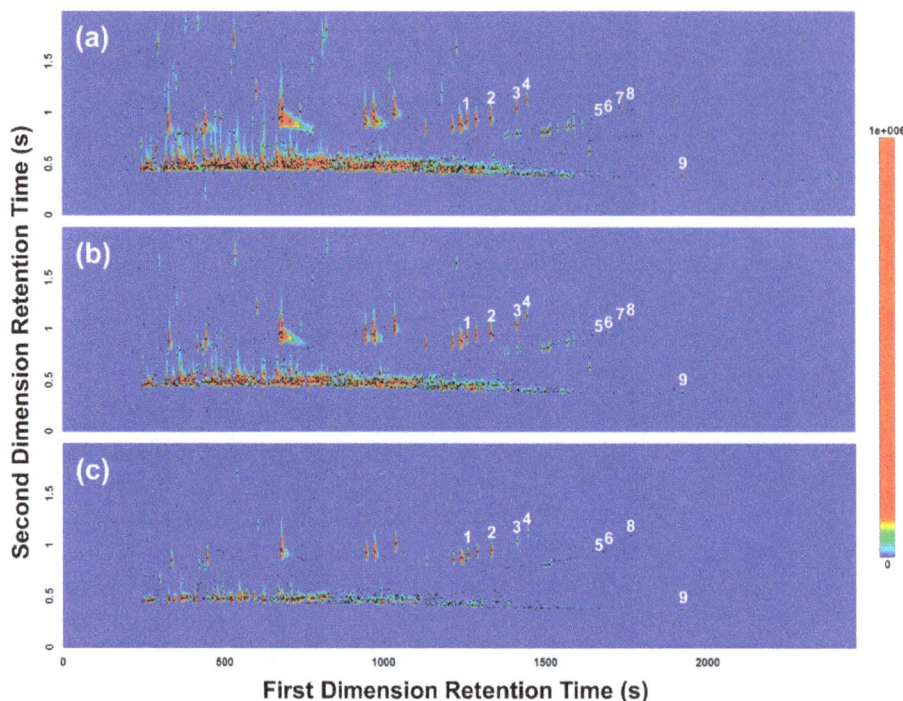

Figure 3. GC×GC-TOFMS total ion current (TIC) contour plots of headspace samples collected from the pig carcass burnt *with* gasoline and analyzed with three different tube desorption/trap desorption split flow conditions: (**a**) 20 mL/min/20 mL/min (split ratio = 15.4:1); (**b**) 20 mL/min/50 mL/min (split ratio = 36.4:1); (**c**) 50 mL/min/100 mL/min (split ratio = 102.0:1). Peak markers (represented by black dots) highlight each VOC detected. Numerals denote ASTM E1618-14 [24] gasoline target compounds detected: (1) 1,3,5-trimethylbenzene; (2) 1,2,4-trimethylbenzene; (3) 1,2,3-trimethylbenzene; (4) indane; (5) 1,2,4,5-tetramethylbenzene; (6) 1,2,3,5-tetramethylbenzene; (7) 5-methylindane; (8) 4-methylindane; and (9) dodecane.

Although a smaller volume of headspace could likely have been sampled in this trial, given the high concentration of gasoline detected following burning, this may not always be the case. Ignitable liquids are consumed or otherwise weathered at an unpredictable rate during the course of a fire and firefighting efforts and, as a result, often leave behind only trace quantities or residues of the original ignitable liquid (hence, ILR) [36]. Given the variability that occurs in every fire, the authors recommend sampling larger volumes of headspace (e.g., 500 mL) and collecting multiple samples if possible (e.g., 4–5). Replicate samples can be used to determine the optimal split ratio, starting with a low to mid-level split ratio, such as 15.4:1, and then increasing or decreasing the split ratio as needed. Additional replicates can be run at the optimal split ratio for sample confirmation or for use with multivariate data analysis, as performed herein. Future studies should, however, consider testing smaller volumes of headspace in addition to varying or entirely removing the headspace re-accumulation step between samples.

3.3. Sample Analysis

The detection and identification of gasoline (or gasoline-interfering contaminants) was facilitated through target compound analysis and comparison with a reference gasoline sample. Target compound analysis was performed using the list of 15 key specific compounds outlined in ASTM E1618-14 [24] for the characterization of gasoline. The gasoline target compounds included (1) 1,3,5-trimethylbenzene; (2) 1,2,4-trimethylbenzene; (3) 1,2,3-trimethylbenzene; (4) indane;

(5) 1,2,4,5-tetramethylbenzene; (6) 1,2,3,5-tetramethylbenzene; (7) 5-methylindane; (8) 4-methylindane; (9) dodecane; (10) 4,7-dimethylindane; (11) 2-methylnaphthalene; (12) 1-methylnaphthalene; (13) ethylnaphthalenes (mixed); (14) 1,3-dimethylnaphthalene; and (15) 2,3-dimethylnaphthalene.

In total, 11 of the 15 gasoline target compounds were detected in the reference gasoline sample (i.e., unleaded 91-octane gasoline) as shown and labelled in Figure 4a. Each of the two empty body bags analyzed (in triplicate) as controls tested negative for the presence of gasoline and were established to be free of gasoline-interfering contaminants, with the exception of dodecane (Figure 4b). Overall, very few volatiles were detected in the body bag control samples, and those that were detected were present at trace levels. Each of the three replicate samples collected from the body bag containing the pig burnt *without* gasoline tested negative for the presence of gasoline (Figure 4c). Several background pyrolysis products were detected, but these compounds did not match the gasoline target compounds with the exception of dodecane, which was detected in all samples analyzed in this trial. Each of the three replicate samples collected from the body bag containing the pig burnt *with* gasoline tentatively tested positive for the presence of gasoline, with 8 gasoline target compounds detected (Figure 4d), matching 8 of the 11 gasoline target compounds detected in the reference gasoline sample (Figure 4a). All gasoline target compounds detected in the experimental samples were established to be well resolved from the background pyrolysis products produced from the burnt pig carcasses (Figure 4d).

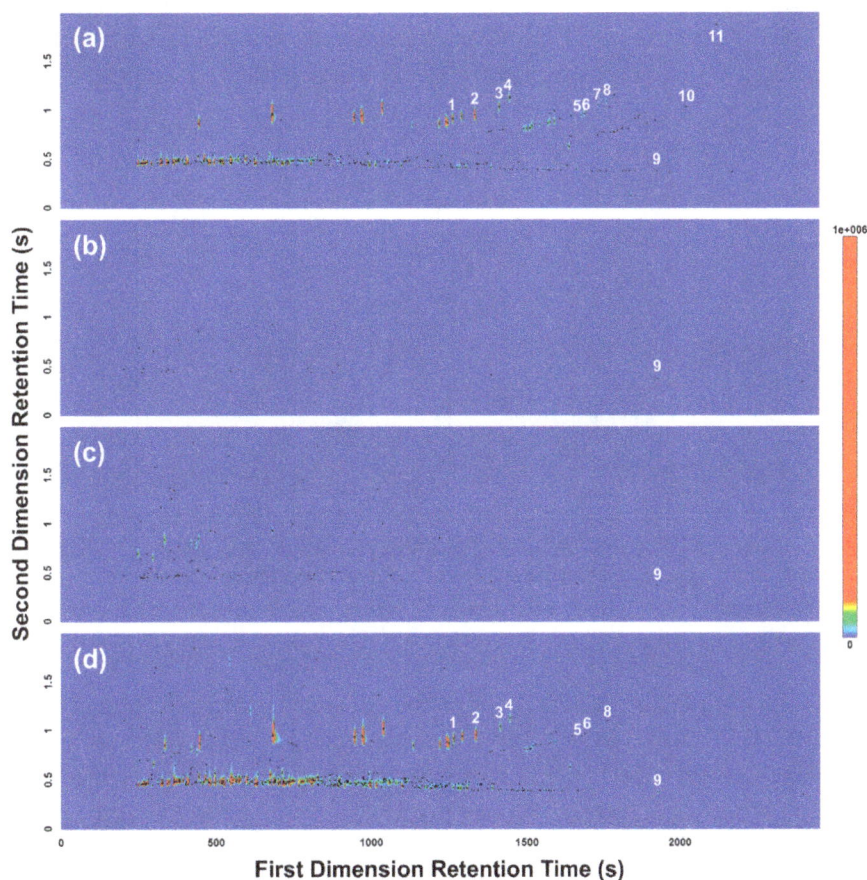

Figure 4. GC×GC-TOFMS TIC contour plots of headspace samples collected and analyzed from (**a**) reference gasoline; (**b**) body bag control; (**c**) pig burnt *without* gasoline; and (**d**) pig burnt *with* gasoline. Peak markers (represented by black dots) highlight each VOC detected. Numerals denote ASTM E1618-14 [24] gasoline target compounds detected: (1) 1,3,5-trimethylbenzene; (2) 1,2,4-trimethylbenzene; (3) 1,2,3-trimethylbenzene; (4) indane; (5) 1,2,4,5-tetramethylbenzene; (6) 1,2,3,5-tetramethylbenzene; (7) 5-methylindane; (8) 4-methylindane; (9) dodecane; (10) 4,7-dimethylindane; and (11) 1-methylnaphthalene.

Following sample collection, the two burnt pig carcasses were utilized in a subsequent study investigating the influence of fire modification on the odor of decomposition [29], as previously described. This study [29] showed that when the remains were left to decompose naturally in an Australian autumn environment, the combustion and pyrolysis products diminished over time, while the gasoline signature was lost entirely by Day 9 (note that no sampling occurred between Day 1 and Day 9 in the referenced study). A more recent study conducted by the authors, further investigating the influence of fire modification on the odor of decomposition during the Australian winter (with more frequent sampling conducted following burning), found that the gasoline signature could be lost within as little as 24 h (data not yet published). This work demonstrates the importance of sampling the remains as soon as possible after discovery, to limit the potential loss of the ignitable liquid signature due to continued weathering and possible microbial degradation, or due to masking by more prevalent decomposition VOCs.

3.4. Multivariate Data Analysis

PCA, a multivariate statistical methodology, was performed on the semi-quantitative data collected in this study as a rapid, objective, and automated/computerized pattern matching technique used to verify the chromatographic results discussed above. Mean centering, variance scaling, and unit vector normalization pre-processing steps were applied to the dataset prior to performing PCA. Figure 5 displays the PCA scores and correlation loadings plots generated using the pre-processed GC×GC-TOFMS peak area data for all analytes of interest detected across the four sample classes: (1) reference gasoline; (2) body bag control; (3) pig burnt *without* gasoline; and (4) pig burnt *with* gasoline.

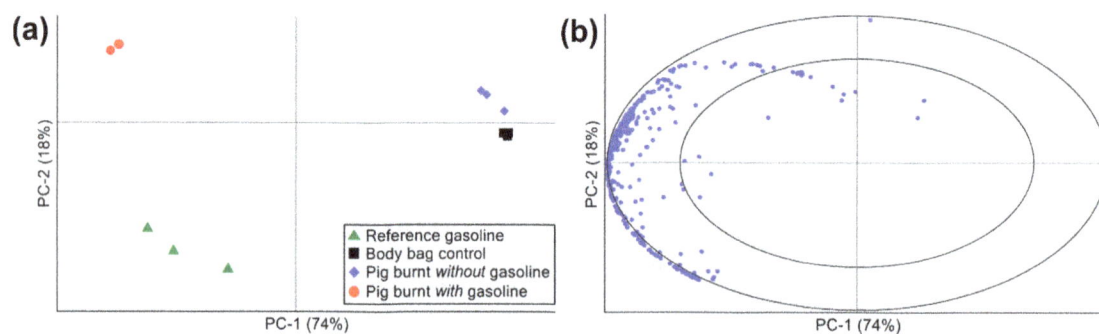

Figure 5. PCA (**a**) scores and (**b**) correlation loadings plots calculated using pre-processed GC×GC-TOFMS peak area data for all analytes of interest detected across the four sample classes: (1) reference gasoline ($n = 3$); (2) body bag control ($n = 6$); (3) pig burnt *without* gasoline ($n = 3$); and (4) pig burnt *with* gasoline ($n = 3$).

Overall, 92% of the explained variation within the dataset was captured by the first two principal components (i.e., PC-1 = 74% and PC-2 = 18%). The reference gasoline and volatile profile of the pig carcass burnt *with* gasoline grouped together on the left side of the scores plot, and were both readily differentiated from the body bag controls and the pig carcass burnt *without* gasoline horizontally along PC-1 (Figure 5a), thus accounting for the largest explained variation within the dataset. The majority of the analytes of interest detected were responsible for this differentiation as observed in the correlation loadings plot (Figure 5b), whereby all but three of the analytes of interest detected grouped on the left side of the plot. These analytes included predominantly alkanes, cycloalkanes, alkenes, and aromatics (e.g., alkylbenzenes, alkylnaphthalenes, and indanes), the majority of which fell within the radius between the ellipses of the correlation loadings plot and, thus, were shown to have high discriminatory power (i.e., explaining greater than 50% of the variance within the dataset, with the inner ellipse representing 50% of the explained variance within the dataset and the outer ellipse (i.e., the unit circle) indicating 100% explained variance). The three analytes appearing on the right side of the correlation loadings plot (tentatively identified as furan, carbon disulfide, and allyl nonyl ester oxalic acid) were detected at low levels across most of the samples (with the exception of the reference gasoline), and are attributed to background VOCs produced by the body bags and/or surrounding sampling environment.

Differentiation along PC-2, accounting for only 18% of the variation in dataset, included differentiation between the body bag controls and the pig burnt *without* gasoline, as well as differentiation between the reference gasoline and the pig burnt *with* gasoline (Figure 5a). Overall, the differentiation observed along PC-2 between the reference gasoline and pig burnt *with* gasoline is small compared to the differentiation observed between these samples and the body bag controls and pig burnt *without* gasoline along PC-1. This minor differentiation observed between the profiles of the reference gasoline and pig burnt *with* gasoline is accounted for by the presence of combustion and pyrolysis products from the pig carcass (e.g., aldehydes, ketones, alcohols, and sulfur-containing compounds) and the consummation/weathering of the gasoline signature (e.g., missing gasoline target compounds). Nevertheless, the grouping and overall similarity observed between the reference gasoline and pig burnt *with* gasoline in Figure 5a supports the chromatographic results above for the positive detection of gasoline in the headspace samples collected from the pig carcass burnt *with* gasoline. The differentiation between the body bag controls and both the reference gasoline and pig burnt *with* gasoline also supports the chromatographic evidence that the body bag is free of gasoline-interfering contaminants.

4. Conclusions

This study demonstrated that dynamic headspace sampling of a body bag, followed by TD-GC×GC-TOFMS analysis, is an integral addition to the fire debris analysts' sample collection and analysis arsenal, allowing for the detection of ILRs in cases of suspected arson involving human remains. This method permits the collection of headspace samples before the remains are removed from the crime scene, limiting the potential for contamination and the loss of volatiles due to potential microbial degradation or further weathering. Overall, the body bag was established to be free of gasoline-interfering contaminants, demonstrating its potential as an alternative sample collection medium for fire debris analysis in cases involving burnt victim remains. However, it is important to note that body bags are not considered vapor-tight and, therefore, should not be used for long-term storage of fire debris evidence. It is recommended that headspace samples be collected immediately after headspace accumulation is complete at the scene of collection, in order to prevent extensive vapor loss and the potential for cross-contamination. Future studies should consider not only sampling different volumes of headspace, but also varying headspace re-accumulation periods, volumes, and types of ignitable liquids, burn times (e.g., allowing fire to extinguish itself without intervention), and time passed before sample collection (e.g., allow time for weathering and/or microbial degradation to occur to reflect a scenario where remains are not discovered immediately after burning).

Author Contributions: Conceptualization, K.D.N. and S.L.F.; Methodology, K.D.N.; Sample analysis, K.D.N.; Data analysis, K.D.N.; Resources, S.L.F.; Writing—Original Draft Preparation, K.D.N. and S.L.F.; Writing—Review & Editing, K.D.N. and S.L.F.; Supervision, S.L.F.

Funding: This research received no external funding.

Acknowledgments: The authors wish to thank Morgan Cook and the Fire & Rescue NSW Fire Investigation & Research Unit for their assistance with carrying out the fires conducted for this trial. The authors also wish to thank all UTS research group members, extended contacts, and laboratory technical staff that contributed to the execution of field work and sample collection/analysis throughout this trial: LaTara Rust, Vitor Taranto, Baree Chilcote, Nicole Cattarossi, Darshil Patel, Katelynn Perrault, Mohammed Shareef, Robert Chatterton, Greg Dalsanto, Ronald Shimmon, and R. Verena Taudte. This work was supported financially by the University of Technology Sydney.

Conflicts of Interest: The authors declare no conflict of interest.

References

1.   Sinkov, N.A.; Sandercock, P.M.L.; Harynuk, J.J. Chemometric classification of casework arson samples based on gasoline content. *Forensic Sci. Int.* **2014**, *235*, 24–31. [CrossRef] [PubMed]

2.   Pahor, K.; Olson, G.; Forbes, S.L. Post-mortem detection of gasoline residues in lung tissue and heart blood of fire victims. *Int. J. Leg. Med.* **2013**, *127*, 923–930. [CrossRef] [PubMed]

3.   Frysinger, G.S.; Gaines, R.B. Forensic analysis of ignitable liquids in fire debris by comprehensive two-dimensional gas chromatography. *J. Forensic Sci.* **2002**, *47*, 471–482. [CrossRef] [PubMed]

4.   Fettig, I.; Krüger, S.; Deubel, J.H.; Werrel, M.; Raspe, T.; Piechotta, C. Evaluation of a headspace solid-phase microextraction method for the analysis of ignitable liquids in fire debris. *J. Forensic Sci.* **2014**, *59*, 743–749. [CrossRef] [PubMed]

5.   Pert, A.D.; Baron, M.G.; Birkett, J.W. Review of analytical techniques for arson residues. *J. Forensic Sci.* **2006**, *51*, 1033–1049. [CrossRef] [PubMed]

6.   Williams, M.R.; Sigman, M. Performance testing of commercial containers for collection and storage of fire debris evidence. *J. Forensic Sci.* **2007**, *52*, 579–585. [CrossRef] [PubMed]

7.   Sandercock, P.M.L. Fire investigation and ignitable liquid residue analysis—A review: 2001–2007. *Forensic Sci. Int.* **2008**, *176*, 93–110. [CrossRef] [PubMed]

8.   Lloyd, J.A.; Edmiston, P.L. Preferential extraction of hydrocarbons from fire debris samples by solid phase microextraction. *J. Forensic Sci.* **2003**, *48*, 130–134. [CrossRef] [PubMed]

9.   Sinkov, N.A.; Johnston, B.M.; Sandercock, P.M.L.; Harynuk, J.J. Automated optimization and construction of chemometric models based on highly variable raw chromatographic data. *Anal. Chim. Acta* **2011**, *697*, 8–15. [CrossRef] [PubMed]

10. Borusiewicz, R. Comparison of new Ampac bags and FireDebrisPAK® bags as packaging for fire debris analysis. *J. Forensic Sci.* **2012**, *57*, 1059–1063. [CrossRef] [PubMed]

11. Borusiewicz, R.; Kowalski, R. Volatile organic compounds in polyethylene bags—A forensic perspective. *Forensic Sci. Int.* **2016**, *266*, 462–468. [CrossRef] [PubMed]

12. Grutters, M.M.P.; Dogger, J.; Hendrikse, J.N. Performance testing of the new AMPAC fire debris bag against three other commercial fire debris bags. *J. Forensic Sci.* **2012**, *57*, 1290–1298. [CrossRef] [PubMed]

13. Belchior, F.; Andrews, S.P. Evaluation of cross-contamination of nylon bags with heavy-loaded gasoline fire debris and with automotive paint thinner. *J. Forensic Sci.* **2016**, *61*, 1622–1631. [CrossRef] [PubMed]

14. Schuberth, J. Post-mortem test for low-boiling arson residues of gasoline by gas chromatography-ion-trap mass spectrometry. *J. Chromatogr. B Biomed. Sci. Appl.* **1994**, *662*, 113–117. [CrossRef]

15. Schuberth, J. A full evaporation headspace technique with capillary GC and ITD: A means for quantitating volatile organic compounds in biological samples. *J. Chromatogr. Sci.* **1996**, *34*, 314–319. [CrossRef] [PubMed]

16. Morinaga, M.; Kashimura, S.; Hara, K.; Hieda, Y.; Kageura, M. The utility of volatile hydrocarbon analysis in cases of carbon monoxide poisoning. *Int. J. Leg. Med.* **1996**, *109*, 75–79. [CrossRef]

17. ASTM E1388-17 Standard practice for static headspace sampling of vapors from fire debris samples. In *Annual Book of ASTM Standards*; ASTM International: West Conshohocken, PA, USA, 2017.

18. ASTM E1412-16 Standard practice for separation of ignitable liquid residues from fire debris samples by passive headspace concentration with activated charcoal. In *Annual Book of ASTM Standards*; ASTM International: West Conshohocken, PA, USA, 2016.

19. Borusiewicz, R.; Zieba-Palus, J. Comparison of the effectiveness of Tenax TA and Carbotrap 300 in concentration of flammable liquids compounds. *J. Forensic Sci.* **2007**, *52*, 70–74. [CrossRef] [PubMed]

20. ASTM E1413-13 Standard practice for separation of ignitable liquid residues from fire debris samples by dynamic headspace concentration. In *Annual Book of ASTM Standards*; ASTM International: West Conshohocken, PA, USA, 2013.

21. Nichols, J.E.; Harries, M.E.; Lovestead, T.M.; Bruno, T.J. Analysis of arson fire debris by low temperature dynamic headspace adsorption porous layer open tubular columns. *J. Chromatogr. A* **2014**, *1334*, 126–138. [CrossRef] [PubMed]

22. Yoshida, H.; Kaneko, T.; Suzuki, S. A solid-phase microextraction method for the detection of ignitable liquids in fire debris. *J. Forensic Sci.* **2008**, *53*, 668–676. [CrossRef] [PubMed]

23. ASTM E2154-15a Standard practice for separation and concentration of ignitable liquid residues from fire debris samples by passive headspace concentration with solid phase microextraction (SPME). In *Annual Book of ASTM Standards*; ASTM International: West Conshohocken, PA, USA, 2015.

24. ASTM E1618-14 Standard test method for ignitable liquid residues in extracts from fire debris samples by gas chromatography-mass spectrometry. In *Annual Book of ASTM Standards*; ASTM International: West Conshohocken, PA, USA, 2014.

25. DeHaan, J.D.; Taormina, E.I.; Brien, D.J. Detection and characterization of volatile organic compounds from burned human and animal remains in fire debris. *Sci. Justice* **2016**, *57*, 118–127. [CrossRef] [PubMed]

26. Nizio, K.D.; Cochran, J.W.; Forbes, S.L. Achieving a near-theoretical maximum in peak capacity gain for the forensic analysis of ignitable liquids using GC×GC-TOFMS. *Separations* **2016**, *3*, 26. [CrossRef]

27. Australian Code for the Care and Use of Animals for Scientific Purposes. 2013. Available online: https://www.nhmrc.gov.au/guidelines-publications/ea28 (accessed on 26 May 2018).

28. U.S. Environmental Protection Agency (EPA). *Compendium Method TO-17: Determination of Volatile Organic Compounds in Ambient Air Using Active Sampling onto Sorbent Tubes*; EPA: Washington, DC, USA, 1999; pp. 1–53.

29. Nizio, K.D.; Forbes, S.L. Preliminary investigation of the influence of fire modification on the odour of decomposition using GC×GC-TOFMS. *Chromatogr. Today* **2017**, *10*, 32–39.

30. Pierce, K.M.; Hoggard, J.C.; Hope, J.L.; Rainey, P.M.; Hoofnagle, A.N.; Jack, R.M.; Wright, B.W.; Synovec, R.E.; Hospital, C.; Point, S.; et al. Fisher ratio method applied to third-order separation data to identify significant chemical components of metabolite extracts. *Anal. Chem.* **2006**, *78*, 5068–5075. [CrossRef] [PubMed]

31. Brokl, M.; Bishop, L.; Wright, C.G.; Liu, C.; McAdam, K.; Focant, J.-F. Multivariate analysis of mainstream tobacco smoke particulate phase by headspace solid-phase micro extraction coupled with comprehensive two-dimensional gas chromatography-time-of-flight mass spectrometry. *J. Chromatogr. A* **2014**, *1370*, 216–229. [CrossRef] [PubMed]

32. Nizio, K.D.; Perrault, K.A.; Troobnikoff, A.N.; Ueland, M.; Shoma, S.; Iredell, J.R.; Middleton, P.G.; Forbes, S.L. In vitro volatile organic compound profiling using GC×GC-TOFMS to differentiate bacteria associated with lung infections: A proof-of-concept study. *J. Breath Res.* **2016**, *10*, 026008. [CrossRef] [PubMed]

33. Turner, D.A.; Goodpaster, J.V. Comparing the effects of weathering and microbial degradation on gasoline using principal components analysis. *J. Forensic Sci.* **2012**, *57*, 64–69. [CrossRef] [PubMed]

34. Glassman, D.M.; Crow, R.M. Standardization model for describing the extent of burn injury to human remains. *J. Forensic Sci.* **1996**, *41*, 152–154. [CrossRef] [PubMed]

35. Bertsch, W. Volatiles from carpet: A source of frequent misinterpretation arson analysis. *J. Chromatogr. A* **1994**, *674*, 329–333. [CrossRef]

36. Sandercock, P.M.L.; Du Pasquier, E. Chemical fingerprinting of gasoline: 2. Comparison of unevaporated and evaporated automotive gasoline samples. *Forensic Sci. Int.* **2004**, *140*, 43–59. [CrossRef] [PubMed]

## 12

A Low-Cost Approach Using Diatomaceous Earth Biosorbent as Alternative SPME Coating for the Determination of PAHs in Water Samples by GC-MS

Naysla Paulo Reinert [1], **Camila M. S. Vieira** [1], **Cristian Berto da Silveira** [2], **Dilma Budziak** [3] and **Eduardo Carasek** [1,*] ⓘ

[1]   Departamento de Química, Universidade Federal de Santa Catarina, Florianópolis 88040-900, SC, Brazil;
      nayslareinert@gmail.com (N.P.R.); camilamaiara.vieira@gmail.com (C.M.S.V.)

[2]   Departamento de Engenharia de Pesca e Ciências Biológicas, Universidade do Estado de Santa Catarina,
      Laguna, Santa Catarina 88790-000, Brazil; cristian.silveira@udesc.br

[3]   Departamento de Ciências Naturais e Sociais, Universidade Federal de Santa Catarina,
      Curitibanos 89520-000, SC, Brazil; dilmabudziak@yahoo.com.br

*   Correspondence: eduardo.carasek@ufsc.br

Abstract: In this study, the use of recycled diatomaceous earth as the extraction phase in solid phase microextraction (SPME) technique for the determination of polycyclic aromatic hydrocarbons (PAHs) in river water samples, with separation/detection performed by gas chromatography-mass spectrometry (GC-MS), is proposed. The optimized extraction conditions are extraction time 70 min at 80 °C with no addition of salt. The limits of quantification were close to 0.5 μg L^{-1} with RSD values lower than 25% (n = 3). The linear working range was 0.5 μg L^{-1} to 25 μg L^{-1} for all analytes. The method was applied to samples collected from the Itajaí River (Santa Catarina, Brazil) and the RSD values for repeatability and reproducibility were lower than 15% and 17%, respectively. The efficiency of the recycled diatomaceous earth fiber was compared with that of commercial fibers and good results were obtained, confirming that this is a promising option to use as the extraction phase in SPME.

Keywords: recycled diatomaceous earth; solid phase microextraction; polycyclic aromatic hydrocarbons; gas chromatography-mass spectrometry

1. Introduction

Water is an extremely valuable natural resource as it is responsible for maintaining biological, geological and chemical cycles [1,2]. Environmental problems caused by anthropogenic activities are continually increasing and gaining attention worldwide [1]. With population growth and increased industrial activities, ever greater amounts of petroleum-based fossil fuels are being consumed [3]. These fuels contain a class of compounds known as polycyclic aromatic hydrocarbons (PAHs).

PAHs are a group of organic compounds composed of multiple aromatic rings [4]. The formation of these molecules is associated with the incomplete combustion of natural organic materials, for instance, due to volcanoes or the incomplete burning of wood in forest fires, and from anthropogenic sources including industrial processes (e.g., refineries), vehicular emissions [5], cane burning [6], and others [7]. According to the International Agency for Research on Cancer (IARC) and the US EPA (the United States Environmental Protection Agency) PAHs are recognized as persistent environmental pollutants with carcinogenic and mutagenic capacity in humans [7,8]. Based on these issues, measures have been taken by governments around the world to monitor the concentrations of compounds that may be harmful to human health, with different standards and regulations being established often aimed at ensuring the quality of drinking water [7]. In Brazil, the Ministry of Health regulates waters for human consumption using benzo[a]pyrene as a marker with maximum permitted values of 0.7 μg L^{-1}.

The determination of these pollutants generally requires a sample preparation procedure to remove matrix interferents, concentrate the analyte and make the extract compatible with the analytical instrumentation. One of the most commonly used sample preparation techniques is solid-phase microextraction (SPME) [9,10].

SPME was proposed by Pawliszyn et al. in 1990 to overcome the drawbacks of traditional sample preparation techniques such as liquid-liquid extraction and solid phase extraction [9,10]. The principle of the technique is the distribution of the analytes between the sample matrix and the sorbent (fiber), combining sampling, isolation and enrichment in a single step [11,12]. SPME fibers are composed of a fused silica or metallic support coated with an extractive phase, for instance, polymethylsiloxane (PDMS), polyacrylate (PA), or other commercially available sorbent [13–16].

In the search for new sorbent materials for SPME, biosorbents have gained prominence in miniaturized techniques because they provide greener, less expensive, renewable, and biodegradable extractive phases. Many of these biosorbents can be found in the environment and consist of macromolecules with different functional groups that can interact with different types of analytes. Our research group has previously used natural sorbents for the determination of organic contaminants using SPME [17,18]. Diatomaceous earth is of particular interest as a new biosorbent since it is discarded in large scale as a waste from breweries, where it is used for the clarification and filtration of organic materials and beers [19].

Diatomaceous earth is obtained from sedimentary rocks, originating from fossilized algae belonging to the class *Bacillariophyta* (diatoms). It is an amorphous mineral, comprised mainly of silica dioxide, of light weight and low molar mass, and its coloration can vary from white to gray. Structurally, diatoms have a hollow cylindrical form of low density and high surface area [20].

In this study the use of diatomaceous earth as an (bio) extractive phase in SPME is explored for the determination of PAHs in river water samples with quantification by gas chromatography coupled to mass spectrometry (GC-MS). The biosorbent was easily adhered onto a NiTi (nitinol) rod using a quick and inexpensive procedure.

2. Materials and Methods

2.1. Reagents and Materials

Analytical standards of PAHs in a mixture containing acenaphthylene, fluorene, phenanthrene, anthracene, pyrene, benzo[a]anthracene, chrysene, benzo[b]fluoranthene, benzo[k]fluoranthene, and benzo[a]pyrene (Bellefonte, PA, USA) were used to prepare stock solutions of 1 mg L^{-1} in acetonitrile purchased from J.T. Baker (Mallinckrodt, NJ, USA). The ionic strength was studied using sodium chloride obtained from Synth (São Paulo, SP, Brazil). The ultrapure water used in the experiments was purified in an ultrapure Mega purity system (Billerica, MA, USA). The fiber was prepared using diatomaceous earth with size less than 200 mesh, nitinol rods (2 cm length and 0.128 mm diameter), epoxy glue acquired from Brascola (São Paulo, SP, Brazil), and a heating block from Dist (Florianópolis, SC, Brazil). SPME extractions were carried out in vials of 40 mL obtained from Supelco (Bellefonte, PA, USA) aided by a thermostatic bath (Lab Companion RW 0525G, Geumcheon-gu, Seoul, Korea) and magnetic stirrers from Dist (Florianópolis, SC, Brazil). Commercial fibers (DVB/Car/PDMS, 50/30 µm; PDMS 100 µm and PDMS/DVB, 65 µm; Supelco, Bellafonte, PA, USA) were used to compare the analyte extraction efficiencies.

2.2. Instrumental and Chromatographic Conditions

An Agilent 7820A gas chromatograph with flame ionization detector (FID) equipped with a split/splitless injector and an Agilent DB-5 capillary column (30 m × 0.25 mm × 0.25 µm; Santa Clara,

CA, USA) was used to optimize the method as well as to compare it with commercial fibers. On the other hand, a Shimadzu GC-MS QP2010 Plus equipped with a split/splitless injector (Kyoto, Japan) containing a Zebron ZB-5MS capillary column (30 m × 0.25 mm × 0.25 μm; Torrance, CA, USA) was used to obtain the analytical parameters of merit. The GC-MS and GC/FID was operated at the same conditions for injection and the columns temperature programs. The injection was performed in splitless mode at 260 °C for 15 min. The column temperature program consisted of maintaining the oven at 80 °C for 1 min and then increasing it 6 °C min^{-1} to 300 °C which was maintained for 10 min. The transfer line temperature, the ion source temperature and the electron impact ionization (EI) mode of the GC-MS were set at 280, 250 °C, and 70 eV, respectively. Helium was used as the carrier gas at a flow rate of 1.0 mL min^{-1}. The PAHs were determined in selected ion monitoring (SIM) mode and the mass/charge (m/z) ratios employed are shown in Table 1. The m/z values in bold were used for the quantitative determination of the analytes.

Table 1. The m/z values used for the determination of PAHs by GC-MS (values in bold were used for the quantification of the analytes).

Analytes	m/z
acenaphthylene	**152**, 153, 151
fluorene	**166**, 165, 167
phenanthrene	**178**, 176, 179
anthracene	**178**, 179, 176
pyrene	**202**, 203, 200
benzo[a]anthracene	**228**, 226, 229
chrysene	**228**, 226, 229
benzo[b]fluoranthene	**252**, 250, 126
benzo[k]fluoranthene	**252**, 250, 126
benzo[a]pyrene	**252**, 250, 126

2.3. Preparation of Diatomaceous Earth Fibers

The diatomaceous earth dust came from the disposal reservoir of a brewery, where this material is used for the filtration and clarification of beer (Santa Catarina, Brazil). Due to its high porosity, the material presents a high degree of saturation with organic matter from the treatment of beer. Thus, a heat treatment is required [20], not only to eliminate the residues originated from the beer filtration but to ensure that all of the organic matter adhered to the material is removed. The diatomaceous earth, after the thermal treatment, was sieved to obtain homogeneous particle size (<200 mesh). The diatomaceous earth was adhered on a 1 cm nitinol wire using epoxy glue. Then, the new fiber was inserted into the heating block at 180 °C for 90 min, resulting in a final phase thickness of approximately 40 μm. The fiber was then conditioned at 240 °C for 90 min in a GC injection port. The fiber lifetime was verified during the study by comparing the responses of the chromatographic areas of the analytes to the optimum extraction condition at a concentration of 5 μg L^{-1}. Fibers were used while the extraction efficiency did not present a reduction greater than 10%.

2.4. Optimization of SPME Procedure

The optimization of the extraction conditions for the diatomaceous earth fiber was performed by multivariate procedures. A central composite design involving 11 experiments with triplicate at the central point was carried out. In the optimization strategy the extraction temperature ranged from 30 to 80 °C and the extraction time from 30 to 117 min. The sodium chloride concentration (0–20% m/v) was also evaluated, but in the univariate form. The extraction procedure consisted of immersing the SPME fiber directly in 25 mL of water sample spiked with 100 μg L^{-1} of each PAH contained in a 40 mL vial and kept under constant magnetic stirring at 1000 rpm. After the extraction, the fiber was immediately inserted into the GC injection port at 240 °C for 15 min for the thermal desorption of the analytes. The analysis was carried out by GC-FID in splitless mode. To obtain the response surface, the geometric mean of the areas of the chromatographic peaks obtained in each extraction using Statistica 8.0 software (Statsoft, USA) was used.

2.5. Comparison of the Extraction Efficiencies Using Diatomaceous Earth and Commercial Fibers

After the optimization of the analytical procedure, the diatomaceous earth was compared to commercial fibers (PDMS and PDMS/DVB) in terms of their efficiency in the extraction of the PAHs studied. The same procedure described at Section 2.4 was carried out but the ultrapure water was spiked with the analytes at a concentration of 5 μg L^{-1}. The extractions were performed using one of the fibers at 80 °C for 70 min. The chromatographic analysis was performed by GC-MS.

2.6. Analytical Figures of Merit of the Method Developed

River water spiked with five concentrations of each analyte ranging from 0.5 to 25.0 μg L^{-1} was prepared to build calibration curves which were used to calculate the linear coefficient of determination (R^2). The lowest concentration on the analytical curve for each analyte which enabled measures with acceptable precision (RSD < 20%) was adopted as limits of quantification (LOQs). The limits of detection (LODs) were obtained dividing the LOQ by 3.3. The precision and the accuracy of the method were evaluated by performing extractions using real water samples spiked with the analytes at 0.5 μg L^{-1}. Precision was calculated as the relative standard deviation (RSD) obtained from spiked river water and accuracy was verified through the relative recovery of the analytes.

3. Results and Discussion

3.1. Characterization of the Diatomaceous Fiber

The diatomaceous earth samples used for the production of SPME fibers belong to the class *Bacillariophyceae centricae* and their color may vary from white to gray. The material consists mainly of silica, SiO$_2$ (87–91%), alumina and ferric oxide [21].

Scanning electron microscopy (SEM) was carried out to characterize the surface morphology of the recycled diatomaceous earth. The images obtained at magnifications of 2000 and 4000× for the surface evaluation are shown in Figure 1 (A and B, respectively). An image of a cross-section of the proposed fiber was obtained at a magnification of 100× (Figure 1C). According to the SEM results, the morphology of the material shows a high porosity which facilitated the physical processes involving the sorption of the analytes.

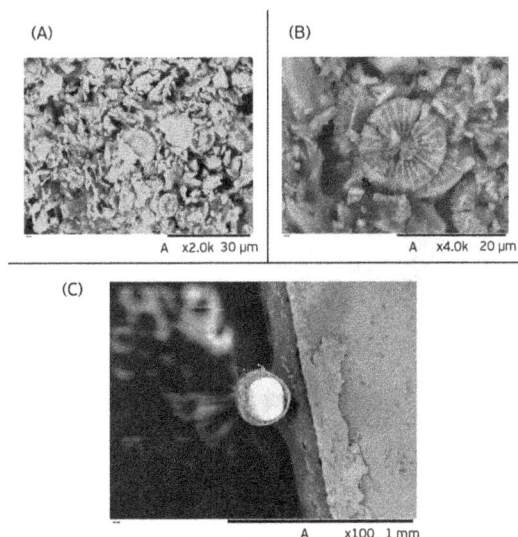

Figure 1. SEM micrographs obtained with the biosorbent fiber at magnifications of (**A**) 2000× and (**B**) 4000×, and a cross-section of the proposed fiber (**C**) at a magnification of 100×.

FTIR spectroscopy was carried out to identify the functional groups in the sorbent. The FTIR spectrum obtained from the material previously conditioned at 240 °C is illustrated in Figure 2. A broad peak at ~3400 cm^{-1} corresponds to the O–H bonds of silanol groups. Two intense peaks between ~1200 and 1080 cm^{-1} were assigned to the asymmetric stretching of the Si–O–Si siloxane groups and one at ~790 cm^{-1} is related to the Si–O–Si vibrations attributed to mesoporous silicas. At ~475 cm^{-1}, a peak related to O–Si–O vibration was present. Lastly, the peak at ~1600 cm^{-1} refers to the angular deformation of the adsorbed water molecules.

Thermogravimetric analysis was conducted to identify if there was any organic material present in the sample and since no mass loss was observed the material can be characterized as thermally stable (data not shown). This result was already expected, since the sorbent comes from inorganic material.

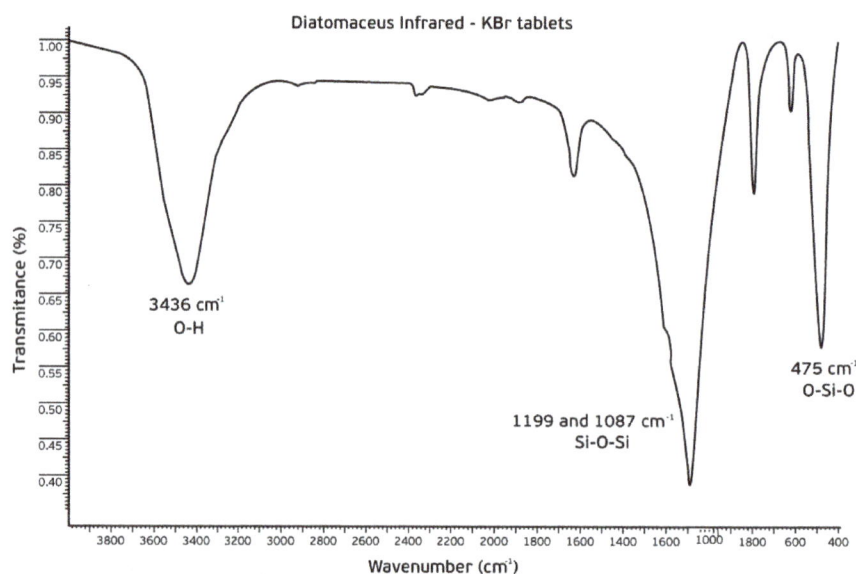

Figure 2. FTIR spectrum of the biosorbent, previously conditioned at 240 °C.

3.2. *Optimization of DI-SPME Extraction Procedure*

The extraction conditions that can influence the SPME efficiency were optimized using the diatomaceous earth fiber. The response used to feed the software Statistica 8.0 was the geometric means of the chromatographic peak areas of the analytes. The response surfaces obtained for the biosorbent fiber are shown in Figure 3.

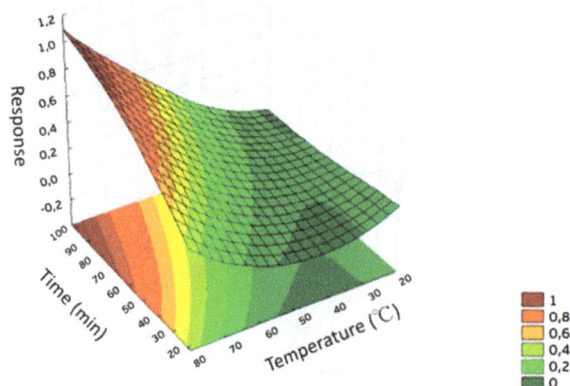

Figure 3. Response surface obtained for the optimization of DI-SPME procedure using biosorbent fiber (diatomaceous earth).

The optimum extraction conditions selected for the proposed fiber were reached using an extraction time of 70 min at 80 °C. The addition of salt was also studied as it is known to lead to the salting-out effect. However, the use of small amounts of salt caused fiber damage and so no salt was added in the extractions.

3.3. Comparison between the Extraction Efficiencies of the Biosorbent and Commercial Coatings

A comparison between the extraction efficiencies using the proposed fiber and commercial fibers (PDMS/DVB and PDMS) was performed. The conditions for the extractions using commercial fibers were optimized (data not shown) as extraction time of 70 min at 80 °C. These values are much closed to those mentioned in the literature (extractions of 60 min at 70 °C) [22–24]. Figure 4 shows this comparison through bar graph using normalized peak area and considering the film thickness of each fiber. The normalization of peak areas for each analyte was made using the highest chromatographic peak areas as 100% for each analyte.

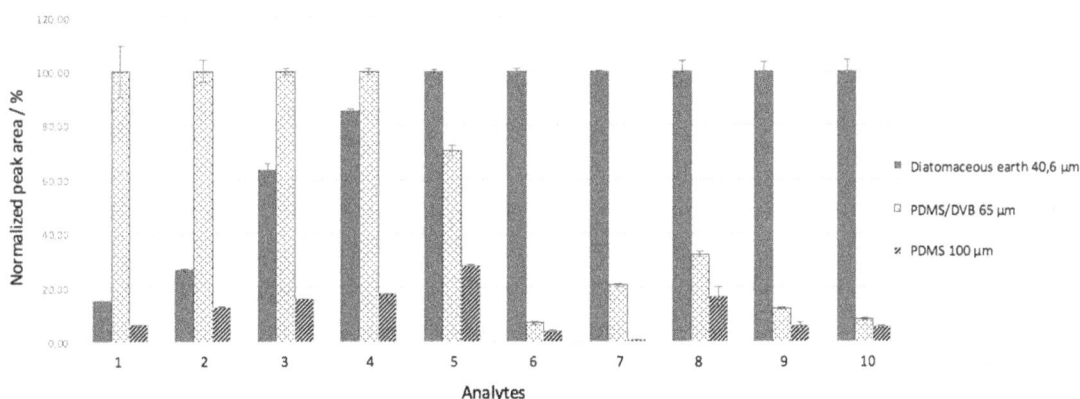

Figure 4. Comparison of extraction efficiencies of the biosorbent fiber, PDMS/DVB and PDMS coatings for determination of PAHs. Analytes: (1) acenaphtlylene; (2) fluorene; (3) phananthrene; (4) anthracene; (5) pyrene; (6) benzo[a]anthracene; (7) chrysene; (8) benzo[b]fluoranthene; (9) benzo[k]fluoranthene; and (10) benzo[a]purene.

It can be observed in Figure 4 that the extraction with the PDMS/DBV coating showed good performance for acenaphthylene, fluorene, phenanthrene, and anthracene, but for the other analytes the results were not as promising. The PDMS fiber gave values below 20%, except for pyrene, and it is not efficient for this application. Taking this into account, the proposed fiber demonstrated very satisfactory performance for PAH extraction when compared to the commercial fibers, with the exception of acenaphthylene, fluorine, and phenanthrene.

In addition, reproducibility studies using two diatomaceous earth fibers were performed and the results showed no significant variation (data not shown). The repeatability obtained with the fibers was estimated comparing the results of the first extraction with those obtained after 115 extractions using the same fiber (data not shown). It was verified that there was no significant loss of extraction efficiency, confirming that the fiber produced with the biosorbent material can be used at least 115 times.

These data demonstrate the high potential for diatomaceous earth fiber as a sorbent candidate for SPME. Moreover, diatomaceous earth is biodegradable, natural, and renewable. In addition, its chemical composition provides numerous possibilities of chemical interaction with a wide range of compounds. Diatomaceous earth has a microporous structure, which facilitates the extraction of the analytes through a physical process (adsorption mechanism).

3.4. Validation Parameters

Table 2 presents some analytical figures of merit obtained in this study. The linear coefficient of determination (R^2) values were >0.95, which indicates a good linear fit. The LOD and LOQ values were satisfactory based on those obtained in other studies.

Table 2. The linear range, linear equation, linearity, and limits of detection and quantification for the method developed using diatomaceous earth coating.

Analyte	LOD (μg L^{-1})	LOQ (μg L^{-1})	Linear Range (μg L^{-1})	Linear Equation	R
Acenaphthylene	0.16	0.49	0.49–25	y = 66,956x − 30,445	0.9890
Fluorene	0.17	0.50	0.50–25	y = 87,809x − 54,517	0.9911
Phenanthrene	0.14	0.42	0.42–25	y = 326,565x − 245,240	0.9777
Anthracene	0.11	0.33	0.33–25	y = 364,057x − 339,574	0.9598
Pyrene	0.15	0.50	0.50–25	y = 979,497x − 935,649	0.9914
benzo[a]anthracene	0.03	0.10	0.10–25	y = 506,040x − 544,597	0.9832
Chrysene	0.14	0.42	0.42–25	y = 691,902x − 796,526	0.9592
benzo[b]fluoranthene	0.06	0.17	0.17–25	y = 158,587x − 50,481	0.9990
benzo[k]fluoranthene	0.11	0.33	0.33–25	y = 431,634x − 806,517	0.9848
benzo[a]pyrene	0.15	0.46	0.46–25	y = 295,450x − 567,387	0.9667

Precision was evaluated in terms of intra-day repeatability ($n = 3$) and inter-day reproducibility ($n = 9$) using samples spiked at the lowest level for each analyte. The results obtained are shown in Table 3. It can be observed that the intra-day and inter-day precision for diatomaceous earth fiber presented values of RSD <15% and <17%, respectively. Relative recovery showed results between 83% and 100%, confirming the accuracy of the method.

Table 3. Relative recovery of analytes and precision (inter- and intra-day) for the extraction of PAHs from spiked river water samples.

Analyte	Spiked Concentration (μg L^{-1})	Relative Recovery (%) ($n = 3$)	RSD, Intra-Day (%) ($n = 3$)	RSD, Inter-Day (%) ($n = 3$)
acenaphthylene	0.5	100	5	10
fluorene	0.5	83	15	10
phenanthrene	0.5	97	10	13
anthracene	0.5	93	13	3
pyrene	0.5	92	2	6
benzo[a]anthracene	0.5	94	2	6
chrysene	0.5	96	2	6
benzo[b]fluoranthene	0.5	90	15	17
benzo[k]fluoranthene	0.5	97	15	17
benzo[a]pyrene	0.5	93	7	17

The selectivity of the proposed method was confirmed by the absence of peaks in the retention time of the target analytes when chromatograms of the extract were obtained from the river water sample without the addition of the analytes. The only exceptions were pyrene, chrysene, and benzo[a]pyrene, but these peaks were not quantifiable. Figure 5 shows the chromatograms obtained for samples of spiked river water (10 μg L^{-1}) and non-spiked river water.

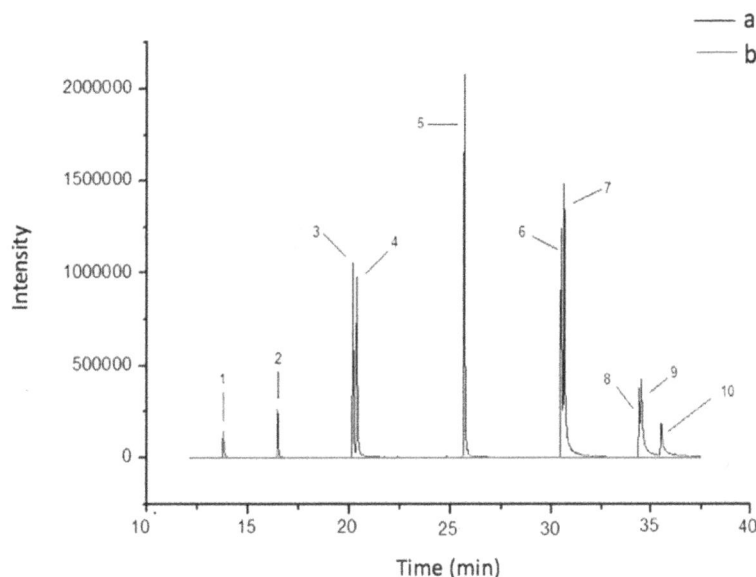

Figure 5. Chromatograms (GC-MS) obtained from a river water sample spiked at 10 µg L^{-1} (**a**) and non-spiked river water sample (**b**). Elution order: (1) acenaphtlylene; (2) fluorene; (3) phananthrene; (4) anthracene; (5) pyrene; (6) benzo[a]anthracene; (7) chrysene; (8) benzo[b]fluoranthene; (9) benzo[k]fluoranthene; and (10) benzo[a]pyrene.

4. Conclusions

In this study, the use of a recycled diatomaceous earth as extractive phase for SPME fiber demonstrated suitable results in comparison to widely used commercial fibers. The production of the biosorbent fiber is simple and the fibers can be reused several times. The separation and detection of the analytes by GC-MS is effective and enables the determination of PAHs in accordance with current Brazilian legislation. The proposed method using the biosorbent achieved good results of parameters of merit. The method is of low cost, because the natural sorbent can be reused in numerous extractions and is widely applicable because the material is easily obtainable.

Author Contributions: All of the authors participated in the same proportion.

Funding: Conselho Nacional de Desenvolvimento Científico e Tecnológico (CNPq), process number 303892/2014-5. Coordenação de Aperfeiçoamento de Pessoal de Nível Superior—Brasil (CAPES) —Finance Code 001.

Acknowledgments: The authors are grateful to the Brazilian governmental agency "Conselho Nacional de Desenvolvimento Científico e Tecnológico (CNPq) and Coordenação de Aperfeiçoamento de Pessoal de Nível Superior" for the financial support which made this research possible.

Conflicts of Interest: The authors declare no conflict of interest.

References

1. Benson, R.; Conerly, O.D.; Sander, W.; Batt, A.L.; Boone, J.S.; Furlong, E.T.; Glassmeyer, S.T.; Kolpin, D.W.; Mash, H.E.; Shenck, K.M.; et al. Human health screening and public health significance of contaminants of emerging concern detected in public water supplies. *Sci. Total Environ.* **2017**, *579*, 1643–1648. [CrossRef] [PubMed]

2. Pal, A.; He, Y.; Jakel, M.; Reinhard, M.; Gin, K.Y. Emerging contaminants of public health significance as water quality indicator compounds in the urban water cycle. *Environ. Int.* **2014**, *71*, 46–62. [CrossRef] [PubMed]

3. Heleno, F.F.; Lima, A.C. Evaluation of analytical methods for BTEX analysis in water using extraction by headspace (HS) and solid phase microextraction (SPME). *Quim. Nova* **2010**, *33*, 329–336. [CrossRef]

4. Hong, W.F.; Jia, H.; Li, Y.F. Polycyclic aromatic hydrocarbons (PAHs) and alkylated PAHs in the coastal seawater, surface sediment and oyster from Dalian, Northeast China. *Ecotoxicol. Environ. Saf.* **2016**, *128*, 11–20. [CrossRef] [PubMed]

5.   Slezakva, K.; Castro, D.; Delerue-Matos, C. Impact of vehicular traffic emissions on particulate-bound PAHs: Levels and associated health risks. *Atmos. Res.* **2013**, *127*, 141–147. [CrossRef]

6.   Cristale, J.; Silva, F.S.; Zocolo, G.J.; Marchi, M.R.R. Influence of sugarcane burning on indoor/outdoor PAH air pollution in Brazil. *Environ. Pollut.* **2012**, *169*, 210–216. [CrossRef] [PubMed]

7.   Dat, N.D.; Chang, M.B. Review on characteristics of PAHs in atmosphere, anthropogenic sources and control technologies. *Sci. Total Environ.* **2017**, *31*, 682–693. [CrossRef] [PubMed]

8.   Siritham, C.; Thammakhet-Buranacha, C. A preconcentrator-separator two-in-one online system for polycyclic aromatic hydrocarbons analysis. *Talanta* **2017**, *15*, 573–582. [CrossRef] [PubMed]

9.   Li, Z.; Ma, R.; Bai, S.; Wang, C.; Wang, Z. A solid phase microextraction fiber coated with graphene-poly9ethylene glycol) composite for the extraction of volatile aromatic compounds from water samples. *Talanta* **2014**, *119*, 498–504. [CrossRef] [PubMed]

10. Laopongsit, W.; Srzednicki, G.; Craske, J. Preliminary study of solid phase micro-extraction (SPME) as a method for detecting insect infestation in wheat grain. *J. Stored Prod. Res.* **2014**, *59*, 88–95. [CrossRef]

11. Lord, H.; Pawliszyn, J. Evolution of solid-phase microextraction technology. *J. Chromatogr. A* **2000**, *885*, 153–193. [CrossRef]

12. Dias, A.N.; Simão, V.; Merib, J.; Carasek, E. Cork as a new (green) coating for solid-phase microextraction: Determination of polycyclic aromatic hydrocarbons in water samples by gas chromatography-mass spectrometry. *Anal. Chim. Acta* **2013**, *772*, 33–39. [CrossRef] [PubMed]

13. Carasek, E.; Merib, J. Membrane-based microextraction techniques in analytical chemistry: A review. *Anal. Chim. Acta* **2015**, *23*, 8–25. [CrossRef] [PubMed]

14. Pawliszyn, J. *Handbook of Solid Phase Microextraction*; Chem. Ind. Press: Beijing, China, 2009.

15. Tsao, Y.U.; Wang, Y.C.; Wu, S.F.; Ding, W.H. Microwave-assisted headspace solid-phase microextraction for the rapid determination of organophosphate esters in aqueous samples by gas chromatography-mass spectrometry. *Talanta* **2011**, *84*, 406–410. [CrossRef] [PubMed]

16. Ahmadi, M.; Elmongy, H.; Madrakian, T.; Abdel-Rehim, M. Nanomaterials as sorbents for sample preparation in bioanalysis: A review. *Anal. Chim. Acta* **2017**, *15*, 1–21. [CrossRef] [PubMed]

17. Do Carmo, S.; Merib, J.; Dias, A.N.; Stolberg, J.; Budziak, D.; Carasek, E. A low-cost biosorbent-based coating for the highly sensitive determination of organochlorine pesticides by solid-phase microextraction and gas chromatography-electron capture detection. *J. Chromatogr. A* **2017**, *1525*, 23–31. [CrossRef] [PubMed]

18. Suterio, N.G.; do Carmo, S.N.; Budziak, D.; Merib, J.; Carasek, E. Use of a Natural Sorbent as Alternative Solid-Phase Microextraction Coating for the Determination of Polycyclic Aromatic Hydrocarbons in Water Samples by Gas Chromatography-Mass Spectrometry. *J. Braz. Chem. Soc.* **2018**, *29*. [CrossRef]

19. Silveira, C.B.; Goulart, M.R. Methodologies for the reuse of the diatomaceous earth residue, from filtration and clarification of beer. *Quim. Nova* **2011**, *34*, 625–629.

20. Souza, G.P.; Filgueira, M. Characterization of natural diatomaceous composite material. *Ceramica* **2003**, *49*, 40–43. [CrossRef]

21. Othmer, K. *Encyclopedia of Chemical Technology*; Wiley: New York, NY, USA, 1993; p. 108.

22. Menezes, H.C.; Paulo, B.P.; Paiva, M.J.N.; Barcelos, S.M.R.; Macedo, D.F.D.; Cardeal, Z.L. Determination of polycyclic aromatic hydrocarbons in artisanal cachaça by DI-CF-SPME–GC/MS. *Microchem. J.* **2015**, *118*, 272–277. [CrossRef]

23. Aguinaga, N.; Campillo, N.; Vinas, P.; Hernández-Córdoba, M. Determination of 16 polycyclic aromatic hydrocarbons in milk and related products using solid-phase microextraction coupled to gas chromatography–mass spectrometry. *Anal. Chim. Acta* **2007**, *23*, 285–290. [CrossRef] [PubMed]

24. Segura, A.; Sánchez, V.H.; Marqués, S.; Molina, L. Insights in the regulation of the degradation of PAHs in Novosphingobium sp. HR1a and utilization of this regulatory system as a tool for the detection of PAHs. *Sci. Total Environ.* **2017**, *590*, 381–393. [CrossRef] [PubMed]

## 13

Modern Instrumental Limits of Identification of Ignitable Liquids in Forensic Fire Debris Analysis

Robin J. Abel [1], Grzegorz Zadora [2,3], P. Mark L. Sandercock [4,†] and James J. Harynuk [1,*] 🔟

[1] Department of Chemistry, University of Alberta, Edmonton, AB T6G 2G2, Canada; rabel@ualberta.ca
[2] Institute of Forensic Research, Westerplatte 9, 31-033 Krakow, Poland; gzadora@ies.krakow.pl
[3] Department of Analytical Chemistry, Institute of Chemistry, The University of Silesia, Szkolna 9, 40-006 Katowice, Poland
[4] Royal Canadian Mounted Police, National Forensic Laboratory Services-Edmonton, 15707-118th Avenue, Edmonton, AB T5V 1B7, Canada
* Correspondence: james.harynuk@ualberta.ca
† The author has retired from Royal Canadian Mounted Police.

Abstract: Forensic fire debris analysis is an important part of fire investigation, and gas chromatography–mass spectrometry (GC-MS) is the accepted standard for detection of ignitable liquids in fire debris. While GC-MS is the dominant technique, comprehensive two-dimensional gas chromatography–mass spectrometry (GC×GC-MS) is gaining popularity. Despite the broad use of these techniques, their sensitivities are poorly characterized for petroleum-based ignitable liquids. Accordingly, we explored the limit of identification (LOI) using the protocols currently applied in accredited forensic labs for two 75% evaporated gasolines and a 25% evaporated diesel as both neat samples and in the presence of interfering pyrolysate typical of fire debris. GC-MSD (mass selective detector (MS)), GC-TOF (time-of-flight (MS)), and GC×GC-TOF were evaluated under matched conditions to determine the volume of ignitable liquid required on-column for correct identification by three experienced forensic examiners performing chromatographic interpretation in accordance with ASTM E1618-14. GC-MSD provided LOIs of ~0.6 pL on-column for both neat gasolines, and ~12.5 pL on-column for neat diesel. In the presence of pyrolysate, the gasoline LOIs increased to ~6.2 pL on-column, while diesel could not be correctly identified at the concentrations tested. For the neat dilutions, GC-TOF generally provided 2× better sensitivity over GC-MSD, while GC×GC-TOF generally resulted in 10× better sensitivity over GC-MSD. In the presence of pyrolysate, GC-TOF was generally equivalent to GC-MSD, while GC×GC-TOF continued to show 10× greater sensitivity relative to GC-MSD. Our findings demonstrate the superior sensitivity of GC×GC-TOF and provide an important approach for interlaboratory benchmarking of modern instrumental performance in fire debris analysis.

Keywords: forensics; trace evidence; fire debris; ignitable liquid; sensitivity; limit of identification; GC-MS; GC×GC-MS

1. Introduction

The purpose of forensic science is to assist the court in assessing the significance of a piece of evidence by interpreting the findings of a scientific examination of exhibit material. One of many considerations when weighing the results of a forensic examination is an understanding of how much target analyte must be present in an exhibit for conclusive identification of its presence (the sensitivity of the technique). Forensic fire debris analysis is the examination of exhibit material for the presence of ignitable liquids, and in this context an understanding of sensitivity is especially important. This is because the most frequently observed ignitable liquids are petroleum products [1,2], but petroleum products are also ubiquitous in our environment due to their widespread use in manufactured goods and their sale as consumer products [3–7]. Ignitable liquids may also be present in an exhibit due to the surrounding environment, since they are composed of volatile organic compounds which can evaporate from one location and condense in another (e.g., the collection of gasoline residues by materials exposed to automotive exhaust) [8–11]. Excessive sensitivity in an analysis can result in the assignment of inappropriately high significance to insignificant amounts of ignitable liquid [5].

Despite the need to limit sensitivity to forensically significant levels for exhibits selected through general fire investigation methods, there remains a long-acknowledged gap between the sensitivity of laboratory methods and accelerant detecting canines used to locate exhibits at the scene [12–15]. This persistent gap has led canine handlers to present the animal's indications at the scene to the court without the necessary laboratory confirmation, which has resulted in vigorous debate within the courts [15,16]. Modern analytical techniques can help close this gap by offering increased sensitivity in cases where canines have been used to select exhibits, but the sensitivity must still be controlled for traditionally selected exhibits. To address both situations, the sensitivity of the analytical process must be understood. Unfortunately, most literature studies into the sensitivity of methods for forensic fire debris analysis exist in outdated literature, assess sensitivity non-quantitatively by relation to other techniques, evaluate less mainstream techniques, or do not assess sensitivity by identification of the ignitable liquid as a whole [10,12,17–20]. Since the techniques and instruments available for fire debris analysis have advanced significantly over the last two decades [21–24], an investigation into the sensitivity offered by modern analytical tools is required.

2. Materials and Methods

2.1. Materials and Reagents

Fuels were purchased from two Edmonton area service stations. One regular gasoline (rated 87 octane) was obtained from an Esso service station in 2009. An aliquot was transferred into a GC vial and manually evaporated under a high purity nitrogen stream (Praxair, Edmonton, AB, Canada) until 75% of the original volume was lost. The vial was then sealed, wrapped with Teflon tape, and stored in a dark and cool area. Before use, its condition was confirmed through examination of its physical characteristics (light yellow color, clear with no resinous deposits or precipitates) and a preliminary GC-MS analysis. A second regular gasoline (rated 87 octane) and a diesel were purchased in March 2015 from a 7-Eleven service station. Aliquots of the second gasoline and diesel were transferred into GC vials, placed into a heated 24-well evaporator (Cole-Parmer, Montreal, QC, Canada), and evaporated with high purity nitrogen gas until 75% of the original gasoline volume and 25% of the original diesel volume were lost.

2.2. Sample Dilution Scheme

Each weathered ignitable liquid was diluted using GC-MS grade SupraSolv™ dichloromethane (EMD Millipore, Burlington, MA, USA) following the scheme in Table 1. Aliquots for dilution were delivered from the sample listed under "Source Vial" using calibrated glass micropipettes (Drummond Scientific, Broomall, PA, USA) into vials pre-filled with diluent delivered from an eVol® calibrated autopipette (SGE, Ringwood, Victoria, Australia). Vials were kept chilled in crushed dry ice (Praxair, Edmonton, AB, Canada) to limit losses of volatile solvent. A second set of dilutions was prepared using dichloromethane doped with 50 µL/mL of an equal mixture of spruce plywood subfloor, black foam underlay, and nylon carpet pyrolysates as diluent. The pyrolysates were generated using an in-house method. The resulting sets of diluted samples (both neat and with pyrolysates) were distributed by 50 µL aliquots into three sets of GC vials with annealed 200 µL glass inserts, one set for each instrumental platform evaluated. Note that low concentration samples were analyzed in triplicate on each platform to ensure the reproducibility of any borderline result during interpretation.

Table 1. Serial dilution scheme used to produce samples for analysis, starting from a vial containing the neat ignitable liquid. The source vial refers to the concentration of the vial being aliquoted. The aliquot from source vial column lists the volume taken for dilution. The volume of standard in aliquot column lists the resulting absolute volume of ignitable liquid contained in the aliquot. The diluent column refers to the amount of solvent or pyrolysate-doped solvent added to the aliquot, and the resulting concentration column refers to the final concentration of the sample produced in each row.

Source Vial Conc. (μL/mL)	Aliquot from Source Vial (μL)	Volume of Standard in Aliquot (μL)	Diluent (μL)	Resulting Conc. (μL/mL)
Neat liquid	50	50	950	50
50	50	2.5	450	5
50	10	0.5	490	1
5	50	0.25	450	0.5
1	50	0.05	450	0.1
0.5	50	0.025	450	0.05
1	10	0.01	390	0.025
0.1	50	0.005	450	0.01
0.05	50	0.0025	450	0.005
0.01	50	0.0005	450	0.001
0.005	50	0.00025	450	0.0005

2.3. General Instrumental Analysis

Chromatographic and mass spectrometric conditions were matched as closely as possible to ensure a fair comparison. All separations were performed on one column configuration: a first length of 5% phenyl capillary column with a nominal length of 30 m (actual length ~26 m), connected to a second length of poly(ethylene glycol) wax capillary column with a measured length of 1.23 m. Both columns had a 0.25 mm inner diameter and a 0.25 μm film thickness, and were connected via an SGE Siltite μUnion (SGE, Ringwood, Victoria, Australia). All chromatography was performed with helium (5.0 grade; Praxair, Edmonton, AB, Canada) as carrier gas under speed-optimized flow conditions (2.0 mL/min based on the column geometry) [25], and temperature programmed under optimum heating rate ($10\,^{\circ}C/t_m$; t_m = column void time, min) [26] for each system. The mass spectrometers were operated in electron impact mode at a potential of 70 eV with freshly cleaned ion sources to ensure optimum ion yield. Mass spectrometers were tuned according to the manufacturers' respective criteria prior to use. During operation, the detector voltages were offset by a magnitude of +200 V relative to the tune voltage, and the mass range was set from 25 m/z to 500 m/z. For both 1D GC methods spectra were acquired at a rate of 10 spectra/s, while the GC×GC method acquired at 200 spectra/s. In all cases, the mass spectrometer filaments were turned off at the start time of the solvent peak and turned back on at the end time of the solvent peak. All sample injections (1 μL) were performed by an Agilent 7683A auto sampler into a split/splitless injector operated at a split ratio of 1:80 and a temperature of 250 °C.

Ignitable liquid standards analyzed by each technique are presented in Table 2 and were analyzed in order of lowest concentration to highest concentration to minimize the risk of sample contamination via carryover. Triplicate injections were performed from three separately prepared vials to avoid concerns of samples concentrating between injections by loss of solvent through punctured vial septa.

Table 2. The scheme used for analysis of the ignitable liquid standards, showing concentration in-vial, the nominal on-column volume after injection, and the number of injections performed at each concentration on each instrument. This scheme was followed for the pure ignitable liquid dilutions, and repeated for the pyrolysate-doped series. N/A represents concentrations not analyzed on those respective instruments.

Ignitable Liquid	Solution Conc. (µL/mL)	Volume on-Column (pL)	Number of Injections by Technique		
			GC-MSD	GC-TOF	GC×GC-TOF
Gasoline 1	0.001	0.0125	N/A	N/A	3
(Esso)	0.005	0.0625	N/A	3	3
75% Evaporated	0.01	0.125	3	3	3
	0.05	0.625	3	3	3
	0.1	1.25	3	3	1
	0.5	6.25	1	1	1
	1	12.5	1	1	1
Gasoline 2	0.001	0.0125	N/A	N/A	3
(7-Eleven)	0.005	0.0625	N/A	3	3
75% Evaporated	0.01	0.125	3	3	3
	0.05	0.625	3	3	3
	0.1	1.25	3	3	1
	0.5	6.25	1	1	1
	1	12.5	1	1	1
Diesel	0.005	0.0625	N/A	N/A	3
(7-Eleven)	0.01	0.125	N/A	3	3
25% Evaporated	0.05	0.625	3	3	3
	0.1	1.25	3	3	3
	0.5	6.25	3	3	3
	1	12.5	1	1	1

2.4. GC-MSD Analysis

Analysis of the ignitable liquid standards in Table 2 was performed using an Agilent 7890A gas chromatograph with a 5975C quadrupole mass spectrometer (Agilent Technologies, Mississauga, ON, Canada). Data acquisition was performed with the Agilent MassHunter Workstation version B.07.00 (build 7.0.7024.0), while data interpretation was performed with ChemStation version E.02.02.1431 (Agilent).

2.5. GC-TOF Analysis

Analysis of the ignitable liquid standards in Table 2 was performed using a Pegasus 4D GC×GC-TOF (LECO, St. Joseph, MI, USA) composed of an Agilent 7890A gas chromatograph with a secondary oven operated at the same temperature as the primary oven, and the stationary quad-jet dual-stage modulator disabled. In this configuration, the Pegasus 4D operates as a 1D GC-TOF. Data acquisition was performed with ChromaTOF® version 4.51.6.0, while data interpretation was performed with ChromaTOF® version 4.71.0.0 (LECO).

2.6. GC×GC-TOF Analysis

Analysis of the ignitable liquid standards in Table 2 was performed using the Pegasus 4D GC×GC-TOF described in Section 2.5, but operated with the secondary oven offset by +5 °C relative to the primary oven, and with the modulator in operation using liquid nitrogen as coolant. The modulation period of 1.3 s was selected with assistance from the LECO GC×GC Column Calculator based on an approximate [1]D peak width of 8 s, and a desired modulation ratio of 4–5. Data acquisition and interpretation were performed as described in Section 2.5.

2.7. Interpretation of Results

Chromatographic interpretation was performed by a forensic examiner with several years of casework experience with the Royal Canadian Mounted Police, and then peer reviewed independently by two other forensic examiners. Collectively, the examiners have 48 years of experience in forensic fire debris casework. Interpretation proceeded from a visual inspection of the total ion chromatogram (TIC), to inspection of the summed extracted ion chromatograms (EICs) for the alkane, cycloalkane, aromatic, indane, and naphthalene compound classes. Each EIC was generated by summation of the ions in Table 3. Classification criteria used for each identification was in accordance with the parameters set out in the ASTM E1618-14 standard [27] and the amplified criteria presented in the Supplementary Materials Sections SI.1 and SI.2.

Table 3. List of characteristic ions which are summed and/or overlaid to generate EICs during examination of sample data for interpretation [6,28].

Class of Compounds	Overlapping Classes [1]	Ions (*m/z*)
Alkanes	Most Organics	43, 57, 71, 85, 99
Cycloalkanes	Alkenes	41, 55, 69, 83, 97
Aromatics	Terpenes	91, 105, 106, 119, 120, 134
Indanes	Tetralins, Styrenes, Aromatics	117, 118, 131, 132, 145, 146
Naphthalenes	Decalins	128, 141, 142, 155, 156, 170

[1] These are other significant compound types that do not belong to the declared class, but that have sufficient abundances of the listed ions that they may appear in the same EIC. It is not an exhaustive list of potentially overlapping classes.

Identification of each peak was achieved by either mass spectral matching to the NIST 11 library (NIST, Gaithersburg, MD, USA) with comparison to known ignitable liquid standards, or by visual inspection relative to known peak groupings accompanied by correct relative ratios of at least three characteristic ions in Table 3 in comparison to known ignitable liquid standards.

3. Results

3.1. General Criteria for Identification of Ignitable Liquids

The ASTM E1618-14 standard [27] provides a good description of the criteria for the classification of ignitable liquids, which is not reproduced here. Other more thorough treatments of ignitable liquid identification are available [29], and forensic examiners also typically establish their own refined criteria based on the peer review process, their shared experience from casework, and from collecting ignitable liquid standards. The resulting amplified criteria used for the identification of gasoline and diesel in this work are detailed in the Supplementary Materials. Please note that, although interpretation of 2D chromatographic data for fire debris analysis has been reported in the literature [30], there are no official ASTM guidelines specific to 2D data at this time. We applied the same general approach described by ASTM E1618-14 for interpreting 1D data when evaluating the 2D data, with the addition of using the structured retention information provided by the ^{2}D separation as an additional point of comparison to reference ignitable liquid standards. A summary of the interpretive results is presented in Table 4.

Table 4. Color-coded results of the interpretation.

pL on Column	Neat Dilutions			Pyrolysate-Doped		
	GC-MSD	GC-TOF	GC×GC	GC-MSD	GC-TOF	GC×GC
Gasoline 1						
0.0125						
0.0625						
0.125						
0.625						
1.25						
6.25						
Gasoline 2						
0.0125						
0.0625						
0.125						
0.625						
1.25					Contested	
6.25						
Diesel						
0.0625						
0.125						
0.625						
1.25						
6.25	Misclass.					
12.5				Misclass.		

Green represents an unambiguous identification of the ignitable liquid, blue represents an identification that has barely passed the necessary threshold, yellow represents indications of an ignitable liquid which suggest its presence but are insufficient for identification, and red represents no ignitable liquid detected. Black entries represent concentrations not analyzed on their respective instruments. The two cells containing "Misclass." represent samples which were not properly classifiable, and where an attempt to classify would have yielded a misclassification as a normal paraffinic product. The one cell containing "Contested" represents the only sample whose identification resulted in disagreement during the peer review.

3.2. Results of Interpretation of GC-MSD Analyses

The full chromatographic data resulting from the GC-MSD analysis are presented on Pages 4–137 of the Supplementary Materials.

3.2.1. GC-MSD Analysis of Gasoline (Sections SI.3 Pages 27–70)

For both Gasoline 1 (Esso) and Gasoline 2 (7-Eleven) at 0.125 pL on-column the TIC appeared to be blank, and the only profile with any sign of peak activity was that of the aromatics with very low indications of toluene, ethylbenzene and the xylenes, and some of the C_3-alkylbenzenes. However, the low abundance leads to a finding of *no ignitable liquid detected*. At 0.625 pL on-column the TIC showed some weak indications of peaks, but the aromatic, indane and naphthalene profiles were all identifiable, leading to the positive identification of both gasolines. Note that the alkane profile eventually became identifiable at 6.25 pL on-column. For the interpretation of the GC-MSD data, example TIC and summed aromatic EIC chromatograms are presented in Figure 1a. EICs displayed as overlaid ion chromatograms are illustrated for alkanes and aromatics in Figure 1b.

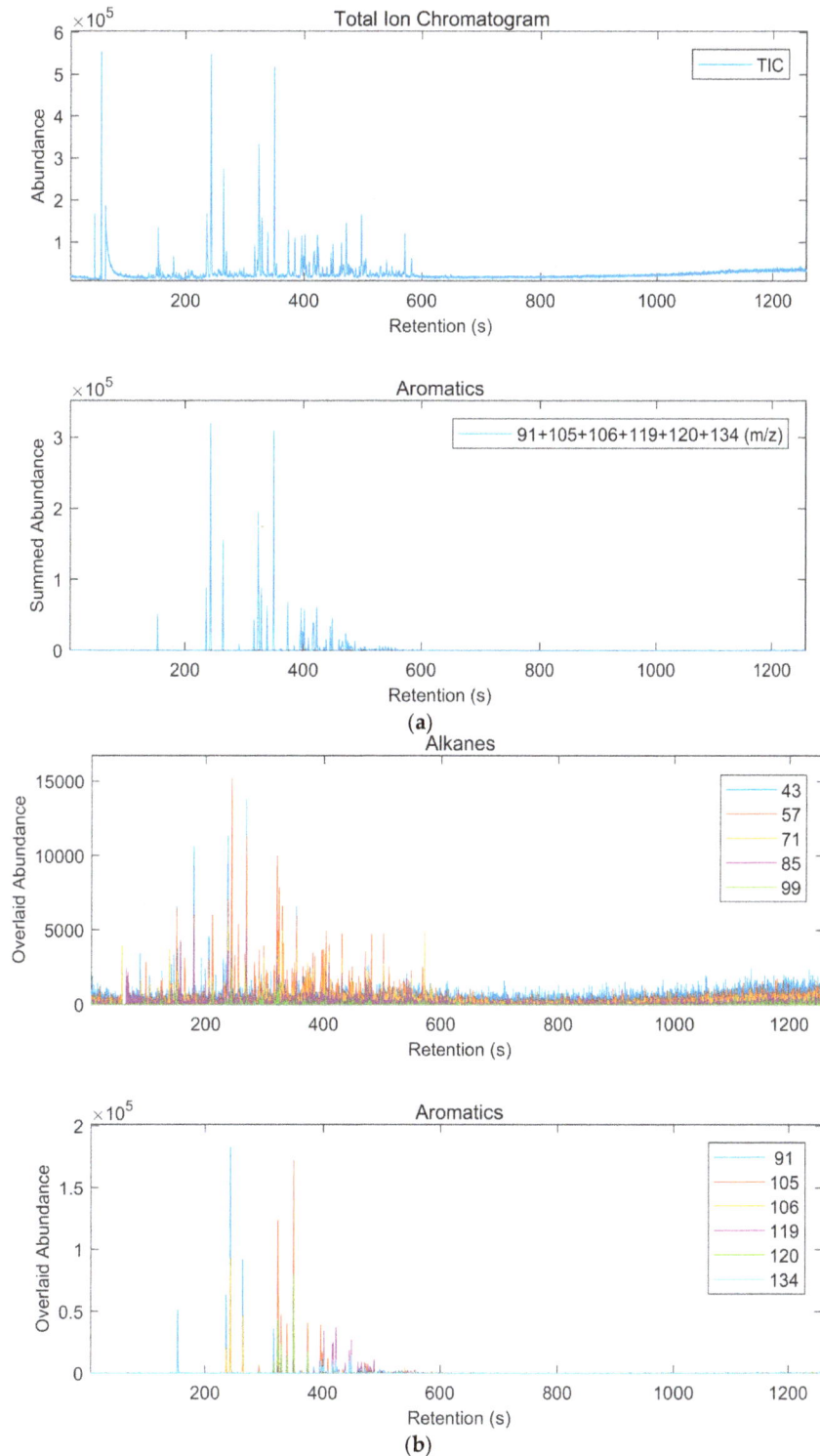

Figure 1. Sample of 1D GC-MSD data for the most abundant gasoline level tested (1.0 µL/mL) as: (**a**) TIC and summed aromatic EIC; and (**b**) alkane and aromatic EICs displayed as overlays of individual ion chromatograms.

3.2.2. GC-MSD Analysis of Gasoline with Pyrolysate (Sections SI.3 Pages 94–137)

In the presence of pyrolysis, the TIC showed predominantly the pyrolysis profile for all samples, so the TIC was not inspected for indications of an ignitable liquid. Both Gasoline 1 and Gasoline 2 had peak activity on all EICs but no identifiable petroleum profiles at 0.125 pL on-column. Although the aromatic profile showed indications of petroleum at 0.625 pL on-column, namely indications of C$_4$-alkylbenzenes not usually observed from pyrolysis, disturbance of the peak ratios in the C$_3$-alkylbenzene range by pyrolysate resulted in an aromatic profile that could not be attributed to petroleum. At 1.25 pL on-column, Gasoline 1 had aromatic, indane and naphthalene profiles suggestive of petroleum, while Gasoline 2 had suggestive aromatic and naphthalene profiles. Despite the indications of petroleum, disturbance of many of the critical peak ratios in the presence of chemically related pyrolysis compounds requires a cautious conclusion that *no ignitable liquid could be identified* at this level for either gasoline. At 6.25 pL on-column, Gasoline 1 showed abundant and competent aromatic, indane and naphthalene profiles resulting in a positive identification of gasoline, while Gasoline 2 showed abundant and competent aromatic, indane and naphthalene profiles as well as a weak but also competent alkane profile, resulting in a positive identification of gasoline. The level of gasoline required on-column in this case is 10-fold higher in the presence of interfering pyrolysate than the amount required when the gasoline is pure. That factor will increase in the presence of more abundant interfering pyrolysate and decrease in cases of weaker interfering pyrolysate or most levels of non-interfering pyrolysate.

3.2.3. GC-MSD Analysis of Diesel (Sections SI.3 Pages 7–26)

For diesel at 1.25 pL on-column, no peaks were visible on the TIC and no indication of peaks were present on the EIC profiles. At 6.25 pL on-column, very weak indications of peaks were visible on the TIC and the *n*-alkanes were identifiable on the alkane EIC. However, no clear presence of branched alkanes could be seen at this level, leading to a situation where a potential misclassification as a heavy normal paraffinic product would be possible under the ASTM E1618-14 guidelines [27]. However, given the very low abundance of the *n*-alkane profile, a cautious finding of *no ignitable liquid could be identified* was given. At 12.5 pL on-column, the alkane EIC showed an identifiable profile including the presence of n-alkane and branched alkanes, along with weak indications of peak activity on the aromatic, indane and naphthalene profiles within the elution range of the Gaussian series of *n*-alkanes between C$_{10}$ (decane) and C$_{19}$ (nonadecane), resulting in a positive identification of a heavy petroleum distillate.

3.2.4. GC-MSD Analysis of Diesel with Pyrolysate (Sections SI.3 Pages 74–93)

In the presence of pyrolysate, the TIC again showed predominantly the pyrolysis profile for all samples, so the TIC was not inspected for indications of an ignitable liquid. At 6.25 pL on-column, no indication of any ignitable liquid was visible, likely due to the interference with the visual appearance of the alkane EIC by pyrolysate. At 12.5 pL on-column, the alkane EIC showed a Gaussian profile of *n*-alkanes but no clear indication of branched alkanes was observed, and pyrolysis contributions were the only apparent peak activity on the aromatic, indane and naphthalene profiles. This leads to a situation where the more abundant *n*-alkanes merit a positive finding but result in a misclassification as a heavy normal paraffinic product. In a forensic lab, such a sample would usually be concentrated and re-analyzed to eliminate any ambiguity prior to issuing a finding, but in this case 12.5 pL on-column was the highest concentration tested for diesel.

3.3. *Results of Interpretation of GC-TOF Analyses*

The full chromatographic data resulting from the GC-TOF analysis are presented in pages 138–313 of the Supplementary Materials.

3.3.1. GC-TOF Analysis of Gasoline (Sections SI.3 Pages 165–220)

For both gasolines at 0.125 pL on-column, the observations were similar to those of the GC-MSD analyses. Gasoline 1 had a slightly more abundant aromatic profile clearly indicative of a petroleum product but resulted in a finding of *no ignitable liquid was identified* since, as detailed in Section 3.2.1, it is inappropriate to identify gasoline from only a weak aromatic profile. For Gasoline 2, the finding was *no ignitable liquid detected*. At 0.625 pL on-column, the observations were again similar to those from the GC-MSD analyses with positive identification of both gasolines, but notably the alkane profile is visible for both gasolines at this level and forms an important part of the identification. For the interpretation of the GC-TOF data, example TIC and summed aromatic EIC chromatograms are presented in Figure 2a. EICs displayed as overlaid ion chromatograms are illustrated for alkanes and aromatics in Figure 2b.

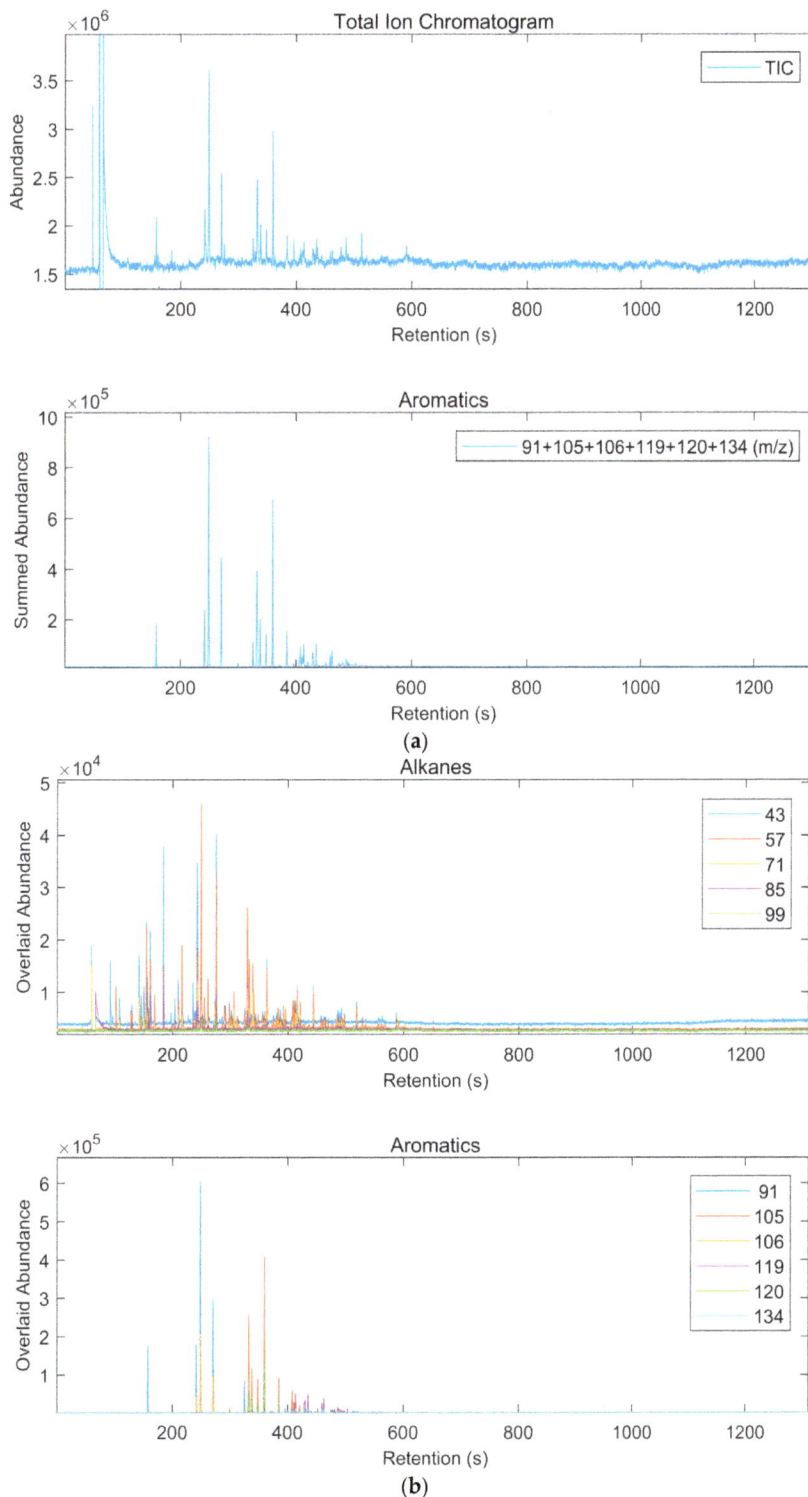

Figure 2. Sample of 1D GC-TOF data for the most abundant gasoline level tested (1.0 μL/mL) as: (**a**) TIC and summed aromatic EIC; and (**b**) alkane and aromatic EICs displayed as overlays of individual ion chromatograms.

3.3.2. GC-TOF Analysis of Gasoline with Pyrolysate (Sections SI.3 Pages 254–313)

In the presence of pyrolysate, Gasoline 1 showed similar results to the GC-MSD analyses with 1.25 pL on-column resulting in a finding of *not identified*, but with positive identification at 6.25 pL based on the alkane, aromatic, indane and naphthalene profiles. Gasoline 2 showed slightly improved results during initial interpretation with a finding of *not identified* at 0.625 pL on-column, and an identification at 1.25 pL based on the aromatic, indane and naphthalene profiles with the presence of light alkanes in ratios comparable to gasoline reference standards. Upon peer review of the interpretation, one reviewer disagreed with the interpretation of Gasoline 2 at 1.25 pL on the basis that the alkanes were not sufficiently comparable to result in a conclusive identification, instead preferring a conservative finding of *not identified*. The original examiner agreed that if the alkane profile is considered insufficient, that the remaining aromatic, indane and naphthalene profiles presented too much interference from pyrolysate to justify conclusive identification of gasoline. Accordingly, the conclusion was downgraded to *not identified* at 1.25 pL on-column, raising the limit of identification to 6.25 pL on-column.

3.3.3. GC-TOF Analysis of Diesel (Sections SI.3 Pages 139–164)

For diesel at 1.25 pL on-column, the observations were similar to the GC-MSD result for 6.25 pL, with the alkane profile showing *n*-alkane peaks but at an abundance too low to confidently identify any ignitable liquid. At 6.25 pL, the observations were similar to that of the GC-MSD result for 12.5 pL, with the alkane profile showing a host of *n*-alkane and branched alkane peaks accompanied by some indications of peaks on the aromatic, indane and naphthalene profiles within in the n-alkane elution range leading to positive identification of a heavy petroleum distillate.

3.3.4. GC-TOF Analysis of Diesel with Pyrolysate (Sections SI.3 Pages 228–253)

In the presence of pyrolysate, diesel showed results somewhat similar to the GC-MSD analyses, except that branched alkanes became identifiable at 6.25 pL via GC-TOF relative to 12.5 pL with GC-MSD. Despite the earlier visibility of branched alkanes at 6.25 pL, they are still insufficient to reach a conclusive identification, resulting in a finding of *no ignitable liquid identified*. However, at 12.5 pL positive identification of a heavy petroleum distillate is achieved based on the alkane profile, supported by indications of peaks on the aromatic, indane and naphthalene profiles within the *n*-alkane elution range.

3.4. Results of Interpretation of GC×GC-TOF Analyses

The full chromatographic data resulting from the GC×GC-TOF analysis are presented in pages 314–413 of the Supplementary Materials.

3.4.1. GC×GC-TOF Analysis of Gasoline (Sections SI.3 Pages 331–360)

For Gasoline 1 at 0.0625 pL on-column, some alkane and aromatic peak activity were visible, suggestive of a petroleum product but insufficient for confirmation, resulting in a finding of *no ignitable liquid identified*. At 0.125 pL, however, competent alkane, aromatic and indane profiles were clearly identifiable, resulting in a positive identification of gasoline. For Gasoline 2 at 0.0625 pL on-column the alkane and aromatic profiles also show peaks indicative of petroleum at a greater abundance than for Gasoline 1, just passing the threshold to result in a positive identification of gasoline. For both Gasoline 1 and Gasoline 2 at 0.125 pL on-column, the alkane, aromatic, indane and naphthalene EICs all show a complete peak profile for gasoline. For the interpretation of the GC×GC-TOF data an example TIC is presented in Figure 3a, and summed alkane and aromatic EICs are presented in Figure 3b. Please note that EICs displayed as overlaid ion chromatograms are not provided for the 2D data, as it was not possible to present them in a meaningful way.

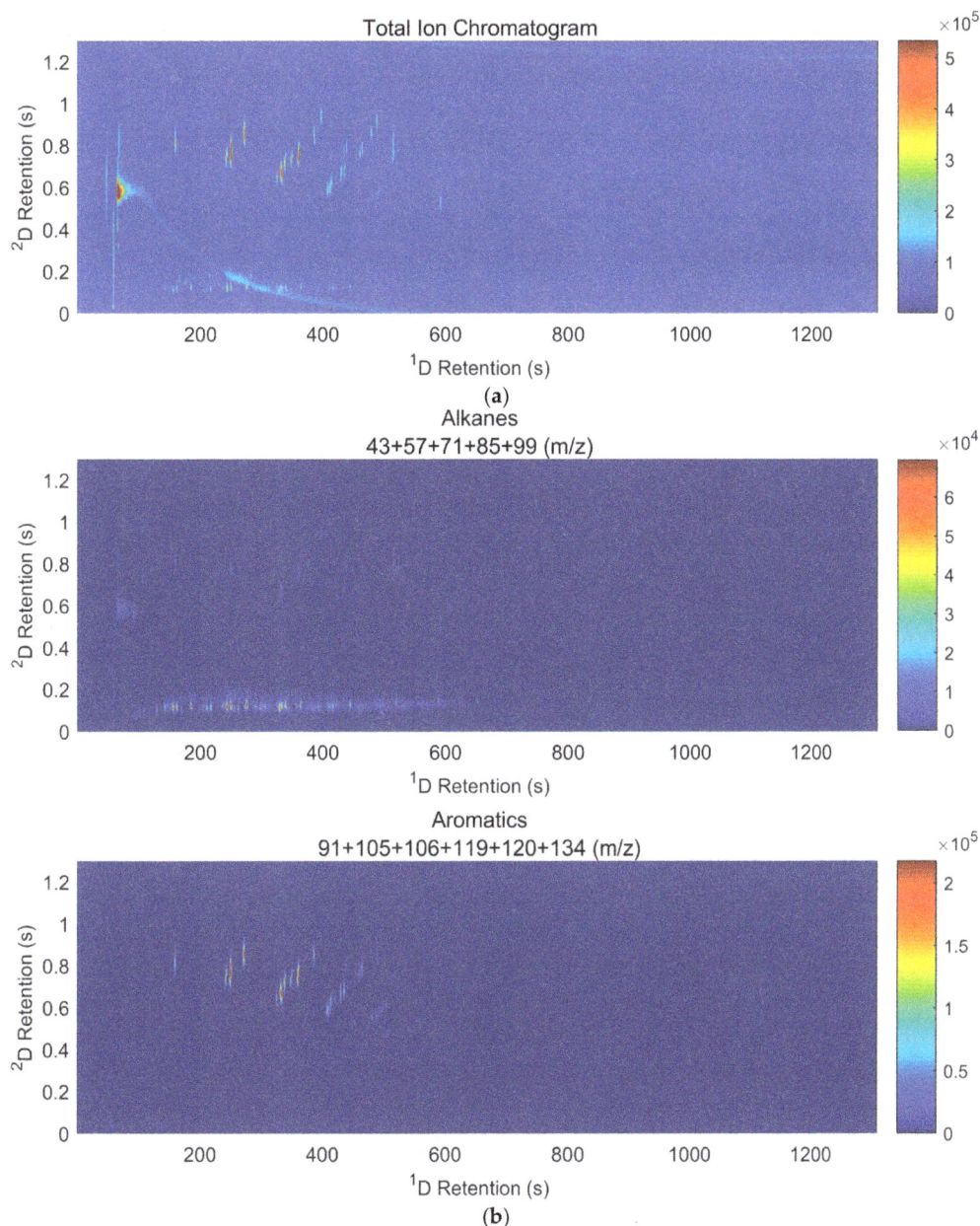

Figure 3. Sample of 2D GC×GC-TOF data for the most abundant gasoline level tested (1.0 μL/mL) as: (**a**) TIC; and (**b**) summed alkane and aromatic EICs. The ^{2}D retention axis has been shifted downwards by 0.55 seconds in all figures to account for the ^{2}D t_m. EICs are not provided as overlays of individual ion chromatograms as it is not possible to present them in a meaningful way for the 2D data.

3.4.2. GC×GC-TOF Analysis of Gasoline with Pyrolysate (Sections SI.3 Pages 382–413)

In the presence of pyrolysate, Gasoline 1 at 0.125 pL on-column showed peak activity on the alkane and aromatic profiles with indications of petroleum, but due to the contribution from pyrolysis to many of the same peaks present in petroleum, only a finding of *no ignitable liquid identified* was possible. At 0.625 pL of Gasoline 1 on-column, competent alkane, aromatic and indane profiles are all present resulting in positive identification of gasoline. For Gasoline 2 at 0.125 pL on-column, peak activity is present on the alkane and aromatic profiles but not indicative of petroleum resulting in a finding of no ignitable liquid detected. At 0.625 pL of Gasoline 2 on-column, the alkane and aromatic EICs show peak profiles indicative of petroleum at low abundances, but just sufficient to positively identify gasoline.

3.4.3. GC×GC-TOF Analysis of Diesel (Sections SI.3 Pages 315–330)

For diesel at 0.625 pL on-column the alkane EIC showed the presence of n-alkanes and branched alkanes at very low levels, but still sufficient to positively identify a heavy petroleum distillate. At 6.25 pL of diesel on-column, the aromatic, indane and naphthalene profiles also showed peak activity indicative of petroleum. Notably, the structured nature of the retention coordinates on the 2D chromatogram allowed clear visualization of the peaks present on these three EICs in a way that was impossible using the 1D techniques.

3.4.4. GC×GC-TOF Analysis of Diesel with Pyrolysate (Sections SI.3 Pages 365–381)

In the presence of pyrolysate, diesel at 1.25 pL on-column showed peak activity on all of the EICs (alkane, aromatic, indane and naphthalene) but the contribution of pyrolysate to many of the same peaks present in the petroleum obscured enough of the critical peak ratios to require a finding of *no ignitable liquid identified*. At 6.25 pL on-column, each of the EIC profiles showed profiles strongly indicative of petroleum resulting in the positive identification of a heavy petroleum distillate. Notably, the high number of isomers from every class of compounds and their easy visualization made confident identification of an HPD especially simple.

4. Discussion

4.1. Notes on Design and Interpretation

Instrumental sensitivity is usually assessed by mass calibration of target substances [31]; however, in the context of forensic examination of fire debris, the concepts of sensitivity and limits of detection require a different treatment. Ignitable liquids are often petroleum products containing hundreds or thousands of individual compounds [1,2]. Additionally, the relative abundances of the constituents of the ignitable liquid can vary considerably from one sample to another due to differences in formulation, weathering, etc. [32–34]. Consequently, the mass of each individual compound in a sample cannot be reliably tracked as a measure of material on-column. Thus, we chose to rely on calibration by total volume of ignitable liquid on-column. Further, the truly important measure for forensic purposes is the limit of identification (LOI), i.e., the concentration at which the ignitable liquid *can be reliably identified*. This identification depends on the combined pattern of numerous peaks [6,27,29], with the threshold for identification determined somewhat by the discretion of a skilled examiner. This makes it difficult to establish a universal rule by which an instrument's absolute sensitivity can be quantified for fire debris analysis. However, among petroleum-based ignitable liquids of varied formulation, there are some relatively stable characteristics (e.g., the C_2-, C_3- and C_4-alkylbenzenes in gasoline, or the dominant n-alkanes and interspersed branched alkanes in middle and heavy distillates) upon which investigators rely [6,29,33,34]. Consequently, our findings may be reasonably extrapolated to most gasolines and diesels in the absence of highly unusual compositions.

When interpreting fire debris data, inspecting the TIC is the first step in assessment as it shows the total overall peak content of the extract [29], but it also maximizes the noise by summing all ion channels even though relatively few of those channels capture the response of any given analyte. EICs are superior to the TIC as they allow filtering peaks by class via summation of only those ions known to be characteristic to a given class of compounds (Table 3) [29,35,36]. As a result, EICs provide a greater signal-to-noise ratio by ignoring the noise in low significance channels, and provide better visualization of individual contributions to an ignitable liquid profile. This improves detectability, but also aids classification and comparison to ignitable liquid standards. For this reason, a description of the EIC profiles is prioritized in the results section. In general, peaks become visible in the EIC long before a sufficient number of compounds are present to indicate the presence of petroleum. For instance, the toluene and p,m-xylene peaks are normally clearly observed in the aromatic EIC well below the limit of identification for gasoline. These peaks alone cannot provide sufficient comparison of within-group peak ratios to positively identify gasoline. For this reason, the term *identifiable* is used when referring to petroleum profiles within the EICs. Additionally, an abundant profile may be present in an EIC far above the threshold required to assess the within-group peak ratios necessary to identify an ignitable liquid, but the presence of even a small amount of pyrolysate may augment enough individual peaks to alter the required ratios. This would also result in a non-identifiable profile even when the remainder of the profile suggests the presence of petroleum.

In fire debris analysis, it is common practice to use a scale of findings starting with *"no ignitable liquid detected"* when no peak profiles indicative of petroleum are observed, which may be a result of their total absence, extremely low abundance, masking by pyrolysate, or a combination thereof. The next level of finding, *"no ignitable liquid identified"*, is reserved for cases where clear indications of petroleum are observed within the EICs, but their abundance is either very low, or there are other factors disqualifying their definitive attribution to a petroleum product such as microbial degradation or interference from pyrolysate. The next level of finding, *"ignitable liquid detected"*, refers to the successful identification of an ignitable liquid sufficient for its classification and uses to be detailed in the conclusion.

4.2. Observations on GC-MSD Results

GC-MSD provided an LOI of 0.625 pL on-column for both gasolines, and 12.5 pL on-column for diesel. Much of the prior work in fire debris analysis referred to an estimated sensitivity of "0.1–0.5 μL" of gasoline [10,12–14,19,37]. This is difficult to place in context as this may refer to the volume left at the scene, the volume captured during exhibit extraction, the volume diluted in a solvent for injection, the un-split volume, or the post-split volume delivered to the GC column. Furthermore, if the chromatographic conditions used to arrive at the given estimate were suboptimal, low peak resolution or excessive band broadening of the analytes would degrade the signal to noise ratio resulting in poorer performance of the instrument. In an attempt to compare our value with the literature, we can consider the most popular extraction method (passive headspace concentration with activated carbon strip) [38] and presume an elution volume of 600 μL CS_2. In this case, a recovery of 0.1 μL would result in a final extract concentration of ~1.7 μL/mL which, when split 1:80 (split ratios in the range of 1:50 to 1:80 are common) would result in a volume of ~2.1 pL delivered on-column. Given our LOI result of 0.625 pL for gasoline, this would suggest an increase in sensitivity of modern instrumentation relative to the instrumentation used in earlier work. In any case, it is impossible to know if this increase in sensitivity represents a real increase in instrumental performance or is the result of a lack of rigor in the historical estimates of sensitivity. It is more important to recognize that the sensitivity offered by fire debris analysis depends on proper collection and preservation of the evidence at the scene, efficient extraction procedures, properly optimized instrumental methods, skilled interpretation of the results, and the nature of the specific ignitable liquid present.

When the contribution of substrate pyrolysate is considered, higher LOIs of 6.25 pL for gasoline and more than 12.5 pL for diesel are found. This increase reflects the confounding effect of pyrolysate, although it is worth noting that interference from coelution of different compounds is largely resolved by using the mass filtering offered by EICs. The real issue of interference is the distortion of peak ratios present in the ignitable liquid due to production of those same compounds by pyrolysis. For example, peaks such as naphthalene, *p,m*-xylene, toluene and some of the C_3-alkylbenzenes are regularly contributed by substrate pyrolysis [23,29,39], and in the case of gasoline these are all peaks of importance [27,29] which will be unavoidably distorted by the background. No amount of additional separation can resolve this particular issue, and both the magnitude and the specific chemistry of the substrate pyrolysis contribution will define the LOI in these cases.

Given the large difference between the LOIs for gasoline and diesel, the nature of the ignitable liquid itself is the most significant uncontrollable factor affecting sensitivity, and to our knowledge these values are the first rigorous estimators reported for the quantitative sensitivity of GC-MSD for gasoline and diesel identification.

4.3. Observations on GC-TOF Results

GC-TOF provided LOIs of 0.625 pL on-column for both gasolines, and 6.25 pL on-column for diesel. The similar performance of GC-MSD and GC-TOF for gasoline is relatively unsurprising given that most of the critical peaks are characterized by a relatively small number of ions, which will be represented reasonably well by the quadrupole relative to the TOF. What is more significant with the TOF is the lower concentrations required for the alkane pattern to be detected and correctly interpreted for both gasoline and diesel. Alkanes generate a wide range of ions during fragmentation, whereas aromatics generate comparatively few ions (Figure 4). This, coupled with differences in ion transmission through the different mass spectrometers, results in improved spectral quality for alkanes at lower absolute masses on-column. This improves trace level identification of both the linear and branched alkanes resulting in a lower LOI for diesel.

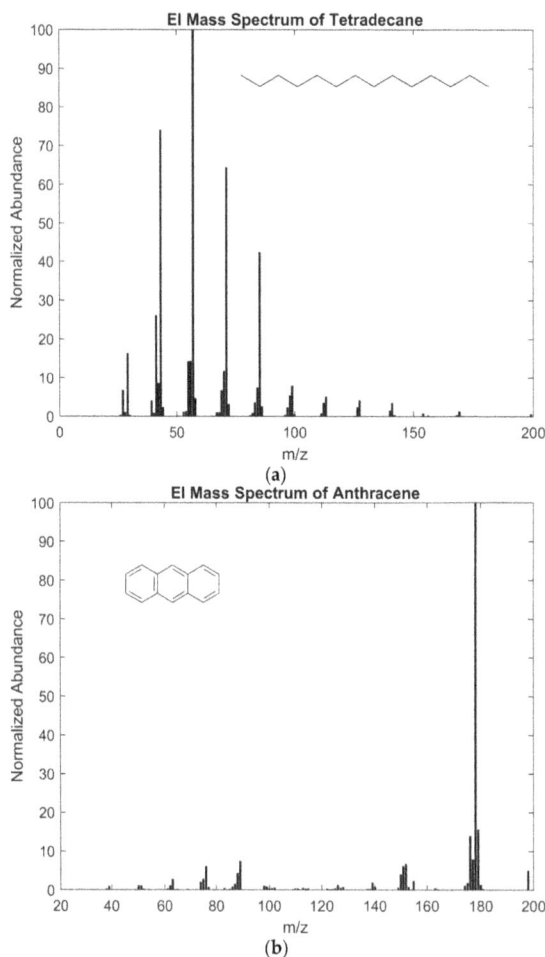

Figure 4. Representative mass spectra from the NIST 11 library for: (**a**) tetradecane (NIST record #229858); and (**b**) anthracene (NIST record #228201).

When the contribution of substrate pyrolysate is considered, the LOIs again increase to 6.25 pL on-column for both gasolines, and to 12.5 pL diesel. These increases are a result of the same factors discussed in Section 4.2, with the greater sensitivity of the TOF towards alkanes accounting for the lower LOI for diesel. The differences in interpretation of the gasolines is not solely attributable to their alkane fraction, but arises from overall differences in their composition as a result of the formulation process at the refinery [32–34].

Overall, the increase in sensitivity offered by the GC-TOF relative to the GC-MSD is modest at best, although the value of more sensitive detection of alkanes should not be understated in the context of analysis of petroleum products.

4.4. Observations on GC×GC-TOF Results

GC×GC-TOF provided LOIs of 0.125 pL on-column for Gasoline 1, 0.0625 pL on-column for Gasoline 2, and 0.625 pL on-column for diesel. These values represent a significant and consistent order of magnitude increase in sensitivity relative to both GC-MSD and GC-TOF resulting from a combination of the focusing effect achieved during modulation onto the second-dimension column, reduced noise due to the improved separation, and greater confidence in peak identification due to the structured retention information displayed in the 2D chromatogram [21].

When the contribution of substrate pyrolysis is considered, the LOIs increase to 0.625 pL on-column for gasoline, and 6.25 pL on-column for diesel, but these represent less of a decrease in sensitivity when compared to the 1D methods. While some have suggested that GC×GC offers better separation of pyrolysis compounds from the ignitable liquid compounds [21], we found this to be of very minor benefit during the interpretation process. The greater benefits offered by GC×GC in the context of fire debris analysis arise from the focusing effect of modulation and most importantly the structured retention information inherent in the chromatogram. Structured retention causes the hosts of heavier isomeric compounds to elute in easily recognized bands ordered by compound type and degree of substitution. This is especially advantageous in the presence of substrate interferences since pyrolysis does not produce the entire array of isomers observed in petroleum. The added sensitivity and structured retention of GC×GC means that compounds normally buried in the noise on 1D chromatograms are easily visualized. In 1D analysis, the interference of pyrolysate distorts peaks and disturbs the ratios within critical peak groupings. However, the ability to see the normally obscured heavy groupings of high isomerism in ignitable liquids allows comparison of richer areas of the chromatogram to ignitable liquid standards (Figure 5). As a result, confident identification can be made even when the critical ratios in classical analysis are heavily perturbed. This effect may alter the classification guidelines currently in use, or even require the creation of a new published standard covering analysis and interpretation of fire debris by 2D instrumentation.

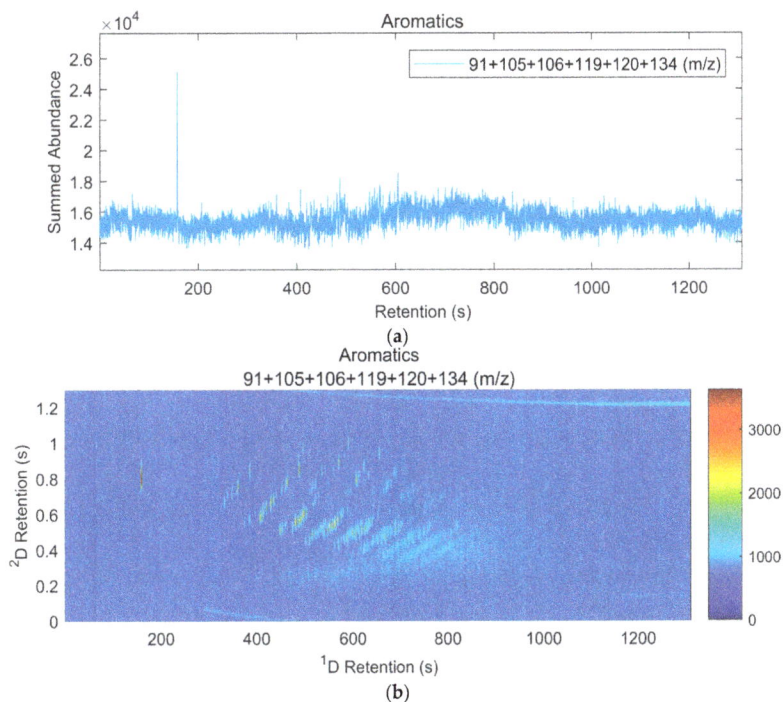

Figure 5. Heavy aromatic compounds illustrating structured retention information and high isomerism as observed on the summed aromatic EIC for the 1.0 μL/mL diesel using: (a) GC-TOF; and (b) GC×GC-TOF. The ^{2}D retention axis has been shifted downwards by 0.55 s in both figures to account for the ^{2}D t_m.

5. Conclusions

We have reported the first quantitative sensitivities of GC-MSD as LOIs (i.e., the concentration on-column at which the ignitable liquid *can be reliably identified*) for gasoline and diesel in the context of fire debris, and compared those sensitivities to GC-TOF and GC×GC-TOF techniques under equivalent conditions. These results will enable future quantitative assessment of the other parameters in fire debris collection and analysis, such as collection at the scene, and extraction efficiency in the lab. This work also provides a solid foundation for future work towards closing the identification gap between the forensic lab and the accelerant-detecting canine. Our results also demonstrate the superiority of GC×GC analysis relative to the classic 1D GC techniques and highlight new reasons for this superiority. However, further study will be required to determine what level of sensitivity is excessive and would result in inappropriately high significance attributed to insignificant amounts of ignitable liquid. We anticipate that our experimental approach will allow other labs to begin validating their instrumental sensitivities, as well as reporting LOIs for other less frequently encountered ignitable liquid classes. Ultimately, we hope to see increased use of GC×GC in fire debris analysis by mainstream forensic labs.

Author Contributions: Conceptualization, R.J.A. and J.J.H.; Data curation, R.J.A.; Formal analysis, R.J.A.; Funding acquisition, J.J.H.; Investigation, R.J.A.; Methodology, R.J.A. and J.J.H.; Peer review of interpretation, G.Z. and P.M.L.S.; Project administration, J.J.H.; Resources, J.J.H.; Supervision, J.J.H.; Writing—original draft, R.J.A. and J.J.H.; and Writing—review and editing, J.J.H., G.Z. and P.M.L.S.

Funding: This research was funded by the Natural Sciences Engineering Research Council (NSERC) Canada through a Discovery Grant to JJH.

Conflicts of Interest: The authors declare no conflict of interest. The funders had no role in the design of the study; in the collection, analyses, or interpretation of data; in the writing of the manuscript, or in the decision to publish the results.

References

1. Jackowski, J.P. The incidence of ignitable liquid residues in fire debris as determined by a sensitive and comprehensive analytical scheme. *J. Forensic Sci.* **1997**, *42*, 828–832. [CrossRef]
2. Sandercock, P.M.L. Passive headspace extraction of ignitable liquids using activated carbon cloth. *Can. Soc. Forensic Sci. J.* **2016**, *49*, 176–188. [CrossRef]
3. Lentini, J.J.; Dolan, J.A.; Cherry, C. The petroleum-laced background. *J. Forensic Sci.* **2000**, *45*, 968–989. [CrossRef] [PubMed]
4. DeHaan, J.D. Fire investigations and the forensic lab: What the lab should be doing, or, it's not about the GC. *CAC News* **2002**, *4*, 14–16.
5. DeHaan, J.D. Our changing world, Part 3: Is more sensitive necessarily more better? and Part 4: A matter of time. *Fire Arson Investig.* **2002**, *52*, 20–23.
6. Almirall, J.R.; Furton, K.G. *Analysis and Interpretation of Fire Scene Evidence*; CRC Press: Boca Raton, FL, USA, 2004; p. 262.
7. Hetzel, S.S.; Moss, R.D. How long after waterproofing a deck can you still isolate an ignitable liquid? *J. Forensic Sci.* **2005**, *50*, 369–376. [CrossRef] [PubMed]
8. Lang, T.; Dixon, B.M. The possible contamination of fire scenes by the use of positive pressure ventilation fans. *Can. Soc. Forensic Sci. J.* **2000**, *33*, 55–60. [CrossRef]
9. Koussaifes, P.M. Evaluation of fire scene contamination by using positive-pressure ventilation fans. *Forensic Sci. Commun.* **2002**, *4*, 4.
10. Armstrong, A.; Babrauskas, V.; Holmes, D.L.; Martin, C.; Powell, R.; Riggs, S.; Young, L.D. The evaluation of the extent of transporting or "tracking" an identifiable ignitable liquid (gasoline) throughout fire scenes during the investigative process. *J. Forensic Sci.* **2004**, *49*, 741–748. [CrossRef]
11. Belchior, F.; Andrews, S.P. Evaluation of cross-contamination of nylon bags with heavy-loaded gasoline fire debris and with automotive paint thinner. *J. Forensic Sci.* **2016**, *61*, 1622–1631. [CrossRef]

12. DeHaan, J.D. Canine accelerant detection teams: Validation and certification. *CAC News* **1994**, *2*, 17–21.

13. Kurz, M.E.; Billard, M.; Rettig, M.; Augustiniak, J.; Lange, J.; Larsen, M.; Warrick, R.; Mohns, T.; Bora, R.; Broadus, K.; et al. Evaluation of canines for accelerant detection at fire scenes. *J. Forensic Sci.* **1994**, *39*, 1528–1536. [CrossRef] [PubMed]

14. Tindall, R.; Lothridge, K. An evaluation of 42 accelerant detection canine teams. *J. Forensic Sci.* **1995**, *40*, 561–564. [CrossRef]

15. Katz, S.R.; Midkiff, C.R. Unconfirmed canine accelerant detection: A reliability issue in court. *J. Forensic Sci.* **1998**, *43*, 329–333. [CrossRef]

16. Ottley, B.L. Beyond the crime laboratory: The admissibility of unconfirmed forensic evidence in arson cases. *N. Engl. J. Crim. Civ. Confin.* **2010**, *36*, 263.

17. Twibell, J.D.; Home, J.M.; Smalldon, K.W. A comparison of the relative sensitivities of the adsorption wire and other methods for the detection of accelerant residues in fire debris. *J. Forensic Sci. Soc.* **1980**, *22*, 155–159. [CrossRef]

18. Loscalzo, P.; DeForest, P.R.; Chao, J. A study to determine the limit of detectability of gasoline vapor from simulated arson residues. *J. Forensic Sci.* **1980**, *25*, 162–167. [CrossRef]

19. Thatcher, P.J. The scientific investigation of fire causes. *Forensic Sci. Prog.* **1986**, *1*, 117–152.

20. Choodum, A.; Nic Daéid, N. Development and validation of an analytical method for hydrocarbon residues using gas chromatography-mass spectrometry. *Anal. Methods* **2011**, *3*, 1136–1142. [CrossRef]

21. Frysinger, G.S.; Gaines, R.B. Forensic analysis of ignitable liquids in fire debris by comprehensive two-dimensional gas chromatography. *J. Forensic Sci.* **2002**, *47*, 471–482.

22. Dolan, J. Recent advances in the applications of forensic science to fire debris analysis. *Anal. Bioanal. Chem.* **2003**, *376*, 1168–1171. [CrossRef] [PubMed]

23. Stauffer, E.; Lentini, J.J. ASTM standards for fire debris analysis: A review. *Forensic Sci. Int.* **2003**, *132*, 63–67. [CrossRef]

24. Taylor, C.M.; Rosenhan, A.K.; Raines, J.M.; Rodriguez, J.M. An arson investigation by using comprehensive two-dimensional gas chromatography-quadrupole mass spectrometry. *J. Forensic Res.* **2012**, *3*, 169.

25. Klee, M.S.; Blumberg, L.M. Theoretical and practical aspects of fast gas chromatography and method translation. *J. Chromatogr. Sci.* **2002**, *40*, 234–247. [CrossRef] [PubMed]

26. Blumberg, L.M.; Klee, M.S. Optimal heating rate in gas chromatography. *J. Microcolumn Sep.* **2000**, *12*, 508–514. [CrossRef]

27. American Society for Testing and Materials. *ASTM E1618-14 Standard Test Method for Ignitable Liquid Residues in Extracts from Fire Debris Samples by Gas Chromatography-Mass Spectrometry*; ASTM International: West Conshohocken, PA, USA, 2014.

28. McLafferty, F.W.; Tureček, F. *Interpretation of Mass Spectra*, 4th ed.; University Science Books: Mill Valley, CA, USA, 1993; p. 371.

29. Stauffer, E.; Dolan, J.A.; Newman, R. *Fire Debris Analysis*, 1st ed.; Elsevier/Academic Press: London, UK, 2008.

30. Sampat, A.; van Daelen, B.; Lopatka, M.; Mol, H.; van der Weg, G.; Vivó-Truyols, G.; Sjerps, M.; Schoenmakers, P.; van Asten, A. Detection and characterization of ignitable liquid residues in forensic fire debris samples by comprehensive two-dimensional gas chromatography. *Separations* **2018**, *5*, 43. [CrossRef]

31. Doong, R.; Chang, S.; Sun, Y. Solid-phase microextraction for determining the distribution of sixteen US Environmental Protection Agency polycyclic aromatic hydrocarbons in water samples. *J. Chromatogr. A* **2000**, *879*, 177–188. [CrossRef]

32. Speight, J.G. *Handbook of Petroleum Analysis*; John Wiley & Sons: New York, NY, USA, 2001; p. 473.

33. Sandercock, P.M.L. A survey of Canadian gasoline (2004). *Can. Soc. Forensic Sci. J.* **2007**, *40*, 105–130. [CrossRef]

34. Sandercock, P.M.L. Survey of Canadian gasoline (Winter 2010). *Can. Soc. Forensic Sci. J.* **2012**, *45*, 64–78. [CrossRef]

35. Bertsch, W.; Zhang, Q.W.; Holzer, G. Using the tools of chromatography, mass spectrometry, and automated data processing in the detection of arson. *J. High Resolut. Chromatogr.* **1990**, *13*, 597–605. [CrossRef]

36. Lentini, J.J. An improved method of obtaining ion profiles from ignitable liquid residue samples. *CAC News* **1995**, *4*, 18.

37. Kurz, M.E.; Schultz, S.; Griffith, J.; Broadus, K.; Sparks, J.; Dabdoub, G.; Brock, J. Effect of background interference on accelerant detection by canines. *J. Forensic Sci.* **1996**, *41*, 868–873. [CrossRef] [PubMed]

## 14

Determination of the Three Main Components of the Grapevine Moth Pest Pheromone in Grape-Related Samples by Headspace-Gas Chromatography-Mass Spectrometry

María del Carmen Alcudia-León *, Mónica Sánchez-Parra, Rafael Lucena and Soledad Cárdenas [iD]

Departamento de Química Analítica, Instituto de Química Fina y Nanoquímica IUIQFN, Universidad de Córdoba, Campus de Rabanales, Edificio Marie Curie (anexo), E-14071 Córdoba, Spain; q32sapam@uco.es (M.S.-P.); q62luror@uco.es (R.L.); qa1caarm@uco.es (S.C.)
* Correspondence: q12caalm@uco.es

Abstract: The grapevine moth (*Lobesia botrana*) is the most significant pest of viticulture. This article reports the development of an analytical method that allows the instrumental determination of the three main pheromone components of the pest ((*E,Z*)-7,9-dodecadien-1-yl acetate, (*E,Z*)-7,9-dodecadien-1-ol and (*Z*)-9-dodecen-1-yl acetate) in grape-related samples (must, table grape and wine grape). The combination of headspace, gas chromatography and mass spectrometry provides limits of detection in the range of 60–420 ng/Kg and precision, expressed as a relative standard deviation, better than 8.5%. This analytical approach is rapid and simple and opens a door to the study of the pest incidence on the final products.

Keywords: *Lobesia botrana*; pheromone; headspace-gas chromatography-mass spectrometry (HS-GC-MS)

---

1. Introduction

The grapevine moth, *Lobesia botrana* Lepidoptera: Tortricidae, is the main grape pest in the Mediterranean basin [1]. Pest damage is mainly caused by larvae feeding on grapes, which leads to fungal colonization of wounds and fruit rot [2]. In Europe, the management of grape wine moth has relied primarily on the timing of conventional insecticides [3–5]. Pesticides in agriculture have a negative public perception as they increase the risk of environmental damage [6–8]. This has generated a demand for less harmful protection strategies as an alternative to traditional pest control [9].

Reliable monitoring systems and selective means of pest control are two prerequisites of integrated pest management. Insect pheromones have achieved a reputation in both areas. During the past few decades, sex pheromones and related attractants for hundreds of insect species, mostly of the order Lepidoptera, have been characterized. As a result of these studies, monitoring traps for numerous agricultural pests have been made available to growers and advisory services.

The main component of the grapevine moth sexual pheromone was identified as (*E,Z*)-7,9-dodecadien-1-yl acetate [10,11]. Two related compounds, (*E,Z*)-7,9-dodecadien-1-ol and (*Z*)-9-dodecen-1-yl acetate, were later discovered in the female sex pheromone gland and found to have a synergistic effect on the male catches [12]. These findings were crucial for the application of pheromone-based control and monitoring techniques. In fact, mating disruption is currently the most successful and widespread technique for controlling the moth in Europe. It is based on the use of a high concentration of the main component of the female pheromone to confuse the male and thus avoid mating [13]. In addition, sex pheromone-baited traps were developed for monitoring the *L. botrana* population, playing an important role in pest detection and treatment timing.

Pheromone determinations have been employed for different purposes [14–16] including the detection of insects [17]. In this sense, our research group has developed some analytical approaches for the detection of the major components of the *Tuta absoluta* pheromone in tomato, water and air [18,19] and the main component of the *Bactrocera oleae* pheromone in olives [20]. The same basic configuration is proposed in this article to identify the three components of the *Lobesia botrana* pheromone in grape-related samples. All of the variables involved in the extraction process have been considered in depth. Once optimized, the method was characterized in terms of linearity, sensitivity and precision, being finally applied to the identification and quantification of the three main components of the pheromone in grape samples and must. To the best of our knowledge, this is the first analytical methodology that achieves the instrumental determination of the presence of the pheromone components of grapevine moth pest in grape-related samples.

2. Materials and Methods

2.1. Reagents and Samples

All chemicals used were of analytical grade or better (purity > 99%). Major components of sex pheromone of grapevine moth (*Lobesia botrana*): (*E,Z*)-7,9-dodecadien-1-ol (component A), (*E,Z*)-7,9-dodecadien-1-yl acetate (component B) and (*Z*)-9-dodecen-1-yl acetate (component C), were purchased from Sociedad Española de Desarrollos Químicos, S.L. (Barcelona, Spain). Stock standard solutions of each analyte were prepared in methanol (Scharlab, Barcelona, Spain) at a concentration of 5 g/L and stored at 4 °C. A working solution containing the three components of pheromone was daily prepared by dilution of the stocks with methanol.

Different samples, including table grapes, wine grapes and must were considered in this study. All the samples were purchased at local markets. Grapes used for making white ("Moscatel" variety) and red ("Cabernet sauvignom" variety) wines in Córdoba (Andalusia, Spain) were included in this study. Unwashed fresh grape samples (stored at 4 °C) were ground and homogenized using a blender (UltraTurraz) for 2 min at ambient temperature. The treated sample was then placed in 100 mL vials and frozen at −18 °C until its analysis.

2.2. HS-GC-MS Analysis

Headspace analyses were performed on a MPS2 32-space headspace autosampler (Gerstel, Mülhein, and der Ruhr, Germany) including a robotic arm and an oven. An automated injector fitted with a 2.5 mL gastight HS-syringe (heated at 80 °C) was used for the introduction of 2.5 mL of homogenized headspace from the vial into the HP6890 gas chromatograph (Agilent, Palo Alto, CA, USA) equipped with an HP5973 mass spectrometric detector based on a quadrupole analyzer and an electron multiplier detector. System control and data acquisition were achieved with an HP1701CA MS ChemStation software.

The gas chromatograph was equipped with an HP5MS fused silica capillary column (30 m × 0.25 mm i.d.) coated with 5% phenylmethyl-polysiloxane (film thickness 0.25 μm; Agilent, Palo Alto, CA, USA). The column oven was initially held at 40 °C for 2 min, raised to 200 °C at a rate of 10 °C/min and finally raised to 300 °C at 50 °C/min and maintained for 2 min. The mass spectrometer detector operated in selection monitoring mode recording the following m/z fragment-ions in a single window: 67, 68 and 79. Electron impact ionization (70 eV) was used for analytes fragmentation. The injector, MS source, and quadrupole temperatures were kept at 270, 230, and 150 °C, respectively. The peak areas were used for quantification of the individual analytes.

2.3. Analytical Procedure

Aliquots of 6 g of ground samples (grapes or must) were accurately weighed in 10 mL headspace vials and 0.6 g of NaCl were added. Then, the vials were hermetically sealed with a silicone septum and placed into the autosampler. The robotic arm took each vial from the tray and transferred it into the oven where samples were heated at 80 °C for 30 min under mechanical stirring (750 rpm) to ensure the equilibration of the analytes between the two phases. Then, 2.5 mL of the gaseous phase were sampled by the gas-tight syringe (heated at 80 °C) and injected into the gas chromatograph/mass spectrometer, where the separation and identification/quantification of the analytes took place.

3. Results and Discussion

3.1. Optimization of Experimental Variables

The aim of this study was to develop a simple, highly efficient and sensitive analytical method for the determination of three grapevine moth pheromone components in grape-related samples. The optimization of the method was done using grape samples spiked with the three components at an individual concentration of 10 µg/Kg. Different variables, which are summarized in Table 1, may affect the efficiency of the extraction procedure and therefore their effects were considered in depth. Table 1 also reflects their initial and optimum values as well as the interval studied. The optimization was performed under a univariate approach.

Table 1. List of the variable involved HS-GC-MS determination of the three sexual pheromone components.

Variable	Initial Value	Interval Studied	Optimal Value
Extraction temperature (°C)		40–80	80
Headspace injected volume (mL)	1	0.5–2.5	2.5
Ionic strength (g/Kg of NaCl)	0	0–300	100
Sample amount (g)	2.5	0.5–7	6
Extraction time (min)	15	2–140	30

In order to study the effect of the temperature on the extraction efficiency, the oven of the headspace module was maintained at different temperatures, namely: 40, 50, 60, 70 and 80 °C. This interval was selected based on the distribution equilibrium of the three analytes between the sample and the vial headspace. The analytical signal increases for the three components with the temperature, as shown in Figure 1. However, temperatures over 80 °C were not assayed in order to avoid excessive vial pressures when samples with high water content are analyzed. An optimal extraction temperature of 80 °C was chosen.

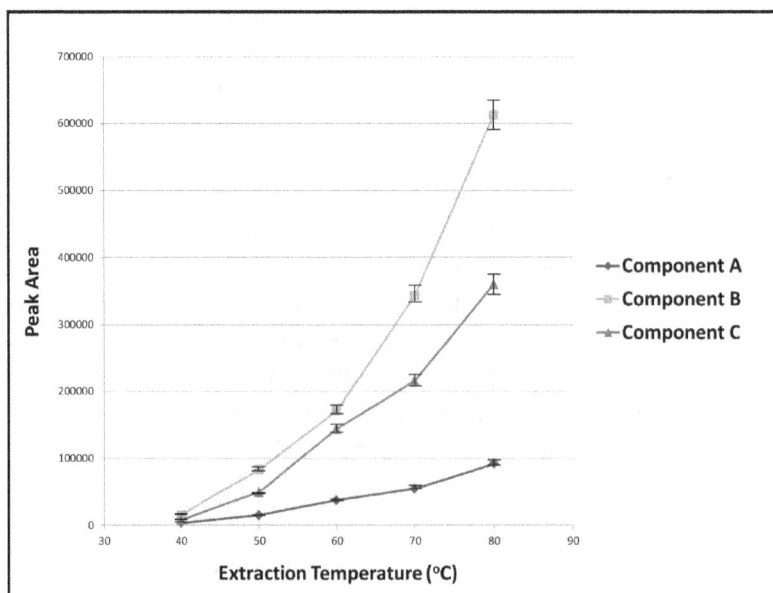

Figure 1. Effect of the extraction temperature on the analysis of the target analytes in grape samples.

The effect of the headspace volume injected in the gas chromatograph was evaluated in the range from 0.5 mL to 2.5 mL. As expected, the peak areas of the analytes increased with the injected volume. The highest peak area, which had a negligible negative effect on the chromatographic resolution, was obtained for 2.5 mL.

The partition coefficient of the analytes between the liquid and the headspace is affected by the ionic strength. In fact, this variable may present two contradictory effects. On the one hand, it can decrease the solubility of the target analytes in water, by the so-called salting-out effect, favoring their transference to the gas phase. On the other hand, ionic strength may increase the viscosity of the sample solution, negatively affecting the transference kinetics. The effect of salt addition on the extraction efficiency of the three analytes from the grape sample was investigated by adding sodium chloride (selected as model electrolyte) to the sample in the range 0–300 g/Kg. The results are shown in Figure 2 and indicated that the extraction of all the analytes is positively affected at lower ionic strength (indicating a prevalence of the salting out effect) whereas at higher concentration (over 100 g/Kg) a decrease in the signals was observed (possibly due to the second effect). Attending to the results, 100 g/Kg of sodium chloride was selected as the optimum value for further studies.

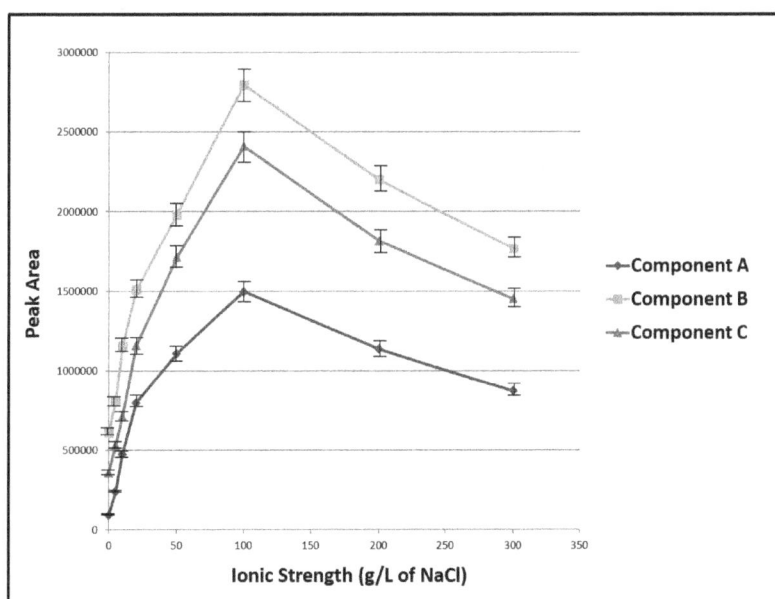

Figure 2. Effect of the ionic strength on the analysis of the target analytes in grape samples.

The sample amount defines the total quantity of analytes present in the vial and the headspace volume. The effect of the sample amount was evaluated at six different levels (0.5, 1, 2.5, 5, 6 and 7 g). The signals for the analytes increased (data not shown) up to 6 g remaining almost constant for higher amounts. Therefore, 6 g was selected as the best value.

Headspace analysis is not an exhaustive procedure because there is a distribution equilibrium of the analytes between the sample and its headspace. Therefore, the incubation time was studied between 2 min and 140 min. The results showed that the signals exhibited a remarkable increase in the range of 2 min to 30 min, reaching a steady-state over this value. Thus, 30 min was selected as the incubation time. This time does not have a great influence on the overall sample throughput since several samples can by extracted simultaneously.

3.2. Analytical Figures of Merit

The analytical performance of the proposed method was studied in order to evaluate its usefulness for the quantitative analysis of the three components of the pheromone in grape samples. The figures of merit are summarized in Table 2. The calibration curves were constructed by analyzing blank grape samples spiked with analytes at different concentrations. Each concentration level was analyzed in triplicate, following the optimized procedure. The method was characterized on the basis of its linearity, sensitivity and precision.

Table 2. Figures of merit of the proposed method for the determination of the target compounds in grape samples.

Analyte	LOD [a]	LOQ [b]	Linear Range [c]	RSD [d]
Component A	0.42	1.39	LOQ-1000	8.5
Component B	0.06	0.21	LOQ-1000	7.2
Component C	0.17	0.55	LOQ-1000	7.9

[a] LOD, limit of detection in µg/kg. [b] LOQ, limit of quantification in µg/kg. [c] Linear Range in µg/kg. [d] RSD; relative standard deviation expressed in percentage ($n = 11$).

The sensitivity of the method was estimated by means of the limits of detection (LOD), calculated for a signal-to-noise ratio (S/N) equal to three, based on the calibration graph parameters. They ranged from 0.06 µg/Kg ((E,Z)-7,9-dodecadien-1-yl-acetate) to 0.42 µg/Kg ((E,Z)-7,9-dodecadien-1-ol). Limits of quantification (LOQ), calculated for a S/N ratio equal to 10, were in the interval of 0.21–1.39 µg/Kg. The linearity was maintained up to 1000 µg/Kg for the three components. The repeatability of the method, expressed in terms of relative standard deviation (RSD, %), was lower than 8.5% (Table 2) calculated from 11 replicate analyses of a blank grape sample spiked with the analytes at a concentration of 1.5 µg/Kg.

Finally, a recovery study was carried out in blank table grape samples spiked with the analytes at a concentration of 1.5 µg/Kg. The samples were left to stand 24 h after spiking to allow potential interaction between the analyte and the sample matrix. The relative recoveries were calculated comparing the spiking level with the concentration obtained after the analysis of the spiked samples. The recovery values ranged from 87 ± 7 to 96 ± 7 which indicates a good accuracy level.

3.3. Analysis of Grape Samples

Once optimized and analytically characterized, the proposed method was applied to the determination of the grapevine moth pheromone components in grape-related samples (six samples of table grapes, two must and two wine grapes). Samples were analyzed in triplicate using three independent aliquots. The analytes were not detected in the six table grape samples as was expected. However, must and wine grape samples resulted to be positive and the results are presented in Table 3.

Table 3. Analysis of different grape samples collected in Córdoba and must samples purchased at local market.

Analyte	White Wine Grape [a] (Moscatel)	Red Wine Grape [a] (Carbenet Sauvignon)	White Must [a]	Red Must [a]
Component A	5.2 ± 0.4	4.5 ± 0.4	Not Detected	Not Detected
Component B	64 ± 5	82 ± 6	94 ± 7	122 ± 9
Component C	21 ± 2	29 ± 2	Not Detected	Not Detected

[a] Concentration (µg/Kg) $\pm$ SD ($n = 3$).

By way of example, Figure 3 shows the chromatogram obtained for the analysis of Moscatel grape sample, where the three components of the pheromone were detected and quantified. In must samples, only Component B was determined.

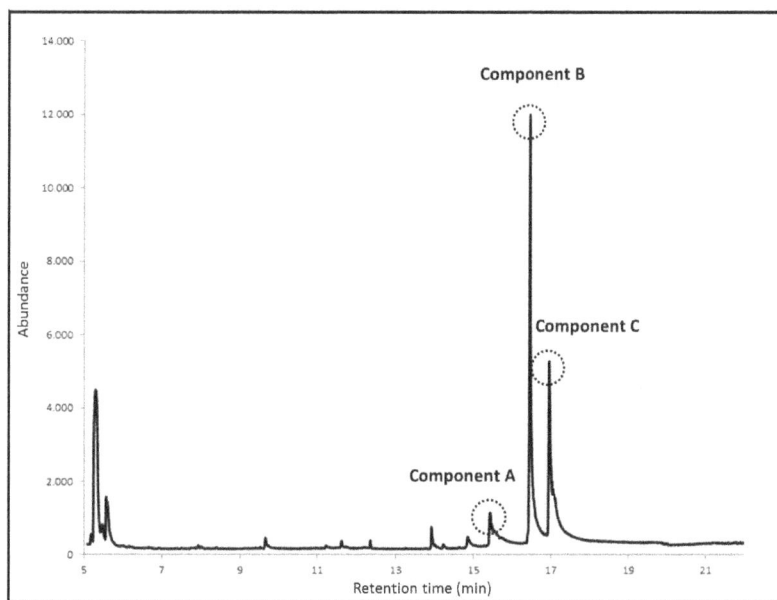

Figure 3. Chromatogram of a Moscatel wine grape sample. Components: (*E,Z*)-7,9-dodecadien-1-ol (**A**), (*E,Z*)-7,9-dodecadien-1-yl acetate (**B**) and (*Z*)-9-dodecen-1-yl acetate (**C**).

4. Conclusions

In this paper, a HS-GC-MS method for the determination of three components of the *Lobesia botrana* pheromone in grape-related samples is proposed. *Lobesia botrana* is an important pest, which can potentially damage crops throughout the growing season. The proposed methodology is completely automated and therefore it can be employed in routine analysis, allowing the determination of the pheromone components at a lower concentration, with precision better than 8.5% in the to ng/Kg level. The usefulness of the proposed approach has been demonstrated by analyzing grape-related samples.

Nonetheless, the proposed method is not able to determine the origin of the pheromone since these molecules can have a natural (pest) or synthetic (trap lure and/or mating disruptant) origin. The potential use of this method for the early detection of the pest would require further studies with a multidisciplinary (chemical and entomological) perspective.

Acknowledgments: Financial support from Spanish Ministry of Science and Innovation (CTQ2014-52939R) is gratefully acknowledged.

Author Contributions: María del Carmen Alcudia-León, Rafael Lucena and Soledad Cárdenas conceived and designed the experiments; María del Carmen Alcudia-León and Mónica Sánchez-Parra performed the experiments; María del Carmen Alcudia-León and Rafael Lucena analyzed the data; María del Carmen Alcudia-León, Rafael Lucena and Soledad Cárdenas wrote the paper.

Conflicts of Interest: The authors declare no conflict of interest.

References

1.　Savopoulou-Soultani, M.; Stavridis, D.G.; Tzanakakes, M.E. Development and reproduction on *Lobesia botrana* on vine and olive inflorescences. *Entomol. Hell.* **1990**, *8*, 29–35.

2.　Vacas, S.; Alfaro, C.; Zarzo, M.; Navarro-Llopis, V.; Primo, J. Effect of sex pheromone emission on the attraction of *Lobesia botrana*. *Etomol. Exp. Appl.* **2011**, *139*, 250–257. [CrossRef]

3.　Caffarelli, V.; Vita, G. Heat accumulation for timing grapevine moth control measures. *Bull. SROP* **1988**, *11*, 24–26.

4.　Oliva, J.; Navarro, S.; Navarro, G.; Camara, M.A.; Baarba, A. Integrated control of grape berry moth (*Lobesia botrana*), powdery mildew (*Uncinula necátor*), downy ildew (*Plasmopara vitícola*) and grapevine sour rot (*Acetobacter* spp.). *Crop Prot.* **1999**, *18*, 581–587. [CrossRef]

5.　Boselli, M.; Scannavini, M. Lotta alla tignoletta della vite in Emilia-Romagna. *Inf. Agrar.* **2001**, *19*, 97–100.

6. Pimentel, D. Environmental and Economic Cost of the Application of Pesticides Primarily in the United States. In *Integrated Pest Management: Innovation-Development Process*; Peshin, R., Dhawan, A., Eds.; Springer: Amsterdam, The Netherlands, 2009; pp. 89–111.

7. McNeil, J.N.; Cotnoir, P.A.; Leroux, T.; Laprade, R.; Schwartz, J.L. A Canadian national survey on the public perception of biological control. *Biocontrol* **2010**, *55*, 445–454. [CrossRef]

8. Gonzalez-Rodriguez, R.M.; Rial-Otero, R.; Cancho-Grande, B.; Gonzalez-Barreiro, C.; Simal-Gandara, J. A review on the fate of pesticides during the processes within the food-production chain. *Crit. Rev. Food Sci.* **2011**, *51*, 99–114. [CrossRef] [PubMed]

9. Witzgall, P.; Kirsch, P.; Cork, A. Sex pheromones and their impact in pest management. *J. Chem. Ecol.* **2010**, *36*, 80–100. [CrossRef] [PubMed]

10. Roelofs, W.L.; Kochansky, J.; Cardé, R.T.; Arn, H.; Rauscher, S. Sex attractant of the grape vine moth, *Lobesia botrana*. *Mitt. Schweiz. Entomol. Ges.* **1973**, *46*, 71–73.

11. Buser, H.R.; Rauscher, S.; Arn, H. Sex pheromone of *Lobesia botrana*: (E,Z)-7,9-Dodecadienyl acetate in the female grape vine moth. *Z. Naturforschung C* **1974**, *29*, 781–783.

12. Arn, H.; Rauscher, S.; Guerin, P.; Buser, H.R. Sex pheromone blends of three tortricid pests in European vineyards. *Agric. Ecosyst. Environ.* **1988**, *21*, 111–117. [CrossRef]

13. Gordon, D.; Zahavi, T.; Anshelevich, L.; Harel, M.; Ovadia, S.; Dunkelblum, E.; Harari, A.R. Mating disruption of *Lobesia botrana* (Lepidoptera: Tortricidae): Effect of pheromone formulations and concentration. *J. Econ. Entomol.* **2005**, *98*, 135–142. [CrossRef] [PubMed]

14. Stewart, M.; Baker, C.F.; Cooney, T. A rapid, sensitive, and selective method for quantitation of lamprey migratory pheromones in river water. *J. Chem. Ecol.* **2011**, *31*, 1203–1207. [CrossRef] [PubMed]

15. Galvan, T.L.; Kells, S.; Hutchison, W.D. Determination of 3-Alkyl-2-methoxypyrazines in Lady Beetle-Infested Wine by Solid-Phase Microextraction Headspace Sampling. *J. Agric. Food Chem.* **2008**, *56*, 1065–1071. [CrossRef] [PubMed]

16. Cudjoe, E.; Wiederkehr, T.B.; Brindle, I.D. Headspace gas chromatography-mass spectrometry: A fast approach to the identification and determination of 2-akyl-3-methoxyperazinepheromones in ladybugs. *Analyst* **2005**, *130*, 152–155. [CrossRef] [PubMed]

17. Eom, Y.; Risticebic, S.; Pawliszyn, J. Simultaneous sampling and analysis of indoor air infested with *Cimex lectularius* L. (Hemiptera: Cimicidae) by solid phase microextraction, thin film microextraction and needle trap device. *Anal. Chim. Acta* **2012**, *716*, 2–10. [CrossRef] [PubMed]

18. Alcudia-León, M.C.; Lucena, R.; Cárdenas, S.; Valcárcel, M. Determination of *Tuta absoluta* pheromones in water and tomato samples by headspace-gas chromatography-mass spectrometry. *Anal. Bioanal. Chem.* **2015**, *407*, 795–802. [CrossRef] [PubMed]

19. Alcudia-León, M.C.; Lucena, R.; Cárdenas, S.; Valcárcel, M.; Kabir, A.; Furton, K.G. Integrated sampling and analysis unit for the determination of sexual pheromones in environmental air using fabric phase sorptive extraction and headspace-gas chromatography-mass spectrometry. *J. Chromatogr. A* **2017**, *1488*, 17–25. [CrossRef] [PubMed]

20. Alcudia-León, M.C.; Cärdenas, S.; Valcárcel, M.; Lucena, R. Green detection of the olive fruit fly pest by the direct determination of its sexual pheromone. *Anal. Methods* **2015**, *7*, 7228–7233. [CrossRef]

## 15

Development of an Automated Method for Selected Aromas of Red Wines from Cold-Hardy Grapes Using Solid-Phase Microextraction and Gas Chromatography-Mass Spectrometry-Olfactometry

Lingshuang Cai [1,2], Somchai Rice [1,3], Jacek A. Koziel [1,3,4,]* (iD) and Murlidhar Dharmadhikari [4]

[1] Department of Agricultural and Biosystems Engineering, Iowa State University, Ames, IA 50011, USA; Lingshuang.Cai@dupont.com (L.C.); somchai@iastate.edu (S.R.)

[2] DuPont Crop Protection, Stine-Haskell Research Center, Newark, DE 19711, USA

[3] Interdepartmental Toxicology Graduate Program, Iowa State University, Ames, IA 50011, USA

[4] Department of Food Science and Human Nutrition, Iowa State University, Ames, IA 50011, USA; murli@iastate.edu

* Correspondence: koziel@iastate.edu

Abstract: The aroma profile of red wine is complex and research focusing on aroma compounds and their links to viticultural and enological practices is needed. Current research is limited to wines made from cold-hardy cultivars (interspecific hybrids of vinifera and native N. American grapes). The objective of this research was to develop a fully automated solid phase microextraction (SPME) method, using tandem gas chromatography-mass spectrometry (GC-MS)-olfactometry for the simultaneous chemical and sensory analysis of volatile/semi-volatile compounds and aroma in cold-hardy red wines. Specifically, the effects of SPME coating selection, extraction time, extraction temperature, incubation time, sample volume, desorption time, and salt addition were studied. The developed method was used to determine the aroma profiles of seven selected red wines originating from four different cold-hardy grape cultivars. Thirty-six aroma compounds were identified from Maréchal Foch, St. Croix, Frontenac, Vincent, and a Maréchal Foch/Frontenac blend. Among these 36 aroma compounds, isoamyl alcohol, ethyl caproate, benzeneethanol, ethyl decanoate, and ethyl caproate are the top five most abundant aroma compounds. Olfactometry helps to identify compounds not identified by MS. The presented method can be useful for grape growers and wine makers for the screening of aroma compounds in a wide variety of wines and can be used to balance desired wine aroma characteristics.

Keywords: cold-hardy grapes; wine; gas chromatography-mass spectrometry (GC-MS); olfactometry; solid phase microextraction (SPME); aroma

1. Introduction

In order to understand the aroma of wine and make a marketable product, it is necessary to separate, identify, and quantify the chemical compounds that impart these aromas. Aromas to note are the primary (varietal aroma), secondary (aromas due to fermentation from yeasts), and bouquet (aromas due to aging and storage) aromas. The pleasant volatiles in wines are due to the presence of higher aliphatic alcohols, ethyl esters, and acetates [1,2]. Wine aroma increased in complexity after malo-lactic fermentation (MLF), which produces changes in the carbonyl compounds [3]. Wines that have undergone MLF can be associated with herbaceous aromas from aliphatic aldehydes [4,5] or buttery aromas from diacetyl [6]. Wine can also have off odors due to volatile sulfur compounds, described as garlic, onion, or cabbage [7]. Vinylphenols have been described as phenolic, medicinal, smoky, spicy, and clove-like [8,9]. An experienced palate can often distinguish the "foxy" characteristic of the main North American vines (*Vitis labrusca* and *Vitis rotundifolia*) from vinifera vines caused by methyl anthranilate [9,10].

Cold climate grapes are newer, and the aroma profiles are less characterized than Vinifera varieties. The identification and quantification of the most aromatic compounds can help the industry maximize the aroma quality in these wines. The varietal flavor profile was used to demonstrate good examples of wine production and the best grape growing and innovative vinification techniques. The first step in the aroma analysis of wines is to extract volatile organic compounds (VOCs). Solid phase microextraction (SPME) coupled with gas chromatography-mass spectrometry (GC-MS) is useful in extracting and pre-concentrating VOCs in wine. The inception of GC-olfactometry (GC-O) in 1964 allowed researchers to link an aroma descriptor to these separated compounds [11]. Although many detectors have been used to identify and quantify aroma compounds from wine, MS is the most widely used [12]. A review and summary of many experimental parameters are available elsewhere [12,13].

In this research, an automated headspace SPME-GCMS-O method was developed and the aroma profiles of seven cold-hardy wine samples were investigated. The chemicals in selected cold-hardy wines were isolated and tentatively identified by matching the mass spectral and aroma character. The grape and wine industry has expanded exponentially in cold climates. Therefore, there is a need to research aroma compounds and their links to grape growing and wine making practices in cold climates. Such information can be used for the monitoring of fruit maturity, developing the best viticultural and wine making practices, and the development of appropriate wine styles specific to cold climates. The flavor and aroma profiling of cold-hardy wine enables the development of high quality and unique wines. Therefore, a method was developed to evaluate the full chemical and sensory aroma profile of wine from cold climate grapes.

2. Materials and Methods

2.1. Samples, Internal Standard, SPME

Seven different red wines were obtained from various wineries in Iowa. Varieties included two Maréchal Foch from separate wineries, two Frontenac from separate wineries, a St. Croix, a Vincent, and a Maréchal Foch/Frontenac blend. Wines were not stored after initial opening for analysis. 3-nonanone (99%), CAS 925-78-0 (Sigma-Aldrich, St. Louis, MO, USA), was used as an internal standard (IS) for the semi-quantification of aroma compounds. 3-nonanone was chosen because the compound is odor-active and not present in these wine samples. The final concentration of IS in wine (0.206 mg/L) was achieved by adding 10 µL of IS in ethanol (82.5 mg/L) to each 4 mL of wine. Each wine sample bottle was opened immediately before each analysis, and triplicate runs were performed for each experiment (n = 3). SPME fibers with seven different coatings were purchased from Sigma-Aldrich (St. Louis, MO, USA). These coatings included: 50/30 µm Divinylbenzene (DVB)/Carboxen (CAR)/Polydimethylsiloxane (PDMS), 100 µm PDMS, 7 µm PDMS, 85 µm Polyacrylate (PA), 65 µm PDMS/DVB, 70 µm Carbowax (CW)/DVB, and 85 µm CAR/PDMS. All SPME fibers were 1 cm in length. Details of SPME fiber cores, coatings, and the internal structure can be found elsewhere [14].

The optimized method used for the automated analysis of wine aroma used a 1 cm 50/30 µm DVB/CAR/PDMS SPME fiber, 10 min extraction time at 50 °C, and 2 min desorption in the heated GC inlet. A 10 min incubation in the heated agitator was used to equilibrate VOCs in the headspace of 4 mL of the wine sample in a 10 mL vial, facilitated by the addition of 2 g of sodium chloride.

2.2. GC-MS-Olfactometry System

The analysis was performed on a standard 6890N GC/5973Network Platform (Agilent Technologies, Santa Clara, CA, USA) and a CTC CombiPal™ autosampler equipped with a heated agitator (Trajan Scientific, Pflugerville, TX, USA). A constant agitation speed of 500 rpm was used throughout this research, so that extraction would only depend on the SPME fiber geometry and diffusion coefficients of the aroma compounds. The instrument was modified after marketing with a Dean's switch for heartcutting, the ability for cryogenic focusing, FID, and the olfactometry port. A detailed schematic of the instrument can be found elsewhere [15]. The GC contains two columns connected in series. The first non-polar column was BPX-5 stationary phase with the following dimensions: 43.5 m length × 0.53 mm ID × 1.0 μm film thickness (SGE Analytical Science, by Trajan, Austin, TX, USA). The second cross-linked polar column was BP-20 (Wax) with the following dimensions: 25 m length × 0.53 mm ID × 1.0 μm film thickness (SGE Analytical Science, by Trajan, Austin, TX, USA). A constant pressure of 5.8 psi was provided at the midpoint between the first and second column using MultiTrax™ V.6.00 (Microanalytics, a Volatile Analysis company, Round Rock, TX, USA) system automation and MSD ChemStation™ E.01.01.335 data acquisition software (Agilent Technologies, Santa Clara, CA, USA). Additional analysis to obtain the total compound chromatogram (TCC) was done using a MassHunter Workstation (Agilent, Santa Clara, CA, USA). Flow from the second analytical column was directed to the single quadrupole mass selective detector and the olfactometry port by fixed restrictor tubing in an open-split interface.

For this research, a full heartcut was utilized from 0.05 to 35.00 min. In other words, the sample flow was first directed through the non-polar column, and then the second polar column, yielding results similar to a mid-polarity GC separation on a long column. Therefore, retention indices were not used for identification in this research, due to the configuration of the two GC capillary columns connected in series.

The following instrument parameters were used: GC inlet temperature, 260 °C; FID, 280 °C; column, 40 °C initial, 3.0 min hold, 7 °C per min ramp, 220 °C final, 11.29 min hold; carrier gas, UHP helium (99.999%) with an inline filter trap. The mass detector was operated in electron ionization (EI) mode with an ionization energy of 70 eV. The mass detector ion source and quadrupole were held at 230 °C and 150 °C, respectively. Full spectrum scans were collected with the mass filter set from m/z 33 to m/z 450. The MS was auto-tuned daily before analysis. The use of a full scan for data acquisition allowed for library search techniques using NIST05 and Wiley 6th edition mass spectral databases.

Olfactometry data was generated using AromaTrax™ V.6.61 software (Microanalytics, a Volatile Analysis company, Grant, AL, USA). Recorded parameters included an aroma descriptor and the perceived intensity. The editable descriptor panel is shown in Figure 1. The area under the peak of each aroma note in the aromagram is calculated as width × intensity × 100, where the width is the length of time that the aroma persisted in minutes. The sum of the areas under the peaks in the aromagram is the total odor, a dimensionless value used to analyze the total aroma detected by the human nose.

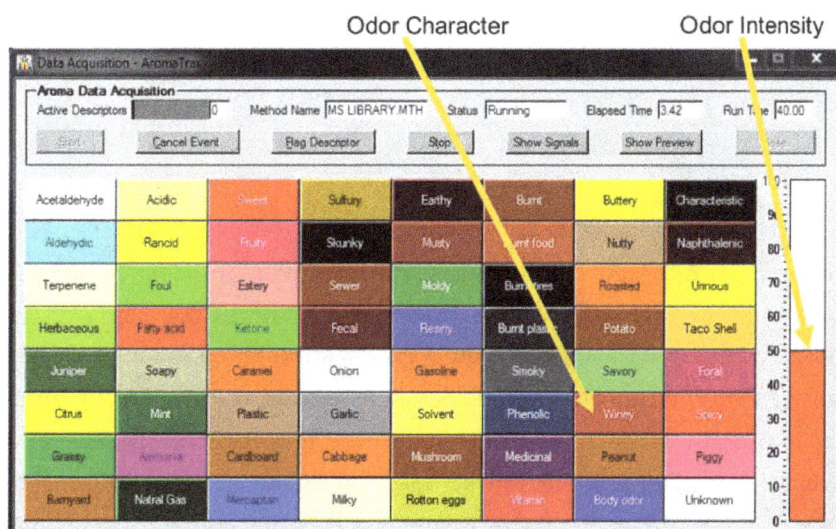

Figure 1. Aroma descriptor panel used to characterize volatiles and perceived intensities.

3. Results

3.1. SPME Optimization

3.1.1. SPME Coating Selection

Seven commercially available SPME coatings were selected for optimization in the extraction of aroma compounds from Iowa red wines (Table 1). The response of the mass spectrometer to volatiles detected throughout the run was used to determine the extraction efficiency of the SPME coating. It was shown that the use of a 50/30 μm DVB/CAR/PDMS SPME coating was most appropriate for the rest of the experiments. In Table 1, a coating with a number lower than 100 indicates a lower extraction efficiency for that analyte when compared to 50/30 μm DVB/CAR/PDMS. These analytes spanned the entire chromatographic run, representing the range of analytes in wine aroma.

Table 1. Optimization of solid phase microextraction (SPME) extraction conditions—Fiber selection.

RT (min)	Compound	50/30 μm DVB/CAR/PDMS	100 μm PDMS	7 μm PDMS	85 μm PA	65 μm PDMS/DVB	70 μm CW/DVB	85 μm CAR/PDMS
4.77	Ethyl isobutyrate	100	58	0	13	75	29	104
5.53	Isobutyl alcohol	100	42	1	95	81	97	83
8.10	Isoamyl alcohol	100	35	0	73	75	66	92
11.00	Ethyl lactate	100	17	0	71	60	79	132
11.75	Ethyl caproate	100	24	0	7	53	17	109
12.92	Acetic acid	100	5	3	124	38	242	158
16.38	Ethyl caprylate	100	67	1	21	98	45	64
18.44	Vitispirane	100	61	0	16	90	38	99
18.34	Diethyl succinate	100	52	0	45	105	56	63
20.42	Ethyl decanoate	100	114	12	56	121	72	53
21.13	Benzenethanol	100	23	0	74	86	67	82
24.03	Ethyl myristate	100	94	32	59	92	66	41
30.37	Ethyl palmitate	100	187	132	135	194	155	17

Bolded numbers indicate when a SPME fiber coating extracted more mass than the 50/30 μm DVB/CAR/PDMS coating. DVB: Divinylbenzene; PA: Polyacrylate; PDMS: Polydimethylsiloxane; CW: Carbowax; CAR: Carboxen.

3.1.2. Extraction Time

Different extraction times were tested using the autosampler. These times were 10 s, 30 s, 1 min, 3 min, 5 min, 10 min, 15 min, 20 min, 30 min, and 60 min. Plots of the mass extracted versus the extraction time varied in shape. The profiles were typically linear or logarithmic, with the exception of acetic acid (Figure 2). Equilibrium was reached for most compounds (i.e., the logarithmic curve had started to flatten out), and was not excessively long, with a figure of about 10 min. Additional SPME fiber sorption capacity limitations were noticed after 10 min, for example, ethyl isobutyrate, ethyl lactate, and acetic acid. An extraction time of 10 min was chosen to avoid these possible interactions due to competitive adsorption and apparent analyte displacement after 10 min.

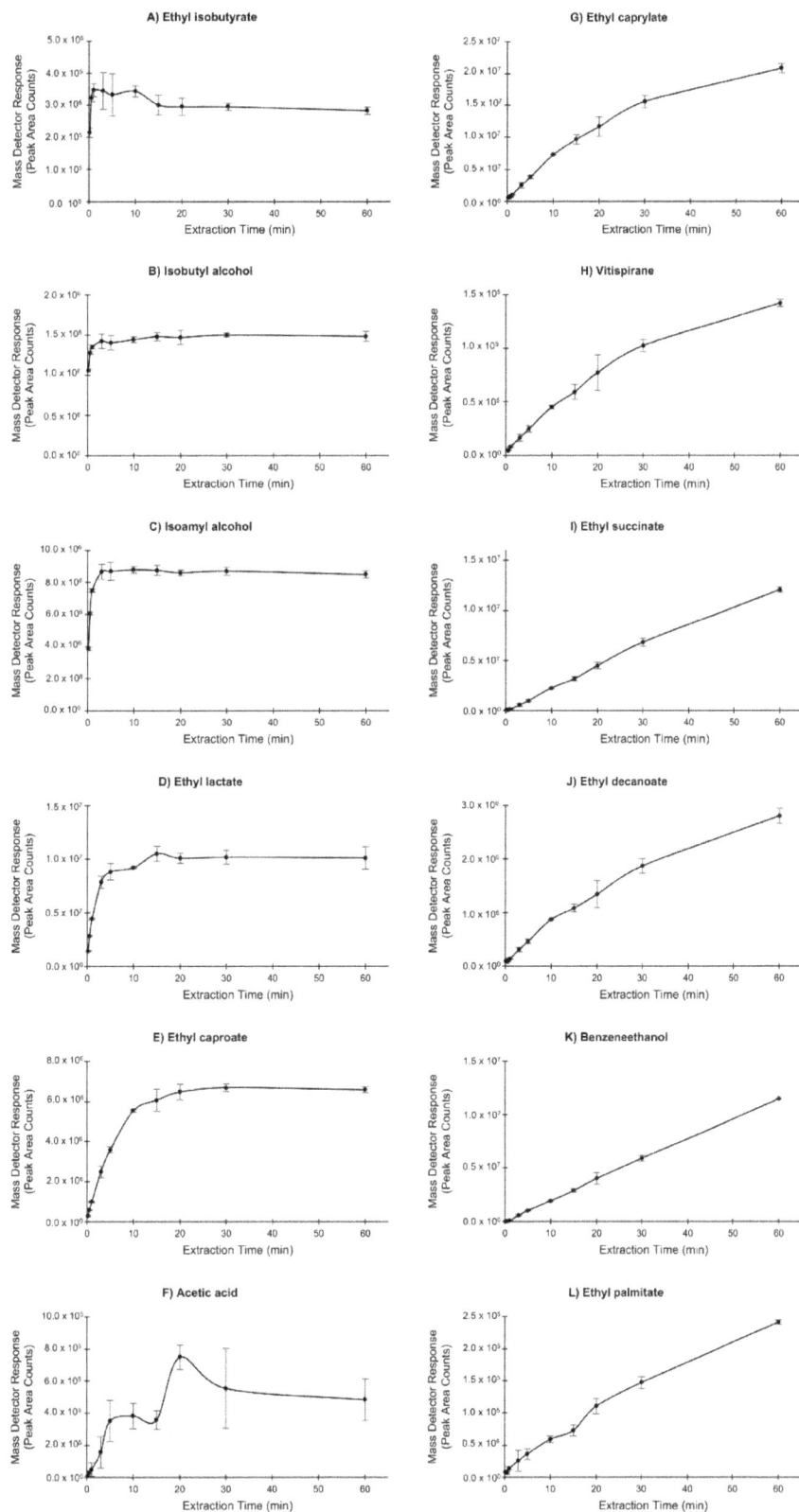

Figure 2. Extraction time profile of selected volatiles extracted with HS-SPME in an Iowa Maréchal Foch. Conditions: 5 min pre-sampling desorption of 50/30 μm DVB/CAR/PDMS SPME fiber; 4 mL wine sample in a 10 mL threaded glass amber vial with PTFE/silicone septa; 10 min incubation at 35 °C; 10 s, 30 s, 1 min, 3 min, 5 min, 10 min, 20 min, 30 min, and 60 min extraction time at 35 °C; agitation speed 500 rpm; 5 min desorption time into GC inlet. 10 min extraction time was chosen for the method.

3.1.3. Extraction Temperature

Extraction times can be shortened by the efficient use of a higher extraction temperature, as seen in Figure 3. Masses extracted of ethyl caprylate, ethyl succinate, ethyl decanoate, benzeneethanol, and ethyl palmitate increased with a higher temperature. Higher temperatures can also decrease the amount of analyte extracted, as observed in ethyl isobutyrate, isoamyl alcohol, and ethyl caproate. Extraction temperatures of 35, 40, 50, 60, 70, and 80 °C were investigated. To efficiently extract the range of volatiles and semi-volatiles, the optimal temperature chosen for this experiment was 50 °C.

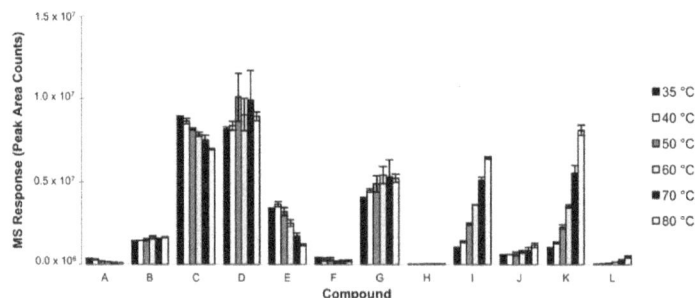

Figure 3. Effects of extraction temperature on selected volatiles extracted with HS-SPME in an Iowa Maréchal Foch. Compound letters correspond to compounds found in Figure 2 (i.e., A—ethyl isobutyrate, B—isobutyl alcohol, L—ethyl palmitate). Conditions: 5 min pre-sampling desorption of 50/30 μm DVB/CAR/PDMS SPME fiber; 4 mL wine sample in a 10 mL threaded glass amber vial with PTFE/silicone septa; 10 min incubation at 35, 40, 50, 60, 70, 80 °C; 10 min extraction time at 35, 40, 50, 60, 70, 80 °C; agitation speed 500 rpm; 5 min desorption time into GC inlet. 50 °C was chosen for the method.

3.1.4. Incubation Time

An extended incubation time allows for volatiles to equilibrate in the headspace before sampling. This was useful in the extraction of less volatile compounds in the wine sample, such as ethyl caprylate or ethyl decanoate (Figure 4). The incubation time did not have a large effect on the extraction efficiency of more volatile compounds such as ethyl isobutyrate, isobutyl alcohol, and isoamyl alcohol, likely due to their abundance in the headspace of the sample. Incubation times of 0, 5 min, 10 min, 15 min, and 20 min were investigated to establish the equilibrium of analytes in the headspace. The incubation time of 10 min was chosen, as determined by the extracted mass of semi-volatiles.

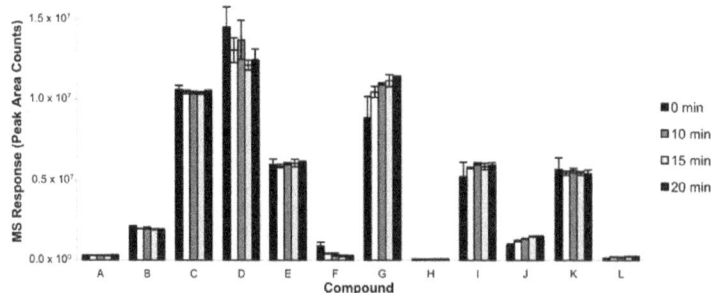

Figure 4. Effects of incubation time on selected volatiles extracted using HS-SPME in an Iowa Maréchal Foch. Compound letters correspond to compounds found in Figure 2 (i.e., A—ethyl isobutyrate, B—isobutyl alcohol, L—ethyl palmitate). Conditions: 5 min pre-sampling desorption of 50/30 μm DVB/CAR/PDMS SPME fiber; 4 mL wine sample in a 10 mL threaded glass amber vial with PTFE/silicone septa; 0, 5 min, 10 min, 15 min, and 20 min incubation at 50 °C; 10 min extraction time at 50 °C; agitation speed 500 rpm; 5 min desorption time into GC inlet. 10 min incubation time was chosen for the method.

3.1.5. Sample Volume

A 10 mL glass amber vial was used with the autosampler. The sample volume was investigated to maximize the mass extracted from the headspace by SPME. A higher volume of wine would yield a greater mass of volatiles in the headspace, up to the equilibrium. This was observed as an increase in the mass extracted was directly proportional to an increase in the sample volume in ethyl caproate, ethyl caprylate, ethyl succinate, ethyl decanoate, benzeneethanol, and ethyl palmitate (Figure 5). From this experiment, a 4 mL sample volume in a 10 mL vial was chosen for the method.

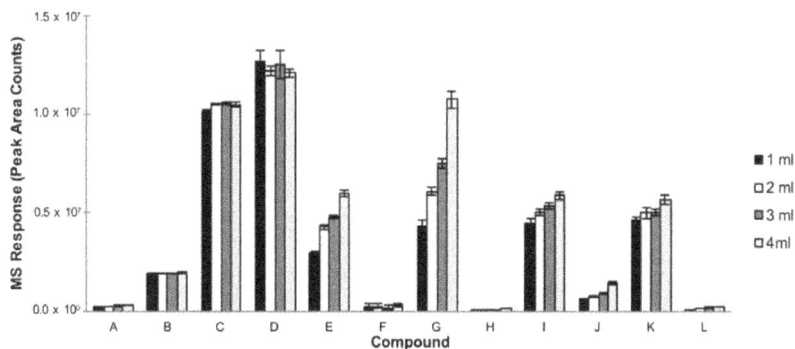

Figure 5. Effects of sample volume on selected volatiles extracted using HS-SPME in an Iowa Maréchal Foch. Compound letters correspond to compounds found in Figure 2 (i.e., A—ethyl isobutyrate, B—isobutyl alcohol, L—ethyl palmitate). Conditions: 5 min pre-sampling desorption of 50/30 μm DVB/CAR/PDMS SPME fiber; 1, 2, 3, and 4 mL wine sample in a 10 mL threaded glass amber vial with PTFE/silicone septa; 10 min incubation at 50 °C; 10 min extraction time at 50 °C; agitation speed 500 rpm; 5 min desorption time into GC inlet. 4 mL sample volume was chosen for the method.

3.1.6. Desorption Time

Optimizing the desorption time maximizes the transfer of analytes into the instrument for analysis. Desorption times of 30 s, 60 s, 120 s, 180 s, 240 s, and 300 s were used to determine the minimum time needed to desorb analytes from the SPME fiber (Figure 6). A desorption time of 120 s was chosen in a 260 °C injector. The inlet pressure was constant and determined by the pressure needed to maintain balance with the midpoint pressure.

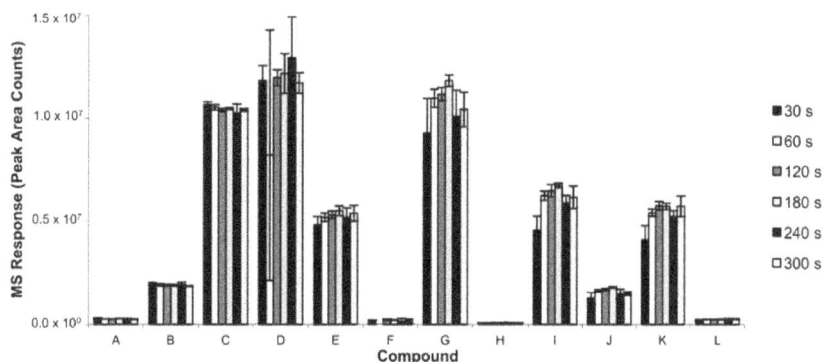

Figure 6. Effects of fiber desorption time on selected volatiles extracted in an Iowa Maréchal Foch. Compound letters correspond to compounds found in Figure 2 (i.e., A—ethyl isobutyrate, B—isobutyl alcohol, L—ethyl palmitate). Conditions: 30 s, 60 s, 120 s, 180 s, 240 s, 300 s pre-sampling desorption of 50/30 μm DVB/CAR/PDMS SPME fiber; 4 mL wine sample in a 10 mL threaded glass amber vial with PTFE/silicone septa; 10 min incubation at 50 °C; 10 min extraction time at 50 °C; agitation speed 500 rpm; 30 s, 60 s, 120 s, 180 s, 240 s, 300 s desorption time into GC inlet. 120 s thermal desorption time was chosen for the method.

3.1.7. Salt Addition

The addition of sodium chloride was used to adjust the ionic strength of the wine sample. This salting-out effect can help drive analytes to the SPME coating with increasing amounts of salt. The addition of 0.5, 1.0, 1.5, 2, and 2.5 g of salt was investigated to maximize the extraction efficiency (Figure 7). For this experiment, a 2.0 g addition of sodium chloride was chosen for the method.

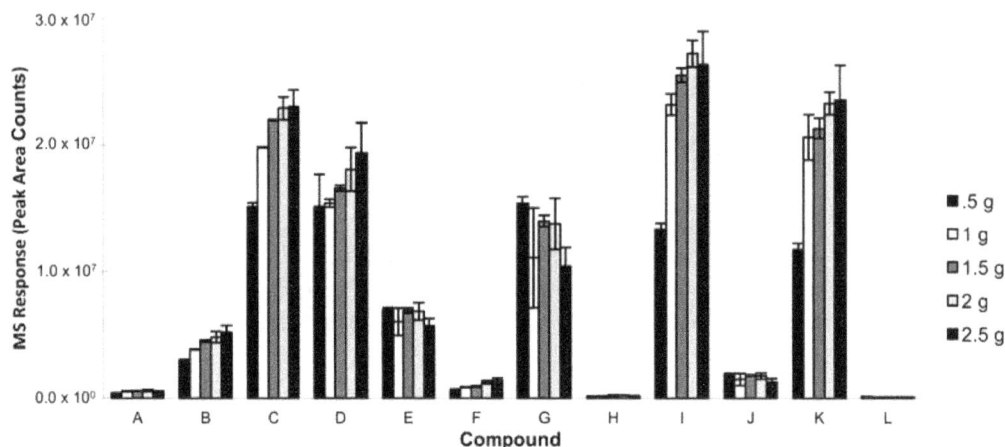

Figure 7. Effects of salt addition on selected volatiles extracted in an Iowa Maréchal Foch. Compound letters correspond to compounds found in Figure 2 (i.e., A—ethyl isobutyrate, B—isobutyl alcohol, L—ethyl palmitate). Conditions: 30, 60, 120, 180, 240, 300 s pre-sampling desorption of 50/30 μm DVB/CAR/PDMS SPME fiber; 4 mL wine sample in a 10 mL threaded glass amber vial with PTFE/silicone septa; 10 min incubation at 50 °C; 10 min extraction time at 50 °C; agitation speed 500 rpm; 30, 60, 120, 180, 240, 300 s desorption time into GC inlet. 120 s thermal desorption time was chosen for the method.

3.2. Analysis of Wine Samples

GC-MS-Olfactometry

Tentative identifications of thirty-six compounds by the mass spectral match of their most significant red wine aroma compounds with the ratio of their compound peak area to the internal standard peak area are listed in Table 2. The molecular weight of the compounds identified by mass spectral match ranged from 60 to 284 amu. The relative quantity of each compound to the corresponding ones in the other wine samples was calculated by the peak area to internal standard ratio. The aroma profile of the seven Iowa red wines varied considerably between samples. These results can reflect the influence of the climate, grape variety, vintage, and different viticultural and enological practices in seven different Iowa red wines. Further research is warranted to link the variables to wine aroma.

Table 2. Tentative identification by mass spectral match of the most significant red wine aroma wine aroma compounds, listed as the volatile compound peak area: internal standard peak area ratio.

#	RT (min)	LRI [A]	Stationary Phase [1-10]	Compound	MW	B Foch	B St. Croix	B Frontenac	B Foch/Frontenac	C Vincent	D Frontenac	D Foch
1	3.58	949 963 955 558 619 575 976	2 2 4 5 8 8 9	2,3-Butanedione	86	1.11	5.18	0.85	0.32	0.94	1.39	1.05
2	3.7	880 710	2 8	Acetal	118	1.24	0.96	8.57	23.1	6.75	4.9	8.08
3	4.08	774	3	Ethyl propanoate	102	1.73	1.05	7.26	19.6	3.42	6.85	5.38
4	4.72	956 746	2 8	Ethyl isobutyrate	116	5.94	1.67	3.99	3.54	6.36	2.3	1
5	5.49	1110 1054 1083 1093 609 616	2 2 4 4 6 8	Isobutyl alcohol	74	14.6	23.8	21.9	27.6	86.1	21.4	28.7
6	6.07	863 788	3 5	Ethyl butanoate	116	1.55	3.43	3.5	2.2	2.63	1.05	3.59
7	6.66	1138 1149 634 653	4 4 5 6	n-Butanol	74	0.9	1.85	1.44	4.07	1.76	4.32	1.68
8	7.22	1060 856 840	2 6 8	Ethyl isovalerate	130	4.29	0.41	2	1.76	0.9	0.38	0.29
9	8.05	1184 719	2 8	Isoamyl alcohol	88	132	355	275	477	475	221	377
10	8.12	1110 860	2 8	Isoamyl acetate	130	32.2	3.89	3.87	4.02	3.89	3.22	0
11	10.03	1250	2	Styrene	104	nd	16.9	3.22	0.05	0.03	nd	0.68

Table 2. *Cont.*

#	RT (min)	LRI [A]	Stationary Phase [1-10]	Compound	MW	[B] Foch	[B] St. Croix	[B] Frontenac	[B] Foch/Frontenac	[C] Vincent	[D] Frontenac	[D] Foch
12	10.91	1312	2	Ethyl lactate	118	13.7	34.1	39.9	34.3	30	67.1	19.8
		1341	4									
		803	8									
13	11.15	1230	2	Ethyl caproate	144	26.4	71	40.4	33	34.7	21.5	64.3
		1060	3									
		1232	4									
		1238	4									
		985	5									
		981	5									
		996	6									
		996	6									
14	11.27	1316	2	n-Hexanol	102	8.64	0	0	19.6	12.6	7.82	51.7
		1330	2									
		1332	2									
		1352	4									
		847	5									
		848	5									
		848	5									
		862	6									
		1354	9									
15	12.81	1400	2	Acetic acid	60	51.4	6.61	58.2	10.6	11.6	13.2	18.7
		1401	2									
		791	3									
		1435	4									
		1442	4									
		1459	4									
		621	5									
		723	8									
16	13.25	1437	2	Furfural	96	nd	nd	nd	10.2	nd	nd	nd
		1438	2									
		1450	2									
		1449	2									
		1447	4									

Table 2. *Cont.*

#	RT (min)	LRI [A]	Stationary Phase [1-10]	Compound	MW	[B] Foch	[B] St. Croix	[B] Frontenac	[B] Foch/Frontenac	[C] Vincent	[D] Frontenac	[D] Foch
17	13.57	1456 1466 1465 802 829 800 800 836 868 830 815	4 4 4 5 5 5 5 6 6 6 8	3-Nonanone (IS)	142	100	100	100	100	100	100	100
18	14.37	1375 1392 1109	1 4 10	Methyl octanoate	158	0.39	1.01	0.32	0.23	0.35	0.18	0.33
19	14.48	1692 941	2 8	1,3-Butanediol	90	1.89	4.03	11	3.38	1.48	8.15	4.43
20	14.81	1518 1516 1509 1454 1520 1482 1502 1086 1515 1496 1513 1538 1530 1522 1516	1 1 2 2 2 2 3 4 4 4 4 4 4 4	Benzaldehyde	106	0.18	0.93	0.95	1.14	0.73	0.12	55.2

Table 2. *Cont.*

#	RT (min)	LRI [A]	Stationary Phase [1-10]	Compound	MW	[B] Foch	[B] St. Croix	[B] Frontenac	[B] Foch/Frontenac	[C] Vincent	[D] Frontenac	[D] Foch
		926	5									
		926	5									
		926	5									
		924	6									
		960	6									
		962	6									
		957	6									
		961	6									
		944	7									
		938	7									
		947	8									
		947	8									
		1540	9									
21	15.02	1501	2	Isobutyric acid	88	2.45	0.28	1.38	0.78	2.46	0.45	0.32
		935	3									
22	15.15			2,3-Butanediol	90	1.41	1.56	2.31	2.03	1.5	2.62	2.35
23	15.72	1258	3	Ethyl caprylate	172	164	398	209	116	187	181	205
		1429	4									
		1466	4									
		1196	6									
		1193	6									
		1195	6									
24	16.83	1631	2	Isovaleric acid	102	8.97	2.58	7.79	5.1	3.73	3.54	3.55
		834	6									
25	17.66	1276	7	Vitispirane	192	0	3.02	0.31	4.5	15.7	3.39	0
26	17.79	1278	5	Ethyl nonanoate	186	0.83	nd	1.68	0	0	0	0.78
27	17.99	1642	2	Diethyl succinate	174	3.62	11.5	18.9	43.8	12.9	5.77	122
		1153	8									
28	19.74	1390	6	Ethyl decanoate	200	54.9	99.5	83.9	29.8	33.3	98.5	34.4
		1391	6									
		1394	6									

Note: # is the chromatographic peak number. RT is the retention time in minutes. A is the GC capillary column linear retention index from the LRI & Odour Database [16]. GC stationary phases are: 1—CP-Wax, 2—CW-20M, 3—DB-1701, 4—DB-Wax, 5—DB1, 6—DB5, 7—HP-1, 8—OV-101, 9—SP-Wax, 10—SPB-1. MW is the molecular weight. B, C, and D stand for the three different Iowa wineries. IS = internal standard.

Table 2. *Cont.*

#	RT (min)	LRI [A]	Stationary Phase [1-10]	Compound	MW	[B] Foch	[B] St. Croix	[B] Frontenac	[B] Foch/Frontenac	[C] Vincent	[D] Frontenac	[D] Foch
29	20.27	1788 1785 1233	2 2 8	Phenethyl acetate	164	0.95	1.56	1.26	1.41	0.95	1.47	1.2
30	20.98	1903	4	Benzeneethanol	122	41.3	146	56	232	111	53.7	134
31	23.06	2007 2007 2100 2013 2075 1183 1256	2 2 2 2 4 6 7	Octanoic acid	144	1.39	1.07	6.63	0.09	1.99	2.22	0.7
32	23.36	1595	6	Ethyl laurate	228	3.3	2.63	6.73	1.62	1.28	8.24	2.72
33	24.76			4-Ethylphenol	122	nd	7.49	2.32	nd	nd	3.4	2.33
34	24.81			8-Pentadecanone	226	1.06	7.49	2.31	2.53	2.33	3.4	2.32
35	26.57			Glycerol	92	nd	nd	26.8	nd	nd	nd	nd
36	26.63	1793	6	Ethyl myristate	256	0.39	0.53	0.48	0.38	0.57	0.84	0.99
37	29.63	1993 1985	6 7	Ethyl palmitate	284	2.35	2.1	5.75	1.92	2.81	4.12	5.14

Note: # is the chromatographic peak number. RT is the retention time in minutes. A is the GC capillary column linear retention index from the LRI & Odour Database [16]. GC stationary phases are: 1—CP-Wax, 2—CW-20M, 3—DB-1701, 4—DB-Wax, 5—DB1, 6—DB5, 7—HP-1, 8—OV-101, 9—SP-Wax, 10—SPB-1. MW is the molecular weight. B, C, and D stand for the three different Iowa wineries. IS = internal standard.

Simultaneous olfactometry, when used with GC-MS, can verify compounds by aroma character. An example of this is highlighted in Figure 8. The chromatographic peak at 19 min was not identified by mass spectral comparison, but it was recorded with an aroma character of burnt food (aromagram peak number 27). An open source aroma database search narrows the possible identification of this compound from hundreds to six: octanol, indole, 3-methyl-1-butanol, ethyldimethylpyrazine, dimethyl sulfone, or furfuryl alcohol [17]. The use of olfactometry, by means of the human nose as a detector, is a very valuable tool for the identification of unknown compounds. The total aroma values, calculated as the sum of the area under the aromagram peaks, are compared in Figure 9 and no significant differences are exhibited between the seven Iowa red wines.

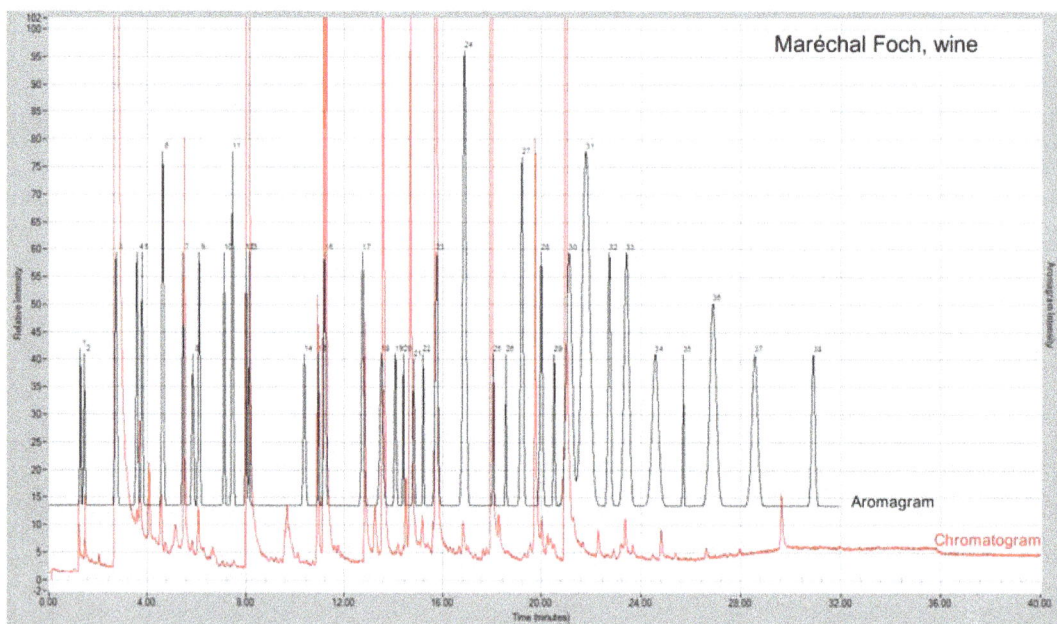

Figure 8. Overlay of total ion chromatogram and aromagram of an Iowa Maréchal Foch wine using the optimized method. Intense aromas (observed as increased peak height, black signal) and responsible chemical compounds (TIC, red signal) were aromagram peak: (#6) sweet, fruity—ethyl isobutyrate (4.71 min); (#11) fruity—ethyl isovalerate (7.21 min); (#24) rancid, sweaty, body odor, burnt—isovaleric acid (16.83 min) (#27) burnt, burnt food—unknown compound (19 min); (#31) sweet, fruity, winey—ethyl laurate (23.26 min). Conditions: 2 min pre-sampling desorption of 50/30 μm DVB/CAR/PDMS SPME fiber; 4 mL wine sample in a 10 mL threaded glass amber vial with PTFE/silicone septa; 10 min incubation at 50 °C; 10 min extraction time at 50 °C; agitation speed 500 rpm; 2 min desorption time into GC inlet.

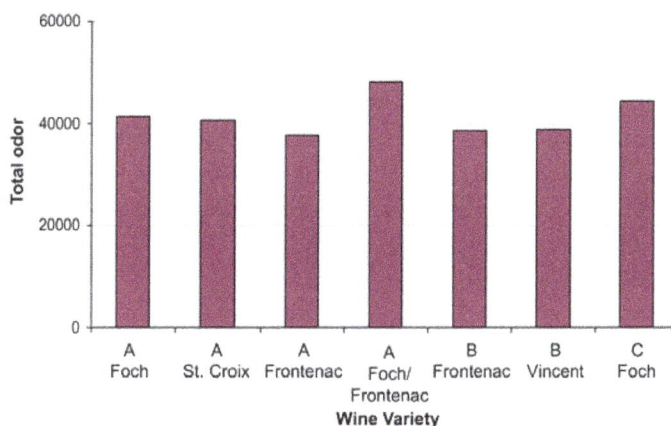

Figure 9. Overlay of total ion chromatogram and aromagram of an Iowa Maréchal Foch wine using the optimized method. Intense aromas (observed as increased peak height, black signal) and responsible chemical compounds (TIC, red signal) were aromagram peaks: (#6) sweet, fruity—ethyl isobutyrate (4.71 min); (#11) fruity—ethyl isovalerate (7.21 min); (#24) rancid, sweaty, body odor, burnt—isovaleric acid (16.83 min) (#27) burnt, burnt food—unknown compound (19 min); (#31) sweet, fruity, winey—ethyl laurate (23.26 min). Conditions: 2 min pre-sampling desorption of 50/30 μm DVB/CAR/PDMS SPME fiber; 4 mL wine sample in a 10 mL threaded glass amber vial with PTFE/silicone septa; 10 min incubation at 50 °C; 10 min extraction time at 50 °C; agitation speed 500 rpm; 2 min desorption time into GC inlet.

4. Discussion

The mixed adsorbent beds of the 50/30 μm DVB/CAR/PDMS coating were best suited for the extraction of wine volatiles, and extracted a greater mass of analytes than the other six fibers. The dual layer Car/PDMS/DVB fiber has been used to overcome the lack of selectivity toward some of the compounds in the single and double-phase fibers [18] and is consistent with previous work by Howard et al. [19]. Even though extraction equilibrium was not reached in some analytes (i.e., the profile is linear in shape at 10 min), precision was assured by using the autosampler to control the mass transfer conditions. The addition of salt can improve the extraction efficiency up to a point, where the target analytes may interact with the salt ions in solution. These interactions will then reduce the extraction efficiency. This phenomenon has been shown to be related to the pKa of the analyte [20]. The total wine aroma is a balance between the heavier aroma of the alcohols, esters, acids, and the unpleasant rancid odors of the aliphatic acids and carbonyls which can be formed during the fermentation process. It should be noted that these 36 compounds were detected in the presence of a highly volatile organic solvent. In Figure 10, ethanol is present at 2.8–3.0 min and is the most abundant compound in the headspace of wine, as expected.

Four esters, including ethyl caproate, ethyl isobutyrate, ethyl isovalerate, and isoamyl acetate, were detected in seven Iowa red wines. In only one previous study, nonanal, (E,Z)-2,6-nonadienal, β-damascenone, ethyl caprylate, and isoamyl acetate had the highest OAVs in Frontenac, Marquette, Maréchal Foch, Sabrevois, and St. Croix wines, using SPME-GCMS(TOF) [21]. Ethyl caproate was previously reported in an analysis of Frontenac and Marquette juice from Quebec using SPME-GCMS [22]. Five compounds in selected Iowa red wines (i.e., isoamyl alcohol, ethyl caproate, benzeneethanol, ethyl decanoate, and ethyl caproate) were also found in Cabernet Sauvignon and Merlot, where ethyl caproate and ethyl caprylate were reported as the most abundant ethyl esters in these vinifera varieties [23]. These compounds have been attributed to yeast metabolism and do not impart any varietal characteristics to wine [24].

A principal components analysis followed by hierarchical clustering analysis is shown in Figure 11. St. Croix (from winery B) and Frontenac (from wineries B and D) are distinguishable by the grape variety and from the other five wine samples. Frontenac from winery B was more significantly associated with (31) octanoic acid than Frontenac from winery D. Maréchal Foch wines (from wineries B and D) were not similar to each other or when blended with Frontenac (from winery B). Maréchal Foch from winery B was associated (8) with ethyl isovalerate, Maréchal Foch from winery D was associated with (20) benzaldehyde, and Maréchal Foch/Frontenac blend from winery B was associated with (16) furfural. Vincent wine from winery C and the Maréchal Foch/Frontenac blend from winery B were similar in wine aroma. It cannot be determined if the difference in aroma is due to the variety or winemaking practices for the Maréchal Foch, Maréchal Foch/Frontenac blend, or Vincent wines.

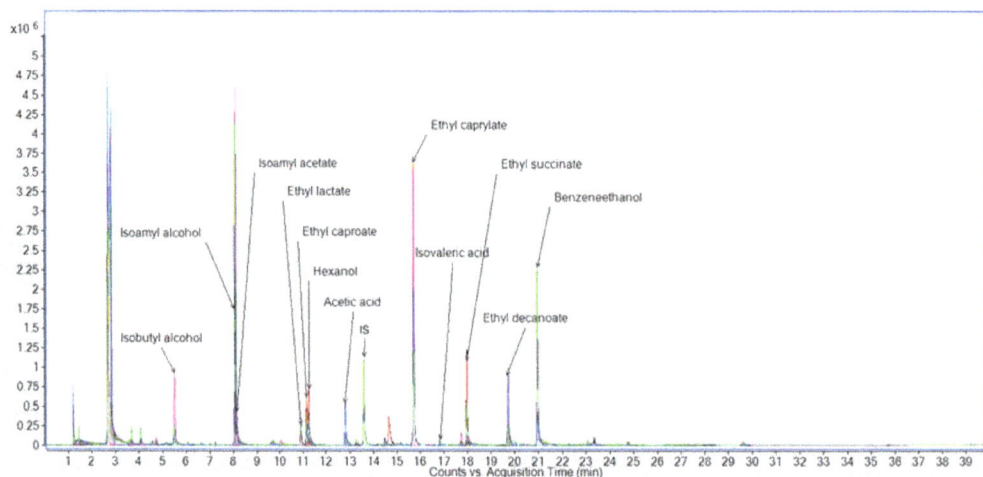

Figure 10. Overlay of total compound chromatograms of seven selected Iowa red wines using the optimized method. Wines included: two bottles of Maréchal Foch, one bottle of St. Croix, two bottles of Frontenac, one bottle of Vincent, and one bottle of Maréchal Foch/Frontenac blend. Conditions: 2 min pre-sampling desorption of 50/30 μm DVB/CAR/PDMS SPME fiber; 4 mL wine sample in a 10 mL threaded glass amber vial with PTFE/silicone septa; 10 min incubation at 50 °C; 10 min extraction time at 50 °C; agitation speed 500 rpm; 2 min desorption time into GC inlet.

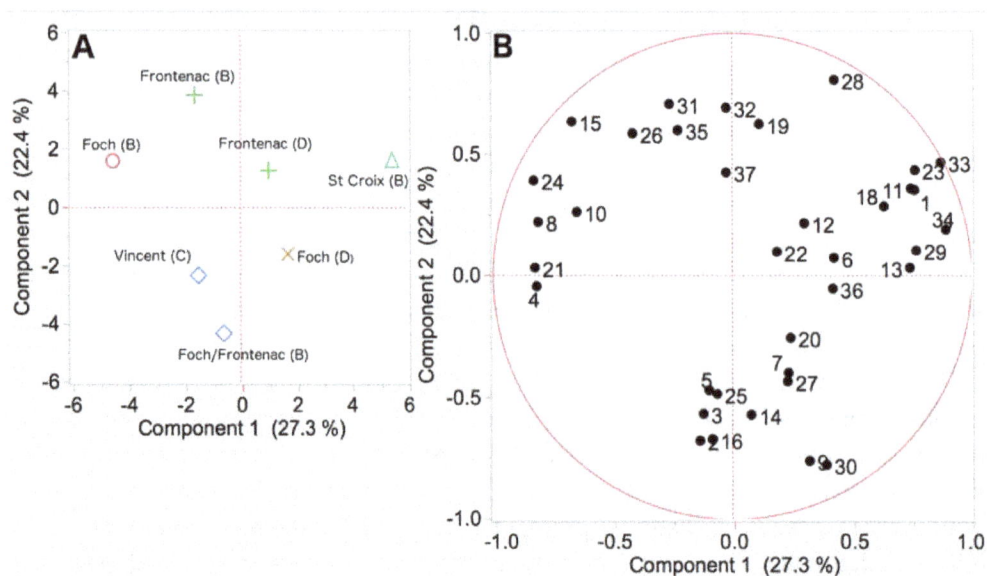

Figure 11. A principal components analysis of wine aroma of seven wines from three different Iowa wineries (**A**) and variables (**B**). Numbered variables refer to peak numbers in Table 2. Internal standard (17) 3-nonanone is not included in (**B**) because there was no variation between the samples.

An automated headspace SPME method, coupled with GCMS-olfactometry was developed to characterize wine aroma. This method was applied to characterize 36 aroma compounds present in seven Iowa red wines from three different wineries. The interactions between the experimental factors were not considered in this research. A multivariate experimental design aimed to determine the main factors followed by a response surface methodology [25] could yield better results. Although a distinct varietal aroma character was not 'pinpointed' for these Iowa red wines, there were notable differences in the aroma profile by grape variety and by the winery. Linking aroma compounds to grape variety and winemaking practices continues to be important in producing high quality Iowa wines.

Acknowledgments: The authors acknowledge funding provided through the Iowa Department of Agricultural and Land Stewardship (07-1273; Project title: Development of a rapid screening method for chemical and aroma analyses of Iowa wines: transresveratrol and aroma-enhancing compounds).

Author Contributions: L.C., J.A.K., and M.D. conceived and designed the experiments; L.C. performed the experiments; L.C. and S.R. analyzed the data; J.A.K. and M.D. contributed reagents/materials/analysis tools; L.C., S.R., J.A.K., and D.M. wrote the paper.

Conflicts of Interest: The authors declare no conflict of interest. The funding sponsors had no role in the design of the study; in the collection, analyses, or interpretation of data; in the writing of the manuscript, and in the decision to publish the results.

References

1. Schreier, P.; Jennings, W.G. Flavor composition of wines: A review. *CRC Crit. Rev. Food Sci. Nutr.* **1979**, *12*, 59–111. [CrossRef] [PubMed]
2. Rapp, A. Wine aroma substances from gas chromatographic analysis. In *Modern Methods of Plant Analysis*; Linskens, H.F., Jackson, J.F., Eds.; Springer: Berlin, Germany, 1988; Volume 6, pp. 29–66. [CrossRef]
3. Sauvageot, F.; Vivier, P. Effects of malolactic fermentation on sensory properties of four burgundy wines. *Am. J. Enol. Vitic.* **1997**, *48*, 187–192.
4. De Revel, G.; Bertrand, A. Dicarbonyl compounds and their reduction products in wine. Identification of wine aldehydes. In *Trends in Flavour Research*; Maarse, H., van der Heij, D.G., Eds.; Elsevier Science B.V.: Amsterdam, The Netherlands, 1994; Volume 35, p. 353. ISBN 0444815872.
5. Allen, M. What level of methoxypyrazines is desired in red wines? The flavour perspective of the classic red wines of Bordeaux. *Aust. Grapegrow. Winemak.* **1995**, *381*, 7–9.
6. Davis, C.R.; Wibowo, D.; Eschenbruch, R.; Lee, T.H.; Fleet, G.H. Practical implications of malolactic fermentation: A review. *Am. J. Enol. Vitic.* **1985**, *36*, 290–301.
7. Rapp, A.; Güntert, M.; Almy, J. Identification and significance of several sulfur-containing compounds in wine. *Am. J. Enol. Vitic.* **1985**, *36*, 219–221.
8. Montedoro, G.; Bertuccioli, M. The flavour of wines, vermouth and fortified wines. In *Food flavours, Part B, The Flavour of Beverages*; Morton, I.D., MacLeod, A.J., Eds.; Elsevier Science Ltd.: Amsterdam, The Netherlands, 1986; p. 171. ISBN 9780444425997.
9. Rapp, A.; Versini, G. Methylanthranilate ("foxy taint") concentrations of hybrid and Vitis vinifera wines. *Vitis* **1996**, *35*, 215–216.
10. Margalit, Y. *Concepts in Wine Chemistry*, 3rd ed.; The Wine Appreciation Guild: San Francisco, CA, USA, 2012; p. 203. ISBN 1-935879-94-4.
11. Fuller, G.H.; Steltenkamp, R.; Tisserand, G.A. The gas chromatograph with human sensor: Perfumer model. *Ann. N. Y. Acad. Sci.* **1964**, *116*, 711–724. [CrossRef] [PubMed]
12. Robinson, A.L.; Boss, P.K.; Solomon, P.S.; Trengrove, R.D.; Heymann, H.; Ebeler, S.E. Origins of grape and wine aroma. Part 2. Chemical and sensory analysis. *Am. J. Enol. Vitic.* **2014**, *65*, 25–42. [CrossRef]
13. Panighel, A.; Flamini, R. Solid phase extraction and solid phase microextraction in grape and wine volatile compounds analysis. *Sample Prep.* **2015**, *2*, 55–65. [CrossRef]
14. Pawliszyn, J. *Handbook of Solid Phase Microextraction*; Chemical Industry Press: Beijing, China, 2009; pp. 90–103. ISBN 978-7-122-04701-4.

15. Zhang, S.; Koziel, J.A.; Cai, L.; Hoff, S.J.; Heathcote, K.Y.; Chen, L.; Jacobson, L.D.; Akdeniz, N.; Hetchler, B.P.; Parker, D.B.; et al. Odor and odorous chemical emissions from animal buildings: Part 5. Simultaneous chemical and sensory analysis with gas chromatography-mass spectrometry-olfactometry. *Trans. ASABE* **2015**, *58*, 1349–1359. [CrossRef]

16. LRI & Odour Database. Available online: http://www.webcitation.org/6rDRATLKY (accessed on 14 June 2017).

17. Flavornet and Human Odor Space. Available online: http://www.webcitation.org/6qr4xTprN (accessed on 30 May 2017).

18. Ferreira, A.C.S.; de Pinho, P.G. Analytical method for determination of some aroma compounds on white wines by solid phase microextraction and gas chromatography. *J. Food Sci.* **2003**, *68*, 2817–2820. [CrossRef]

19. Howard, K.L.; Mike, J.H.; Riesen, R. Validation of a solid-phase microextraction method for headspace analysis of wine aroma compounds. *Am. J. Enol. Vitic.* **2005**, *56*, 37–45.

20. Hall, B.J.; Brodbelt, J.S. Determination of barbiturates by solid-phase microextraction (SPME) and ion trap gas chromatography-mass spectrometry. *J. Chromatogr. A* **1997**, *777*, 275–282. [CrossRef]

21. Slegers, A.; Angers, P.; Ouellet, E.; Truchon, T.; Pedneault, K. Volatile compounds from grape skin, juice, and wine from five interspecific hybrid grape cultivars grown in Quebec (Canada) for wine production. *Molecules* **2015**, *20*, 10980–11016. [CrossRef] [PubMed]

22. Pedneault, K.; Martine, D.; Angers, P. Flavor of cold-hardy grapes: Impact of berry maturity and environmental conditions. *J. Agric. Food Chem.* **2013**, *61*, 10418–10438. [CrossRef] [PubMed]

23. Cheng, G.; Liu, Y.; Yue, T.X.; Zhang, Z.W. Comparison between aroma compounds in wines from four Vitis vinifera grape varieties grown in different shoot positions. *Food Sci. Technol. Camp.* **2015**, *35*, 237–246. [CrossRef]

24. Jackson, R.S. *Wine Science Principles and Applications*, 3rd ed.; Elsevier: Amsterdam, The Netherlands, 2008; p. 662. ISBN 978-0-12-373646-8.

25. Gunst, R.F. Response surface methodology: Process and product optimization using designed experiments. *Technometrics* **1996**, *38*, 284–286. [CrossRef]

## 16

Postmortem Internal Gas Reservoir Monitoring Using GC×GC-HRTOF-MS

Pierre-Hugues Stefanuto [1,*], Katelynn A. Perrault [1], Silke Grabherr [2], Vincent Varlet [3] and Jean-François Focant [1]

[1] Organic and Biological Analytical Chemistry Group, Chemistry Department, University of Liège, Allée du 6 Août 11 (Bât B6c), Quartier Agora, 4000 Liège (Sart-Tilman), Belgium; katelynn.perrault@ulg.ac.be (K.A.P.); jf.focant@ulg.ac.be (J.-F.F.)

[2] Forensic Anthropology and Imaging Unit, University Center of Legal Medicine, Chemin de la Vulliette 4, CH-1000 Lausanne 25, Switzerland; silke.grabherr@chuv.ch

[3] Forensic Toxicology and Chemistry Unit, University Center of Legal Medicine, Chemin de la Vulliette 4, CH-1000 Lausanne 25, Switzerland; vincent.varlet@chuv.ch

* Correspondence: phstefanuto@ulg.ac.be

Academic Editor: Shari Forbes

Abstract: Forensic investigations often require postmortem examination of a body. However, the collection of evidence during autopsy is often destructive, meaning that the body can no longer be examined in its original state. In order to obtain an internal image of the body, whole body postmortem computed tomography (PMCT) has proven to be a valuable non-destructive tool and is currently used in medicolegal centers. PMCT can also be used to visually locate gas reservoirs inside a cadaver, which upon analysis can provide useful information regarding very volatile compounds that are produced after death. However, the non-targeted profiling of all potential volatile organic compounds (VOCs) present in these reservoirs has never been attempted. The aim of this study was to investigate the VOC profile of these reservoirs and to evaluate potential uses of such information to document circumstances surrounding death, cause of death and body taphonomy. Comprehensive two-dimensional gas chromatography coupled to time-of-flight high-resolution mass spectrometry (GC×GC-HRTOF-MS) was used for VOC measurements. This study demonstrated that the chemical composition of VOCs within the gas reservoirs differed between locations within a single body but also between individuals. In the future, this work could be expanded to investigate a novel, non-destructive cadaver screening approach prior to full autopsy procedures.

Keywords: human decomposition; forensic chemistry; comprehensive two-dimensional gas chromatography coupled to time-of-flight high-resolution mass spectrometry (GC×GC-HRTOF-MS); volatile organic compounds; multivariate statistics

1. Introduction

Postmortem imaging techniques are commonly used prior to an autopsy for forensic investigations. These techniques allow some information to be collected and saved prior to their possible destruction and loss due to the invasiveness of the autopsy procedure. Whole body postmortem computed tomography (PMCT) is often used as the imaging technique in postmortem examination. It allows the visualization of internal characteristics of the body and can serve to detect the precise localization of gas reservoirs present inside the body and further allow their collection [1–5]. Routinely, the occurrence and the distribution of postmortem gases are used in medicolegal laboratories, often to provide information about the cause of death [5,6]. However, the full chemical composition of these reservoirs has never been analyzed [5]. This lack of chemical profiling could lead to the misinterpretation of the origin of these gases. For example, the chemical profile of this gas could potentially discriminate between putrefactive gas buildup and the occurrence of a gas embolism [6]. In 2015, Varlet et al. demonstrated the possibility to chemically differentiate alteration gases and gas embolisms using gaseous compositions (e.g., O_2, N_2, CO_2, CH_4, and H_2S) [5]. In order to obtain more information regarding the cause of death or the postmortem interval (PMI), a complete screening of the volatile organic compounds (VOCs) present in these gas reservoirs would be extremely valuable.

Comprehensive two-dimensional gas chromatography (GC×GC) techniques have already been reported to be of high efficacy for the monitoring of decomposition VOCs from different matrices such as soils [7–11] and whole body headspace [12–19]. An increase in separation power is required to efficiently separate the hundreds of VOCs emitted by decomposing remains. Moreover, the hyphenation of GC×GC with mass spectrometry also provides identification information based on signal deconvolution and compound-specific fragmentation patterns. When high-resolution time of flight mass spectrometry (HRTOF-MS) is used, the high mass measurement accuracy additionally contributes to better tentative identifications of compounds present in the mix of decomposition volatiles [11].

Considering PMCT gas reservoir sampling in combination with the separation power of GC×GC and the mass measurement accuracy of HRTOF-MS is a pioneering approach for comprehensive screening of the chemical composition of internal gas reservoirs within a dead body. For the present study, internal gases were sampled under laser guidance and assisted by postmortem multi-detector computed tomography (MDCT) following an established procedure [4,5,20]. Solid-phase microextraction (SPME) was applied to concentrate and to introduce the VOCs into the chromatographic system. GC×GC-HRTOF-MS was then used in order to obtain good separation and robust identification of the compounds of interest. The primary objective was to demonstrate the ability of SPME GC×GC-HRTOF-MS to detect low-level VOCs from internal gases. The combination of various statistical approaches was then investigated to provide information regarding the decomposition status of organs within a body.

2. Materials and Methods

2.1. Sampling

2.1.1. Internal Gas Sampling

First, the location of each intracadaveric gaseous reservoir was revealed during PMCT. In each reservoir, a needle equipped with a closed three-way tap was introduced through the body and into the reservoir, which was facilitated by radiological guidance using three-dimensional coordinates in biopsy mode. A second control computed tomography scan (CT-Scan) was then performed to check the position of the needles in each reservoir. Finally, the three-way tap inserted in each reservoir was slowly opened to withdraw gas into a Luer-lock polytetrafluoroethylene (PTFE) syringe. The three-way taps were then closed and the syringe, tap and needle from each reservoir were removed from the body. The gas samples were then individually transferred from the syringe into a 20 mL headspace vial containing water [4,21,22], where water was expelled to compensate for the introduction of gas volume. A residual water quantity was left in each vial to permit complete seal. Vials were inverted and stored at +4 °C prior to analysis. Preliminary studies were conducted on gases and have assured sample stability over a minimum period of two months while refrigerated [5].

The sampling and storage of gas is complex and even more so in the specific case of forensic evidence. This procedure for collecting gas reservoirs has already been implemented during the medico-legal investigation of the bodies [4,21,22]. Thus, the samples for this study were collected in the same manner in order to allow several analyses to be performed on the same sample. The partitioning of the most polar compounds into the water phase was expected; however, this effect was not characterized within this study. The partitioning between the aqueous and the gas phase can be an issue if the biomarkers have high solubility in water, reducing their presence in the gas phase compared to less polar compounds. A further evaluation of this effect would be necessary after the validation of marker compounds.

2.1.2. VOC Sampling

Once internal gases were trapped in headspace vials, VOCs were sampled using solid phase micro-extraction (SPME) which was an ideal sampling approach given the limited gas volume of each sample. The fiber was coated with 100 μm of polydimethylpolysiloxane (PDMS) (23 ga., Supelco, Belfonte, PA, USA). The PDMS fiber was chosen for this initial screening for its hydrophobic properties, due to high water content within the vials. The fiber was exposed to the gas sample for 15 min at room temperature (i.e., 20 °C). The main practical limitation for SPME was the minimal space (2 cm) available to expose the fiber in the gaseous volume above the water, which was limited by the volume of gas in each reservoir available to be sampled into the vial. Following this proof of concept, a deeper investigation into the equilibrium exchange between the gas and the water phase would provide additional information about VOC concentrations (e.g., using different SPME fibers, SPME in immersion, multiple SPME extraction, or by other chromatographic methods). All gas samplings were performed with ethical agreement (Ethics Committee of Canton de Vaud, Protocol 181/12). Information regarding each cadaver can be found in Table 1.

Table 1. Information regarding the human bodies analyzed in this study.

Body #	Gender	Age	Internal Gas Cavities Location	Cause of Death	Last Time Seen Alive
Body 1	M	32	Pectoral muscle, Left lung, Right lung Pericardium, Abdominal cavity	Suicide by neuroleptics and solvent abuse	2 weeks
Body 2	M	67	Abdominal cavity, Heart	Undetermined	Minimum 1 week
Body 3	F	75	Heart, Abdominal cavity, thorax cavity (2 times)	Undetermined (known cardiac pathologies)	N/A (minimum several weeks)
Body 4	M	68	Subcutaneous, Abdominal cavity, Pericardium, Jugular, Pectoral muscle	Undetermined (known cardiac pathologies)	10 days

2.2. Sample Analysis and Data Processing

The analytical method was based on previously used techniques for whole body decomposition VOC analysis and are described briefly herein [11]. The instrument used was an Agilent 7890A gas chromatograph (Agilent Technologies, Palo Alto, CA, USA) coupled with an AccuTOF™ GCv 4G HRTOF-MS (JEOL Ltd., Tokyo, Japan). A Rxi-5Sil MS first dimension (^{1}D) column (30 m × 0.250 mm ID × 0.25 μm df, Restek Corporation, Bellefonte, PA, USA) and a Rxi-17Sil MS second dimension (^{2}D) column (2 m × 0.250 mm ID × 0.25 μm df, Restek Corporation) were used. Modulation between columns was performed using a ZX2 dual-stage thermal loop modulator (Zoex Corporation, Houston, TX, USA). The cold jets were cooled by the ZX2 system to −90 °C and hot jets were maintained at 200 °C by a thermal auxiliary. The ^{1}D GC oven was ramped from 35 °C to 240 °C at a rate of 5 °C/min and held for an additional 5 min. No ^{2}D oven was used. The modulation period (PM) was 5 s with an 800 ms hot pulse duration. The helium carrier gas (high purity ALPHAGAZ™, Air Liquide, Liège, Belgium) was held at a constant flow rate of 1.0 mL/min. The HRTOF-MS was operated in electron ionization (EI) mode with an ionizing voltage of 70 V. An acquisition rate of 50 Hz was used with a mass range of 30–400 m/z and an acquisition delay of 1 min. The plate voltage was 2150 V with a sampling interval of 0.25 μs. Data were acquired using MassCenter version 2.6.2b (JEOL Ltd.). Instrument tuning was performed using perfluorokerosene (PFK) (Tokyo Chemical Industry Co. Ltd., Tokyo, Japan) and the mass resolution was above 7000 at m/z 293. Data was analysed in GC Image 2.5 HR (Zoex Corporation) using the GC Project and Image Investigator features. Compound identification was made by comparison to the 2014 National Institute of Standards and Technology (NIST) library [15].

The data were first exported in Microsoft Excel (Microsoft Corporation, Redmond WA, USA) for the initial matrix creation. Hierarchical Cluster Analysis (HCA) and heat map visualization were performed in the open source software for statistics, R (The R Foundation for Statistical Computing, version 3.1.3, Vienna, Austria).

Not all bodies could be included into the statistical part of the study. For the HCA, body 2 was not used since only two gas reservoirs were available. The clustering analysis of two samples is irrelevant. For the heat map processing, body 3 was not used because the analytical conditions were slightly different from the other samples.

3. Results and Discussion

3.1. Chromatographic Considerations and Data Processing

The chromatogram displayed in Figure 1 was acquired on a classical column set (i.e., 5% phenylsiloxane—50% phenylsiloxane). Further tests were also conducted using an Rxi-624Sil MS (30 m × 0.25 mm ID × 1.4 μm df)—Stabilwax (2 m × 0.25 mm ID × 0.5 μm df) (Restek Corporation) combination, as already applied in previous decomposition studies [11]. The peak shape and the artifact level improved when using the Rxi-624Sil MS—Stabilwax column set. However, since a second sampling was not possible due to restricted access to samples, the classical column set was used for this proof-of-concept study, and future studies will utilize other column combinations that may be more beneficial to chromatographic qualities. To overpass the chromatographic issues in this data set (e.g., tailing, siloxane background, etc.), a robust alignment process was necessary.

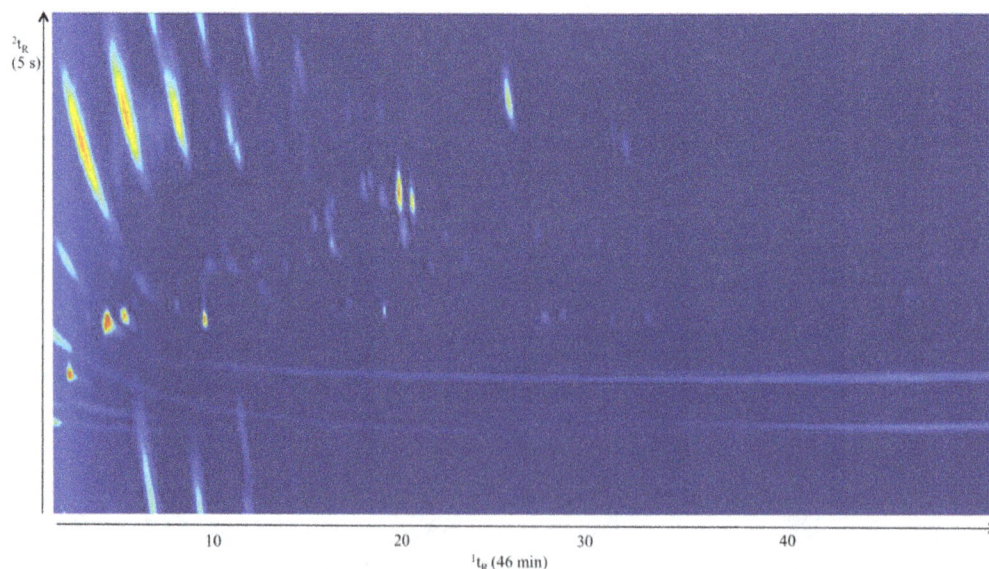

Figure 1. SPME GC×GC-HRTOF-MS total ion current (TIC) chromatogram (30 m × 0.25 mm ID × 0.25 μm df Rxi-5Sil MS as ^{1}D and 2.0 m × 0.25 mm ID × 0.25 μm df Rxi-17Sil MS as ^{2}D) of the pericardium cavity gas sample collected from body 1 (see Table 1).

The first step of data processing was to pre-process each chromatogram for baseline correction, peak detection, and phase shift in order to improve chromatogram alignment. Following these steps, the processed images were used and summed to generate a cumulative image and a feature template from the aligned chromatograms (Figure 2). Every feature designated in the feature template was manually reviewed in order to check the assignment and peak shape for each compound and remove chromatographic artifacts from the template. The resulting template was then applied to all of the samples and the generated data matrix was exported as a *.csv file. This data processing approach was conducted for every comparison. In the absence of these template control steps, there would have been a number of unfortunate outcomes including: multiple peak identifications for a single chromatographic peak, peak misidentification, and assignment of non-relevant peaks as components of internal gas.

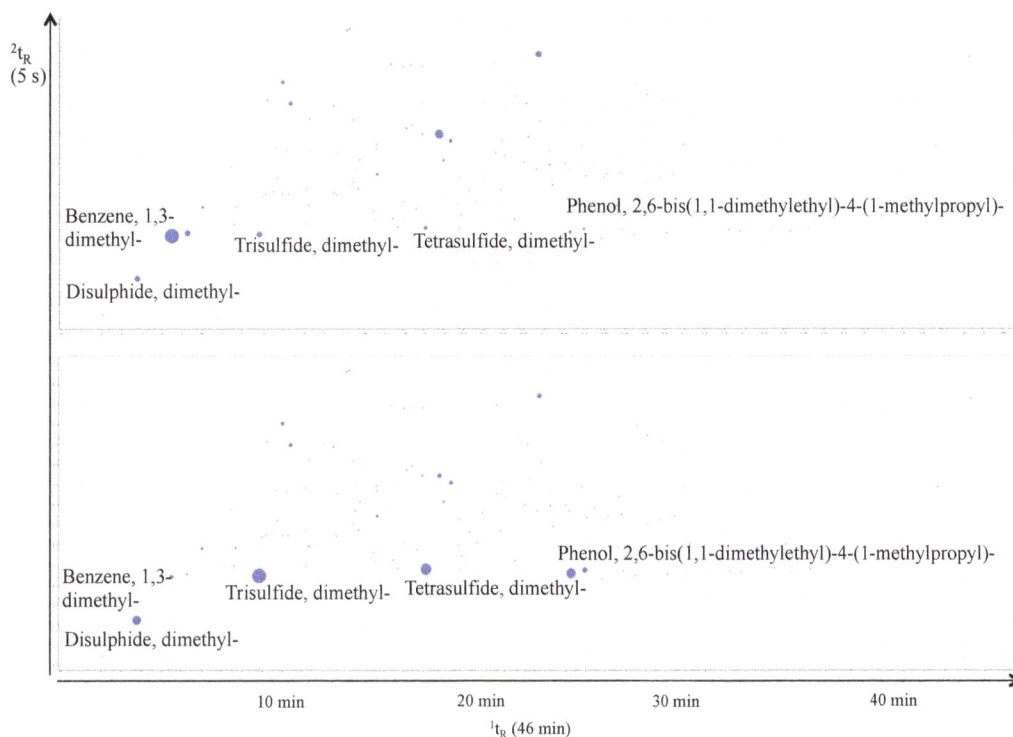

Figure 2. Apex plot of the cumulative image used in the inter-body comparison. On the top, the size of the blobs represents the area of the peak. On the bottom one, the size is proportional to the Fisher Ratio (FR) value. The most intense peaks are not always the most variable.

3.2. Intra-Body Analysis

First, the intra-body variation was investigated considering gas samples collected from various cavities in same individuals. HCA was performed to highlight possible differences between gas reservoir locations in a single body. As illustrated in the corresponding dendrograms (Figure 3), the arrangement after hierarchical clustering underlines the difference between gas samples while showing apparent similarities between VOC profiles issued from samples. For all HCA analyses conducted herein, peak area values were transformed using z-score normalization, and Euclidean distances were calculated based on these values. For each body, the data set contained between 60 and 100 compounds. Figure 3 illustrates two examples of the typical output obtained using this method. For the cluster tree of body 1, the two most closely related VOC profiles resulted from samples originating from the lungs. This is consistent with the hypothesis of tissue-specific degradation processes reported in previous studies [17,23]. Following the lung samples, the pectoral muscle, the abdominal cavity, and the pericardium exhibited decreasing similarities, based on the Euclidean distances. Similar results were also obtained for body 3, whereby the thoracic cavity samples were the most comparable to each other, although the abdominal cavity and the heart were less similar to these. For body 4, a different trend was observed using HCA. The pectoral muscle and jugular were the most similar to each other, followed by the subcutaneous gluteal tissue and pericardium samples. The abdominal cavity appeared to be the most different sample in the case of body 4. Based on these preliminary results, the similarity of samples did not follow the same trends across all bodies investigated. Some of the sampling locations (i.e., organ containing a gas reservoir) were comparable in both cases; however, they did not cluster in the same way on the HCA dendrogram. Due to the fact that there are a variety of extraneous factors that can influence decomposition VOC profiles, it is hypothesized that the differences in trends could be resulting from various postmortem factors (e.g., cause of death, postmortem interval, location of death [24–27]) or antemortem factors (e.g., health conditions, origin, lifestyle, gender, height, weight, diet, medications taken) which can both influence the thanatomicrobiome responsible for alteration gas generation [28]. In the case of body 4, the body was in a fresher state of decomposition compared to body 1 and body 3 (Table 1); therefore, it is suspected that one of the major contributing factors to the difference in trends within each body was likely the PMI.

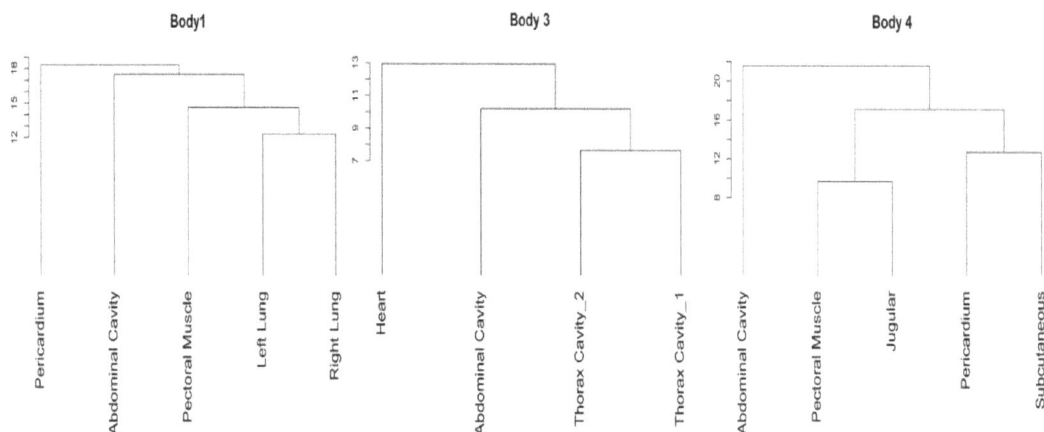

Figure 3. Hierarchical Cluster Analysis (HCA) dendrogram based on the Euclidian distances of z-score normalized peak areas for three different cadavers. Segregation based on full volatile organic compound (VOC) profiles measured in each gas reservoirs.

Because the peak table corresponding to the template chromatogram listed several sulfide compounds at significant relative levels, the clustering behavior observed in Figure 3 was investigated based on the level of these sulfide compounds. As illustrated in Figure 4, a relationship appeared to exist between the level of sulfide VOCs and the clusters of bodies observed on the HCA dendrograms (Figure 3). For body 1, the pericardium sample contained high levels of dimethyl trisulfide and dimethyl tetrasulfide (Figure 4). The abdominal cavity, the pectoral muscle, and the lungs exhibited decreasing levels of sulfides respectively (Figure 4). For Body 4, the first two clusters (i.e., pericardium/subcutaneous gluteal muscle and pectoral muscle/jugular vein) (Figure 3) displayed similar levels of sulfide compounds (Figure 4), especially dimethyl disulfide, dimethyl trisulfide and dimethyl tetrasulfide. The abdominal cavity differed in comparison to these first two clusters (Figure 3) based on higher levels of these three compounds. Similar results were obtained for body 3. Perrault et al. [29] have also discussed the importance of the polysulfide compounds in the monitoring of decomposition VOCs. Indeed, the presence of these compounds at different levels throughout the decomposition promoted them as potential PMI indicators [29]. The mass spectral identification of the sulfide compounds was facilitated by the specific isotopic pattern obtained from the sulfur component on the corresponding mass spectrum. Scripting on such mass spectral feature can easily be implemented for rapid identification of sulfides. Moreover, these compounds could be interesting for the development of routine forensic screening methods, since sulfur-specific detectors are also commercially available and come with reduced cost and higher portability compared to mass spectrometers [30].

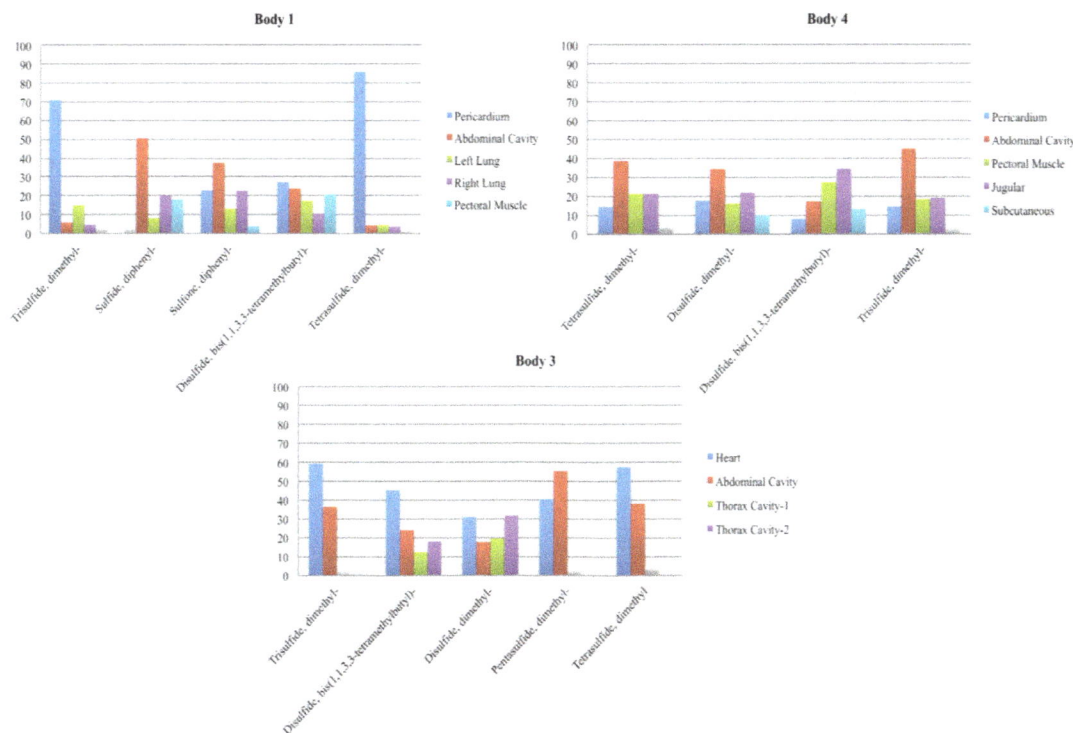

Figure 4. Relative levels of major sulfide compounds based on normalized relative peak area for each compound.

3.3. Inter-Body Analysis

The next stage of this study was to compare the inter- and intra-body differences. Data were reprocessed grouping samples by bodies rather than by sampling localization. Each body was defined as a class of samples. For all compounds remaining after data alignment, Fisher Ratios (FR) were calculated [31]. These FR values were compared to the F critical value (F_{crit}) defined regarding the number of classes, degrees of freedom, and a confidence interval of 99%. Compounds with a FR exceeding F_{crit} were retained and defined as being compounds of interest for the differentiation of bodies. This data analysis approach has been previously reported to be efficient for handling complex multivariate GC×GC data [13,32]. This approach allows data reduction without loss of information, therefore leading to extraction of the most pertinent variables and removal of artifacts that do not discriminate the variables under examination. This data reduction step is necessary due to the fact that GC×GC allows the detection of a large number of compounds, generating high dimensionality matrices that are not ideal for classical statistics. Out of the 102 analytes present in the template list, the process extracted a set of 37 analytes that were the most specific amongst the different classes.

HCA was performed on the resulting matrix of analytes and results can be visualized in the tree-structured graph (left) and heat map (right) of Figure 5. In the heat map, each of the 37 analytes is displayed along the x-axis and the amount of each of them is coded based on a color scale. The y-axis lists the different bodies. The dendrogram on the left clusters samples originating from the same body together. Based on Figure 5, it appeared that the inter-body variation exceeded the organ-to-organ variation. Two samples originating from two different areas within the same body were therefore more similar to each other than two samples collected from the same location in two different bodies. The abundance of compounds was lowest in body 4, with the exception of A27 which had the highest concentration in this body compared with other samples collected from different bodies. The variation observed in body 1 and body 2 could potentially be related to the different postmortem intervals of the bodies, since they were in different states of decomposition. This observation could indicate that internal VOC monitoring may have potential in estimating PMI in the future. Regardless of the position of the internal gas reservoir, the chemical composition appeared to be related to the state of decomposition. Due to the limited number of cadavers in this study and limited knowledge regarding the exact postmortem interval of each body, further statistical correlation between PMI and chemical composition was not possible. However, future studies will endeavor to provide more details regarding this point and will also attempt to elucidate specific VOC compounds that may fluctuate in a reproducible pattern to allow PMI estimation using this procedure.

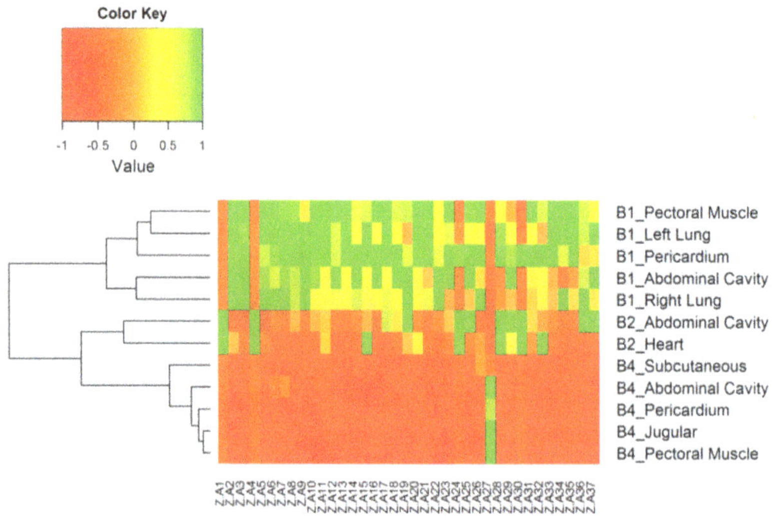

Figure 5. Hierarchical cluster analysis (HCA) heat map representation using the Euclidian distance between the z-score for each compound of interest detected from the gas reservoirs within three cadavers. The compound labels along the x-axis are listed in Table 1a,b (e.g., Z-A18 indicates the z-score based on compound A18, p-Cymene). B1: Body 1, B2: Body 2, B3: Body 3.

The 37 compounds of the heat map in Figure 5 are listed in Table 2a,b. All of these compounds of interest were first identified using mass spectral library searches. From these 37 compounds, 17 identifications were further supported by high-resolution mass spectral information with mass accuracy values lower than 30 ppm (Table 2a). To calculate the mass accuracy, the drift mass was corrected based on the siloxane peak (m/z 207.0329), taken directly beside the corresponding compound peak. The identified decomposition-specific VOCs were polysulfide compounds (dimethyl trisulfide and dimethyl tetrasulfide) [33], different isomers of xylene and cymene, as well as phenolic compounds [8,33,34].

The HRTOF-MS used in this study was operated using an electron ionization source, which does not provide the molecular ion for less stable compounds (e.g., long hydrocarbon chains, oxygenated compounds). For some of the compounds, either the parent ion or ions resulting from the loss of small masses (e.g., CH_3, H_2O) can be of sufficient specificity for mass error calculation, and these can be used to support library identifications. However, for less stable compounds, soft ionization techniques would be much more beneficial in the future to be able to calculate the chemical formula using the molecular ion. Table 2b lists the compounds that were identified by mass spectral library searching but for which identities were not further supported by high-resolution mass spectral formula calculation. For some of them, no compound names are therefore listed in Table 2b because the library identification was clearly incorrect as the compound position on the GC×GC space was wrong. For example, A27 was 'identified' as a branched cyclohexane; however, the second dimension retention time (^2t_R = 3.3 s) was too long for this type of compound. A formula simulation, run using the highest mass of the spectra, did not provide a more reliable identification, and therefore these compounds were represented separately from those in Table 2a. Nevertheless, some compounds (e.g., A10 and A37) had sufficient matches with libraries that they could be considered as acceptable for a proof-of-concept study, although such identities will require further validation in the future.

Table 2. (a) Compounds with identification supported by high-resolution mass spectral information. (b) Compound identifications not supported by high-resolution mass spectral information.

(a)

Label	Compound Name	^1t_R (min)	^2t_R (s)	FR	Match	Reverse	Probability	Formula	Exact Mass (amu)	Corrected Mass (amu)	Measured Mass (amu)	Accuracy (ppm)
A1	Cyclic octaatomic sulfur	35.6	1.7	294.1	548	552	90.66	S_8	223.8045	223.7791	223.7777	6.06
A2	Azulene	16.5	2.1	143.8	932	932	53.88	$C_{10}H_8$	128.0626	128.0432	128.0458	−20.68
A3	Benzene, 1-methyl-3-(1-methylethyl)-	11.0	2.4	127.1	899	899	14.34	$C_{10}H_{14}$	134.1096	134.0901	134.0913	−8.93
A4	Hexathiepane	34.0	1.7	119.7	758	796	98.87	CH_2S_6	205.8481	205.8226	205.8233	−3.24
A5	Benzene, 1-ethyl-2,4-dimethyl-	11.3	2.4	80.7	882	890	17.39	$C_{10}H_{14}$	134.1096	134.0901	134.0919	−13.41
A6	Benzene, 1,3-dimethyl-	5.5	1.9	55.0	832	862	23.85	C_8H_{10}	106.0783	106.0650	106.0639	10.39
A7	Benzene, 1,4-dimethyl-	6.3	1.9	54.2	877	881	25.07	C_8H_{10}	106.0783	106.0750	106.0718	30.19
A9	Benzene, 1,2,3,5-tetramethyl-	12.2	2.5	45.4	856	880	15.1	$C_{10}H_{14}$	134.1096	134.0901	134.0927	−19.38
A13	1H-Indene, 2-butyl-5-hexyloctahydro-	23.8	4.1	24.2	672	676	7.27	$C_{19}H_{36}$	249.2582	249.2328	249.2282	18.36
A14	Phenol, 2,6-bis(1,1-dimethylethyl)-4-(1-methylpropyl)-	25.0	2.8	24.0	787	794	7.53	$C_{18}H_{30}O$	262.2297	262.2042	262.2002	15.32
A18	p-Cymene	13.5	2.5	21.0	581	634	10.41	$C_{10}H_{14}$	134.1096	134.0901	134.0915	−10.43
A19	p-Isopropenylphenol	19.8	2.0	20.8	892	905	60.64	$C_9H_{10}O$	134.0732	134.0537	134.0551	−10.32
A22	Phenol, 4,4′-(1-methylethylidene)bis-	41.7	2.2	16.9	884	964	82.47	$C_{15}H_{16}O_2$	228.1150	228.0956	228.0940	6.94
A24	Trisulfide, dimethyl-	9.8	1.9	16.8	725	741	88.51	$C_2H_6S_3$	125.9632	125.9377	125.9341	28.73
A28	Benzene, 1,3-bis(1,1-dimethylethyl)-	14.9	2.9	12.2	914	917	76.52	$C_{14}H_{22}$	190.1722	190.1527	190.1496	16.31
A30	Tetrasulfide, dimethyl-	18.1	2.0	10.5	742	742	55.43	$C_2H_6S_4$	157.9352	157.9320	157.9306	8.80
A32	Phenol, 2-(1,1-dimethylethyl)-5-methyl-	19.8	2.3	9.6	885	885	26.9	$C_{11}H_{16}O$	164.1201	164.1007	164.1000	4.06

(b)

Label	Compound Name	^1t_R (min)	^2t_R (s)	FR	Match	Reverse	Probability	Formula
A8	Sulfurous acid, cyclohexylmethyl isobutyl ester	28.4	3.8	46.6	630	803	16.3	$C_{11}H_{22}O_3S$
A10	2-Bromobenzothiazole	27.8	2.7	36.1	812	892	70.5	C_7H_4BrNS
A11	1,2,3-Trimethyl-cyclopent-2-enecarboxaldehyde	17.1	3.2	27.2	619	724	10.3	$C_9H_{14}O$
A12	Stiripentol	29.8	2.6	26.1	661	673	39.91	$C_{14}H_{18}O_3$
A15	1,2-Diazaspiro[4.4]nonen-3-carboxylic acid, 6,6,9,9-tetramethyl-, methyl ester	17.4	3.3	22.1	605	690	9.55	$C_{13}H_{22}$
A16	Phthalic acid, methyl phenyl ester	27.0	2.6	22.0	748	836	5.57	$C_{10}H_{10}$
A17	2-Dodecen-1-yl(-)succinic anhydride	18.8	3.2	21.3	576	671	17.82	$C_{16}H_{26}O_3$
A20	Neophytadiene	16.9	4.1	20.7	676	715	7.4	$C_{20}H_{38}$
A21	3-Tetradecyn-1-ol	20.9	2.7	20.5	687	692	24.49	$C_{14}H_{26}O$
A23	Vinyl caprylate	23.9	3.3	16.9	601	623	8.39	$C_{10}H_{18}O_2$
A25	Disulfide, bis(1,1,3,3-tetramethylbutyl)	16.4	4.0	15.5	760	766	46.1	$C_{16}H_{34}S_2$
A26	Benzoic acid, 3,5-dimethyl-, (2,4-dimethylphenyl)methyl ester	15.5	2.2	13.7	761	761	27.59	$C_{18}H_{20}O_2$
A27	Cyclohexane, 2,4-diisopropyl-1,1-dimethyl-	18.5	3.3	12.4	530	678	8.35	$C_{12}H_{24}$
A29	Vinyl caprylate	15.7	3.0	11.2	714	725	25.61	$C_{10}H_{18}O_2$
A31	Neophytadiene	11.4	3.5	10.2	717	774	4.96	$C_{20}H_{38}$
A33	4-imidazolidinone, 3-ethyl-5-[2-(3-ethyl-2(3H)-benzoxazolylidene)ethylidene]-1-phenyl-2-thioxo-	14.2	4.4	9.6	492	770	41.69	$C_{22}H_{21}N_3O_2S$
A34	4-Hepten-3-one, 5-methyl-. (E)-	19.3	3.1	9.3	599	665	5.96	$C_8H_{14}O$
A35	Sulfurous acid, cyclohexylmethyl isobutyl ester	29.8	3.6	9.3	630	801	16.6	$C_{11}H_{22}O_3S$
A36	Succinic acid, 2-(2-chlorophenoxy)ethyl ethyl ester	31.0	1.9	8.9	612	626	34.43	$C_{14}H_{17}ClO_5$
A37	2,4,6-Tris(1,1-dimethylethyl)-4-methylcyclohexa-2,5-dien-1-one	22.8	2.6	8.7	803	857	31.2	$C_{19}H_{32}O$

4. Conclusions

This proof-of-concept study demonstrates the potential of monitoring volatile organic compounds from human intracadaveric gas reservoirs. The precision of PMCT scanning for gas reservoirs within a body combined with the analytical power of SPME GC×GC-HRTOF-MS allowed a non-destructive, non-targeted screening of these volatile samples. The chemical composition of these gas reservoirs may have the potential to provide important information about the decomposition state of a body, and could therefore lead to potential enhancement of PMI estimation based on the use of additional taphonomic markers that would be valuable during early stage decomposition. A trend to different volatile patterns was also observed for each organ, indicating a possible specific decomposition pathway for each of them. The sulfide compounds appeared to be an important class of compounds for ascertaining these differences. An inter-body comparison demonstrated that the variation between individuals exceeds the variation between samples collected from a single individual. This observation is a major point of interest since it could help to develop new techniques for PMI estimation during early decomposition, where few non-invasive techniques are available. Future work will aim to increase the number of individuals studied in order to identify key VOCs that are related to the state of decomposition. A larger data set will also assist with further supporting tentative identifications of some compounds made herein. The purpose of performing this non-targeted analysis will aid in identifying several compounds that can further be investigated in targeted studies. In this case, a quantification method would also be valuable, especially for the sulfide compounds that appear to have potential as decomposition stage markers.

Acknowledgments: We would like to thank JEOL (Europe) B.V. (Niew-Vennep, The Nederlands) for strong instrumental and technical support. We also thank Restek® Corporation (Bellefonte, PA, USA) and Sigma Aldrich® (St. Louis, MO, USA) for providing us with GC phases, SPME fibers, and various GC consumables.

Author Contributions: J.-F.F., P.-H.S., S.G. and V.V. conceived and designed the experiments; P.-H.S., K.A.P. and V.V. performed the experiments; P.-H.S. analyzed the data and wrote the paper; J.-F.F., K.A.P., S.G. and V.V. performed manuscript corrections and edits.

Conflicts of Interest: The authors declare no conflict of interest.

Abbreviations

The following abbreviations are used in this manuscript:

MDCP	multi-detector computed tomography
VOCs	volatile organic compounds
GC×GC	comprehensive two-dimensional gas chromatography
HRTOF-MS	time of flight high-resolution mass spectrometry
PMI	postmortem interval
SPME	solid phase microextraction
PDMS	polydimethylsiloxane
^{1}D	first dimension
^{2}D	second dimension
1D GC	one-dimensional gas chromatography
PM	modulation period
EI	electron ionization
PFK	perfluorokerozene
HCA	hierarchical cluster analysis
TIC	total ion current
FR	Fisher ratio
F_{crit}	F critical
^1t_R	1st dimension retention time
^2t_R	2nd dimension retention time

References

1. Jacobsen, C.; Lynnerup, N. Craniocerebral trauma—Congruence between post-mortem computed tomography diagnoses and autopsy results. *Forensic Sci. Int.* **2010**, *194*, 9–14. [CrossRef] [PubMed]

2.   Roberts, I.S.D.; Benamore, R.E.; Benbow, E.W.; Lee, S.H.; Harris, J.N.; Jackson, A.; Mallett, S.; Patankar, T.; Peebles, C.; Roobottom, C.; et al. Post-mortem imaging as an alternative to autopsy in the diagnosis of adult deaths: A validation study. *Lancet* **2012**, *379*, 136–142. [CrossRef]

3.   Kasahara, S.; Makino, Y.; Hayakawa, M.; Yajima, D.; Ito, H.; Iwase, H. Diagnosable and non-diagnosable causes of death by postmortem computed tomography: A review of 339 forensic cases. *Legal Med.* **2012**, *14*, 239–245. [CrossRef] [PubMed]

4.   Varlet, V.; De Croutte, E.L.; Augsburger, M.; Mangin, P. A new approach for the carbon monoxide (CO) exposure diagnosis: measurement of total CO in human blood versus carboxyhemoglobin (HbCO). *J. Forensic Sci.* **2013**, *58*, 1041–1046. [CrossRef] [PubMed]

5.   Varlet, V.; Giuliani, N.; Palmiere, C.; Maujean, G.; Augsburger, M. Hydrogen sulfide measurement by headspace-gas chromatography-mass spectrometry (HS-GC-MS): Application to gaseous samples and gas dissolved in muscle. *J. Anal. Toxicol.* **2015**, *39*, 52–57. [CrossRef] [PubMed]

6.   Banaschak, S.; Janßen, K.; Becker, K.; Friedrich, K.; Rothschild, M.A. Fatal postpartum air embolism due to uterine inversion and atonic hemorrhage. *Int. J. Legal Med.* **2013**, *128*, 147–150. [CrossRef] [PubMed]

7.   Brasseur, C.; Dekeirsschieter, J.; Schotsmans, E.M.J.; de Koning, S.; Wilson, A.S.; Haubruge, E.; Focant, J.-F. Comprehensive two-dimensional gas chromatography-time-of-flight mass spectrometry for the forensic study of cadaveric volatile organic compounds released in soil by buried decaying pig carcasses. *J. Chromatogr.* **2012**, *1255*, 163–170. [CrossRef] [PubMed]

8.   Perrault, K.; Stuart, B.; Forbes, S. A longitudinal study of decomposition odour in soil using sorbent tubes and solid phase microextraction. *Chromatography* **2014**, *1*, 120–140. [CrossRef]

9.   Forbes, S.L.; Perrault, K.A. Decomposition odour profiling in the air and soil surrounding vertebrate carrion. *PLoS ONE* **2014**, *9*. [CrossRef] [PubMed]

10.   Perrault, K.A.; Stefanuto, P.-H.; Stuart, B.H.; Rai, T.; Focant, J.-F.; Forbes, S.L. Reducing variation in decomposition odour profiling using comprehensive two-dimensional gas chromatography. *J. Sep. Sci.* **2014**, *38*, 73–80. [CrossRef] [PubMed]

11.   Stefanuto, P.-H.; Perrault, K.A.; Lloyd, R.M.; Stuart, B.; Rai, T.; Forbes, S.L.; Focant, J.-F. Exploring new dimensions in cadaveric decomposition odour analysis. *Anal. Methods* **2015**, *7*, 2287–2294. [CrossRef]

12.   Dekeirsschieter, J.; Stefanuto, P.-H.; Brasseur, C.; Haubruge, E.; Focant, J.-F. Enhanced characterization of the smell of death by comprehensive two-dimensional gas chromatography-time-of-flight mass spectrometry (GC×GC–TOFMS). *PLoS ONE* **2012**, *7*. [CrossRef] [PubMed]

13.   Stefanuto, P.-H.; Perrault, K.A.; Stadler, S.; Pesesse, R.; LeBlanc, H.N.; Forbes, S.L.; Focant, J.-F. GC×GC–TOFMS and supervised multivariate approaches to study human cadaveric decomposition olfactive signatures. *Anal. Bioanal. Chem.* **2015**, *407*, 4767–4778. [CrossRef] [PubMed]

14.   Stefanuto, P.-H.; Perrault, K.; Stadler, S.; Pesesse, R.; Brokl, M.; Forbes, S.; Focant, J.-F. Reading cadaveric decomposition chemistry with a new pair of glasses. *Chem. Plus Chem.* **2014**, *79*, 786–789. [CrossRef]

15.   Stadler, S.; Stefanuto, P.-H.; Brokl, M.; Forbes, S.L.; Focant, J.-F. Characterization of volatile organic compounds from human analogue decomposition using thermal desorption coupled to comprehensive two-dimensional gas chromatography-time-of-flight mass spectrometry. *Anal. Chem.* **2013**, *85*, 998–1005. [CrossRef] [PubMed]

16.   Focant, J.-F.; Stefanuto, P.-H.; Brasseur, C.; Dekeirsschieter, J.; Haubruge, E.; Schotsmans, E.; Wilson, A.; Stadler, S.; Forbes, S. Forensic cadaveric decomposition profiling by GC×GC–TOFMS analysis of VOCs. *KazNU Bull. Chem. Ser.* **2013**, *72*, 177–186. [CrossRef]

17.   Rosier, E.; Loix, S.; Develter, W.; Van de Voorde, W.; Tytgat, J.; Cuypers, E. The search for a volatile human specific marker in the decomposition process. *PLoS ONE* **2015**, *10*. [CrossRef] [PubMed]

18.   Vass, A.A.; Barshick, S.-A.; Sega, G.; Caton, J.; Skeen, J.T.; Love, J.C.; Synstelien, J.A. Decomposition chemistry of human remains: A new methodology for determining the postmortem interval. *J. Forensic Sci.* **2002**, *47*, 542–553. [PubMed]

19.   Statheropoulos, M.; Spiliopoulou, C.; Agapiou, A. A study of volatile organic compounds evolved from the decaying human body. *Forensic Sci. Int.* **2005**, *153*, 147–155. [CrossRef] [PubMed]

20.   Varlet, V.; Smith, F.; Giuliani, N.; Egger, C.; Rinaldi, A.; Dominguez, A.; Chevallier, C.; Bruguier, C.; Augsburger, M.; Mangin, P.; et al. When gas analysis assists with postmortem imaging to diagnose causes of death. *Forensic Sci. Int.* **2015**, *251*, 1–10. [CrossRef] [PubMed]

21.   Varlet, V.; Smith, F.; de Froidmont, S.; Dominguez, A.; Rinaldi, A.; Augsburger, M.; Mangin, P.; Grabherr, S.

Innovative method for carbon dioxide determination in human postmortem cardiac gas samples using headspace-gas chromatography–mass spectrometry and stable labeled isotope as internal standard. *Anal. Chim. Acta* **2013**, *784*, 42–46. [CrossRef] [PubMed]

22. Varlet, V.; Bruguier, C.; Grabherr, S.; Augsburger, M.; Mangin, P.; Uldin, T. Gas analysis of exhumed cadavers buried for 30 years: A case report about long time alteration. *Int. J. Legal Med.* **2014**, *128*, 719–724. [CrossRef] [PubMed]

23. Hoffman, E.; Curran, A.; Dulgerian, N. Characterization of the volatile organic compounds present in the headspace of decomposing human remains. *Forensic Sci. Int.* **2009**, *186*, 6–13. [CrossRef] [PubMed]

24. Dekeirsschieter, J.; Verheggen, F.J.; Gohy, M.; Hubrecht, F.; Bourguignon, L.; Lognay, G.; Haubruge, E. Cadaveric volatile organic compounds released by decaying pig carcasses (*Sus domesticus* L.) in different biotopes. *Forensic Sci. Int.* **2009**, *189*, 46–53. [CrossRef] [PubMed]

25. Kasper, J.; Mumm, R.; Ruther, J. The composition of carcass volatile profiles in relation to storage time and climate conditions. *Forensic Sci. Int.* **2012**, *223*, 64–71. [CrossRef] [PubMed]

26. Vass, A.A. Odor mortis. *Forensic Sci. Int.* **2012**, *222*, 234–241. [CrossRef] [PubMed]

27. Perrault, K.A.; Rai, T.; Stuart, B.H.; Forbes, S.L. Seasonal comparison of carrion volatiles in decomposition soil using comprehensive two-dimensional gas chromatography—Time of flight mass spectrometry. *Anal. Methods* **2014**, *7*, 690–698. [CrossRef]

28. Can, I.; Javan, G.T.; Pozhitkov, A.E.; Noble, P.A. Distinctive thanatomicrobiome signatures found in the blood and internal organs of humans. *J. Microbiol. Methods* **2010**, *106*, 1–7. [CrossRef] [PubMed]

29. Perrault, K.A.; Nizio, K.D.; Forbes, S.L. A comparison of one-dimensional and comprehensive two-dimensional gas chromatography for decomposition odour profiling using inter-year replicate field trials. *Chromatographia* **2015**, *78*, 1057–1070. [CrossRef]

30. Furne, J.; Majerus, G.; Lenton, P.; Springfield, J.; Levitt, D.G.; Levitt, M.D. Comparison of volatile sulfur compound concentrations measured with a sulfide detector vs. gas chromatography. *J. Dent. Res.* **2002**, *81*, 140–143. [CrossRef] [PubMed]

31. Pierce, K.M.; Hoggard, J.C.; Hope, J.L.; Rainey, P.M.; Hoofnagle, A.N.; Jack, R.M.; Wright, B.W.; Synovec, R.E. Fisher ratio method applied to third-order separation data to identify significant chemical components of metabolite extracts. *Anal. Chem.* **2006**, *78*, 5068–5075. [CrossRef] [PubMed]

32. Brokl, M.; Bishop, L.; Wright, C.G.; Liu, C.; McAdam, K.; Focant, J.-F. Multivariate analysis of mainstream tobacco smoke particulate phase by headspace solid-phase micro extraction coupled with comprehensive two-dimensional gas chromatography-time-of-flight mass spectrometry. *J. Chromatogr.* **2014**, *1370*, 216–229. [CrossRef] [PubMed]

33. Armstrong, P.; Nizio, K.D.; Perrault, K.A.; Forbes, S.L. Establishing the volatile profile of pig carcasses as analogues for human decomposition during the early postmortem period. *Heliyon* **2016**, *2*. [CrossRef] [PubMed]

34. Swann, L.M.; Forbes, S.L.; Lewis, S.W. Analytical separations of mammalian decomposition products for forensic science: A review. *Anal. Chim. Acta* **2010**, *682*, 9–22. [CrossRef] [PubMed]

## 17

GC×GC-TOFMS for the Analysis of Metabolites Produced by Termites (*Reticulitermes flavipes*) Bred on Different Carbon Sources

Catherine Brasseur [1,2], Julien Bauwens [3], Cédric Tarayre [4], Catherine Millet [5], Christel Mattéotti [5], Philippe Thonart [4], Jacqueline Destain [4], Frédéric Francis [3], Eric Haubruge [3], Daniel Portetelle [5], Micheline Vandenbol [5], Edwin De Pauw [2] and Jean-François Focant [1,*]

[1]   Organic and Biological Analytical Chemistry, Department of Chemistry, University of Liège, Sart-Tilman, Liège B-4000, Belgium; cbrasseur@alumni.ulg.ac.be
[2]   Mass Spectrometry Laboratory, Department of Chemistry, University of Liège, Sart-Tilman, Liège B-4000, Belgium; e.depauw@ulg.ac.be
[3]   Department of Functional and Evolutionary Entomology, Gembloux Agro-Bio Tech, University of Liège, Gembloux B-5030, Belgium; julien.bauwens@ulg.ac.be (J.B.); frederic.francis@ulg.ac.be (F.F.); e.haubruge@ulg.ac.be (E.H.)
[4]   Department of Bio-Industries, Gembloux Agro-Bio Tech, University of Liège, Gembloux B-5030, Belgium; cedric.tarayre@ulg.ac.be (C.T.); p.thonart@ulg.ac.be (P.T.); j.destain@ulg.ac.be (J.D.)
[5]   Department of Microbiology and Genomics, Gembloux Agro-Bio Tech, University of Liège, Gembloux B-5030, Belgium; cmillet@alumni.ulg.ac.be (C.M.); cmattéotti@alumni.ulg.ac.be (C.M.); daniel.portetelle@ulg.ac.be (D.P.); m.vandenbol@ulg.ac.be (M.V.)
*   Correspondence: jf.focant@ulg.ac.be

Academic Editor: Shari Forbes

Abstract: More and more studies are dedicated to termites and their symbionts, to better understand how they efficiently produce energy from lignocellulose. In that context, a powerful analytical method was developed to perform the detection, separation and identification of compounds in the 1 µL fluid volume of the gut of the termite *Reticulitermes flavipes*. Comprehensive two-dimensional gas chromatography (GC×GC) coupled to time-of-flight mass spectrometry (TOFMS) was tested with three different column combinations: (1) low-polar/mid-polar; (2) polar/low-polar and (3) mid-polar/low-polar. The column set (3) offered the best separation and was chosen for further analysis and comparison study. Metabolites were detected in the samples, including amino acids, sugars, amines and organic acids. Samples collected from termites fed for 30 days on Avicel cellulose or xylan powder diets were analyzed and compared with the wood diet. Principal component analysis (PCA) of metabolite profiles demonstrated a separation of different clusters corresponding to the three different diets, with a similar trend for diets containing cellulose. The Analysis of variance (ANOVA) (one way-ANOVA and Tukey's test) was used to compare compound levels between these three different diets. Significant differences were observed, including higher levels of aromatic derivatives in the wood diet and higher levels of sugar alcohols in the xylan diet. A higher accumulation of uric acid was observed with the artificial diets (cellulose and xylan), likely to be related to the nitrogen deficiency. The present study highlighted the capability of adaptation of the termite system to non-optimal carbon sources and the subsequent modification of the metabolite profile. These results demonstrate the potential interest to investigate metabolite profiling with state-of-the-art separation science tools, in order to extract information that could be integrated with other omics data to provide more insight into the termite-symbiont digestion system.

Keywords: termite; comprehensive two-dimensional gas chromatography-time-of-flight mass spectrometry (GC×GC-TOFMS); lignocellulose; xylan; reticulitermes flavipes

1. Introduction

Termites are important bioreactors on the planet as they efficiently convert cellulose into glucose, the omnipresent source of energy [1]. They have always intrigued the scientific community, especially for their ability to thrive on lignocellulosic material. Lignocellulose is a complex association of cellulose, hemicellulose and lignin. These compounds are closely linked and recalcitrant to the break-up into bio-convertible substrates, making their commercial utilization as a biomass resource difficult [2]. Termites have an efficient strategy to convert lignocellulose, with the aid of their associated microbial symbionts. A broad range of microorganisms are present in termites, including bacteria, archaea, protists (only present in some families of termites, called the group of "lower termites") and fungi [3,4]. These make termites a very interesting source of enzymes, although the isolation and cultivation of the symbionts remain a difficult step [5]. The majority of termite gut microbiota has still escaped proper cultivation, probably because termites offer a very specific internal environment, both oxic and anoxic with microenvironments and substrate gradients [6]. Such conditions are difficult to reproduce in batch culture, which limit the possibility of material amplification for identification and characterization.

With the revolution of "omics" technologies, it is interesting to apply advanced studies on termites such as metagenomics, transcriptomics or proteomics [7–9]. These technologies offer the possibility to improve the screening for new enzymes and to study the system as a whole. This is important to enhance the understanding of the termite-microbiota symbiotic association [10], with the need to reproduce their efficient lignocellulolytic system at a larger scale [11]. Metabolomics is the study of small molecule profiles related to biochemical processes. To our knowledge, there is only one recent study on comprehensive metabolomic profiling of termites [12]. It used ^{13}C-labelled cellulose and two-dimensional nuclear magnetic resonance (2D-NMR) to follow cellulose digestion in the dampwood termite *Hodotermopsis sjostedti*. In the present study, we propose a different approach to analyze metabolites in the termite *Reticulitermes flavipes*, using comprehensive two-dimensional gas chromatography coupled to time-of-flight mass spectrometry (GC×GC-TOFMS). The main advantages of this technique for metabolite study have already been demonstrated [13,14]. The great peak capacity, provided by the combination of two different chromatographic columns, as well as the sensitivity, are two important criteria for the analysis of the complex termite samples. In the GC×GC system used for our study, the sensitivity is enhanced by the cryogenic zone compression (CZC) process, implemented by the use of a modulator at the interface between the first dimension (^{1}D) and the second dimension (^{2}D) columns (dimensions) [15]. Sensitivity is a key aspect to challenge the detection of compounds in the very low volume of material available from the internal volume gut of *R. flavipes*, estimated at 1 μL [16]. *R. flavipes* is a widespread and well-studied termite species [17]. These termites belong to the subterranean species, and are smaller in size compared to drywood and dampwood species [18]. Previous studies successfully analyzed metabolites produced by cockroaches (closely related to termites) or by cellulolytic cultivated microorganisms, using high-performance liquid chromatography coupled to infrared spectroscopy (HPLC-IR), or gas chromatography coupled to time-of-flight mass spectrometry (GC-TOFMS), respectively [19,20]. The development of a sensitive and powerful analytical method is of interest to perform similar studies on termites, especially for the smaller species.

For the implementation of an efficient GC×GC-TOFMS method, different column sets were tested, in order to obtain the best separation of the compounds and specifically provide more resolution for the polar compounds. The different phase combinations included low-polar, mid-polar and highly polar columns. The highly polar column was one of the recently introduced ionic liquid (IL) capillary GC columns. These IL columns present an attractive alternative to polysiloxane phases and offer unique selectivity for polar compounds. They are of high potential for metabolomics studies but have not yet been tested much [21].

For the present study, *R. flavipes* was grown on poplar wood, mainly composed of cellulose (45%), hemicelluloses (25%) and lignin (20%) [22]. Artificial diets were tested with microcrystalline cellulose (Avicel) and xylan powders. Cellulose is a polymer of β,1-4 glucose. Xylans are hemicelluloses that consist of a backbone of xylose, often acetylated or branched with arabinose and acidic sugars with various compositions. Statistical analyses were used to identify trends and specific compounds produced, related to the influence of the diet. The purpose was to identify the effect of the diet on the metabolites detected and evaluate how the termite's system adapts depending on the carbon source.

2. Materials and Methods

2.1. Termites, Culture Conditions and Artificial Diets

Reticulitermes flavipes (*ex santonensis* De Feytaud [23]) were collected on Oleron Island, France. The culture was maintained in a laboratory on wet wood (pine wood gradually replaced by poplar wood) at 27 °C and 70% humidity. Only worker-caste termites were collected for experiments. For artificial diets, termites were placed in tubes containing sterile disks prepared with 20% carbon source, 1.5% agar, 0.06% β-sisterol and water. Carbon sources used in this study were microcrystalline cellulose (Avicel) and xylan from beech wood. All substrates were purchased from Sigma-Aldrich (St. Louis, MO, USA). Termites were collected after 30 days of feeding.

2.2. Chemicals

Methanol, chloroform and ethanol were analytical grade and purchased from VWR International (Leuven, Belgium). Water was purified with a Milli-Q® filtration system (Merck Chemicals, Overijse, Belgium). Methoxyamine and pyridine were purchased from Sigma-Aldrich. *N*-methyl-*N*-trimethylsilyltrifluoroacetamide (MSTFA) with 1% trimethylsilyl chloride (TMCS) was purchased from Pierce, Thermo Scientific (Bellefonte, PA, USA). Addition of TMCS aids derivatization of amides, secondary amines, and hindered hydroxyls not derivatized by MSTFA alone. Standards of carbohydrates were purchased from Sigma-Aldrich.

2.3. Sample Preparation and Derivatization

Samples of five guts were collected in triplicate from each diet (each gut included fore-mid- and hind parts). Samples were homogenized using a piston pellet (Eppendorf, Nijmegen, The Netherlands) in 200 μL of a mixture of methanol/water (*v/v*: 50/50). 300 μL of a cold mixture of methanol/water/chloroform (*v/v/v*: 1/1/1) were added and the samples were vortexed. Centrifugation was performed at 4 °C (relative centrifugal force (rcf) of 16,100 for 10 min). The supernatant was transferred into a GC vial and evaporated to dry under a nitrogen stream. The extract was kept at −20 °C until derivatization and GC×GC-TOFMS analysis.

The extract was derivatized in two-steps with 25 μL of a solution of 30 mg/mL methoxyamine hydrochloride in pyridine, incubated at 70 °C during 60 min, then 75 μL of MSTFA + 1% TMCS, were incubated at 30 °C during 2 h. A blank procedure of extraction and derivatization was included. For carbohydrate standard analysis, 5 mM solutions were prepared in a mixture of ethanol/water. The solutions were evaporated to dryness and derivatized with the same procedure used for the extracts.

2.4. GC×GC-TOFMS Analysis

Analyses were performed with a DANI MasterTOF GC-MS system (DANI Instruments SpA, Milano, Italy) equipped with a ZX1-LN$_2$ Cooled Loop Modulation GC×GC System (Zoex Corp., Houston, TX, USA). The different column sets tested were (1) ^{1}D low-polarity Crossbond® (5% diphenyl/95% dimethyl polysiloxane phase (Rtx®-5 (Restek Corp., Bellefonte, PA, USA), 30 m × 0.18 mm, 0.20 μm) and ^{2}D mid-polarity Crossbond® 50% diphenyl/50% dimethyl polysiloxane phase (Rxi®-17 (Restek Corp.), 2 m × 0.10 mm, 0.10 μm); (2) ^{1}D high-polarity ionic liquid non-bonded 1,12-di(tripropylphosphonium)dodecane bis(trifluoromethylsulfonyl)imide trifluoromethylsulfonate phase (SLB®-IL-61 (Sigma-Aldrich, St. Louis, MO, USA), 30 m × 0.25 mm, 0.20 μm) and ^{2}D Rtx®-5 (2 m × 0.10 mm, 0.10 μm); and (3) ^{1}D mid-polarity Crossbond® trifluoropropylmethyl polysiloxane phase (Rtx®-200 (Restek Corp.), 30 m × 0.25 mm, 0.25 μm) and ^{2}D low-polarity

Crossbond® 1,4-bis(dimethylesiloxy)phenylene dimethyl polysiloxane phase (Rxi®-5Sil MS (Restek Corp.), 2 m × 0.10 mm, 0.10 μm). The modulation period (P_M) was set at 4 or 6 s, with a hot pulse duration of 600 ms. The GC oven temperature programs were: for column combination (1) 70 °C for 2 min, 5 °C/min to 300 °C held for 5 min, column combination (2) 50 °C for 5 min, 5 °C/min to 280 °C held for 10 min, and for column combination (3) 60 °C held for 1 min, 20 °C/min to 120 °C then 5 °C/min to 280 °C held for 5 min. A volume of 1 μL of the final extract was injected into a split/splitless injector held at 250 °C and used in splitless mode. Helium was used as the carrier gas at a constant flow rate of 1.0 mL/min. The MS transfer line temperature was held at 280 °C. The temperature of the ion source was set to 200 °C. The TOFMS was operated in Electron Ionization (EI) mode at 70 eV and tuned to 1000 mass resolution. The collected mass range was 35–550 amu with an acquisition rate of 100 spectra/s.

2.5. Data Processing and Statistical Analyses

The DANI software Master Lab 2.03 (DANI Instruments, Milan, Italy) was used for data acquisition. The data were exported in computable document format (CDF) and processed with GC Image™ 2.4 (Zoex Corp., Houston, TX, USA). Automated peak finding was performed with a defined volume threshold. Further processes were applied to remove the chromatographic noise using Computer Language for Chemical Identification (CLIC expression). Comparative analyses of the samples were performed using the Image Investigator tool of GC Image™ (Zoex Corp.), for advanced analysis of multiple chromatograms. The triplicated samples from each diet were compared and examined for statistical characteristics and trends. Multivariate analysis was performed on the data using principal component analysis (PCA) (The Unscrambler® 10.3, CAMO Software Inc., Woodbridge, NJ, USA). The matrix was set with the termite samples on different diets as objects, and the variables were the Percent Response of the compounds detected (ratio of volume to sum of all volumes as percent). This projection method was used to show a potential structure within the data related to the diet. The levels of each compound detected in the samples were also compared using analysis of variance (one way-ANOVA). As there were three groups (three diets), the F-test was used to compare the between-group variance with the within-group variance. F-values were considered with 5% significance level. Then, the Tukey's test was used to compare all possible pairs of means. The Tukey's test was chosen over the Fisher test for the pair-wise comparison as it includes a correction for multiple testing and has a higher power. The identification of compounds was carried out by mass spectral data comparison against the 2011 National Institute of Standards and Technology (NIST) library. Only compounds with a library match score > 700 were named and reported as metabolites. For carbohydrates, standards were analyzed to confirm the library identification.

3. Results and Discussion

3.1. Test of Different Column Sets

A sample from termites cultured on wood was analyzed with a ^{1}D low-polar/^{2}D mid-polar column combination (Column set 1). The two-dimensional chromatogram plot (Figure 1a) showed that a limited separation was achieved without taking advantage of most of the peak capacity available from the 2D system. It also highlighted the large dynamic range of compounds in such samples with overloading issues. To optimize the separation of polar compounds, a reverse column set was tested (Column set 2). The ionic liquid phase was chosen for its particularly high polarity while keeping more thermal stability (up to 290 °C) than classical more common polyethylene glycol/wax column phases. With this column set, the peak occupation in the second dimension was somehow improved (Figure 1b). However, the column bleed was very important in the chromatogram. This rendered data difficult to be processed as the proper peak detection was not achievable anymore, resulting in essentially interferences and background noise produced from the derivatization reagents and column bleed to be found, despite fine tuning of peak finding parameters. A different reverse column set (Column set 3) was tested with a less polar trifluoropropyl methyl polysiloxane stationary phase as ^{1}D. This particular crossbond phase exhibits low bleed and provides an interesting selectivity for compounds with intermediate polarity, including organic acids and nitro-containing compounds. Amongst the three sets, Column set 3 (Figure 1c) offered the best separation and chromatographic fingerprint. As the volumes of samples were very small (internal termite volume gut < 1 μL), the sample preparation was limited to liquid extraction without an additional purification step, in order to extract as many compounds as possible, and avoid loss of material. The main challenges for the analysis of these complex samples were the large dynamic range to consider and the resulting overloading of some compounds and interferences. Column set 3 offered the best fingerprint for data processing and was chosen for the diet comparison study.

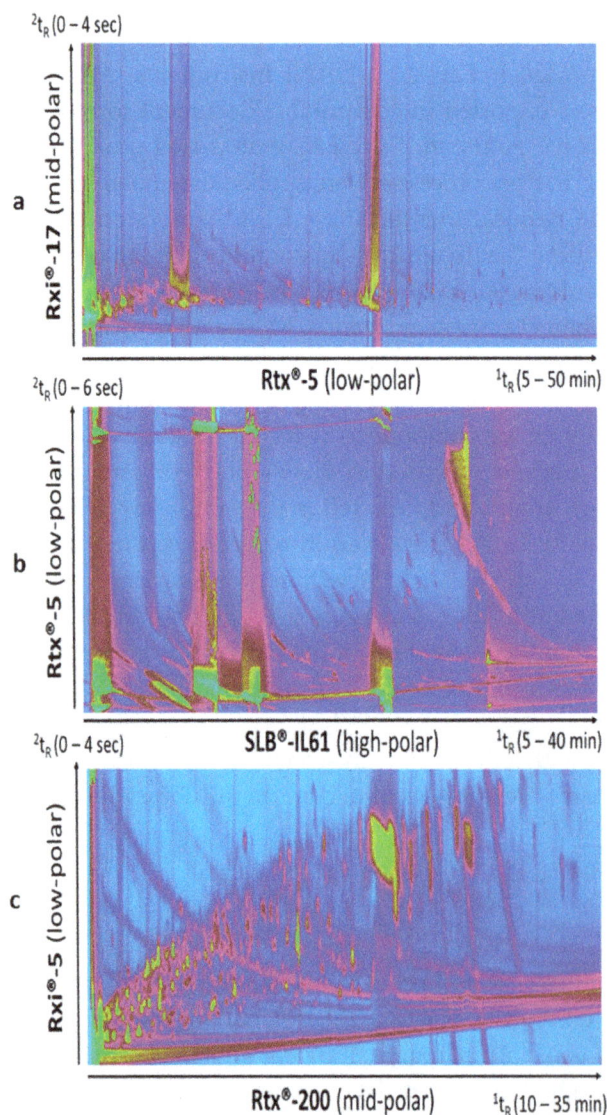

Figure 1. Two-dimensional gas chromatography-time-of-flight mass spectrometry (GC×GC-TOFMS) chromatograms obtained from analyses of a rough metabolite extraction of the gut of the termite Reticulitermes santonensis. Column sets 1, 2, and 3 were used for Figure 1a–c, respectively. Samples were derivatized with the N-Methyl-N-(trimethylsilyl)trifluoroacetamide (MSTFA) silylation reagent. Column set 3 (^{1}D mi-polarity fluoropropyl × ^{2}D low polarity 5% phenyl) provided the best occupation of the two-dimensional capacity and the best chromatographic fingerprint for data processing and comparison study. (**a**): Rtx®-5 (low-polar); (**b**): SLB®-IL61 (high-polar); (**c**): Rtx®-200 (mid-polar).

3.2. *Filtering Out Column Bleed and Noise in Derivatized Samples*

Peak tables generated after processing the samples included peaks from column bleed and derivatization reagents. The manual selection of the regions of chromatograms to exclude from the peak detection processing was difficult and required extensive care in order to avoid the exclusion of compounds of interest. As depicted in Figure 2, the noise and column bleed was not only located at the baseline, but they also crossed the entire chromatogram. Because samples were derivatized with trimethylsilyl groups (TMS), filtering column bleed related signals was not possible using a "sil" keyword-based approach through the list of compound names obtained from the library search identification. Instead, the "trifluoro" keyword was used to remove noise peaks corresponding to the derivatizing reagent MSTFA. Additional background was present and its origin was difficult to identify. The mass spectra probably resulted from a mixture of column bleed and pyridine derivatives. Simple scripts (CLIC expression) were used to recognize feature of these unwanted peaks in the mass spectra, based on two mass spectral specific fragments expressing the highest intensities. The expressions "(Ordinal (59) < = 2) & (Ordinal (137) < = 2)" and "(Ordinal (137) < = 2) & (Ordinal (215) < = 2)" were able to specifically select most of the unwanted peaks. It was efficient to quickly and simply clean up the peak table. Additionally, "(Ordinal (294) < = 2) & (Ordinal (309) < = 2) was used to remove other unwanted peaks identified as silanol trimethyl- rhenium complex. Specific selection resulting from each command and expression is shown in Figure 2. The mass spectra of the unwanted peaks are available in supporting information (Figure S1).

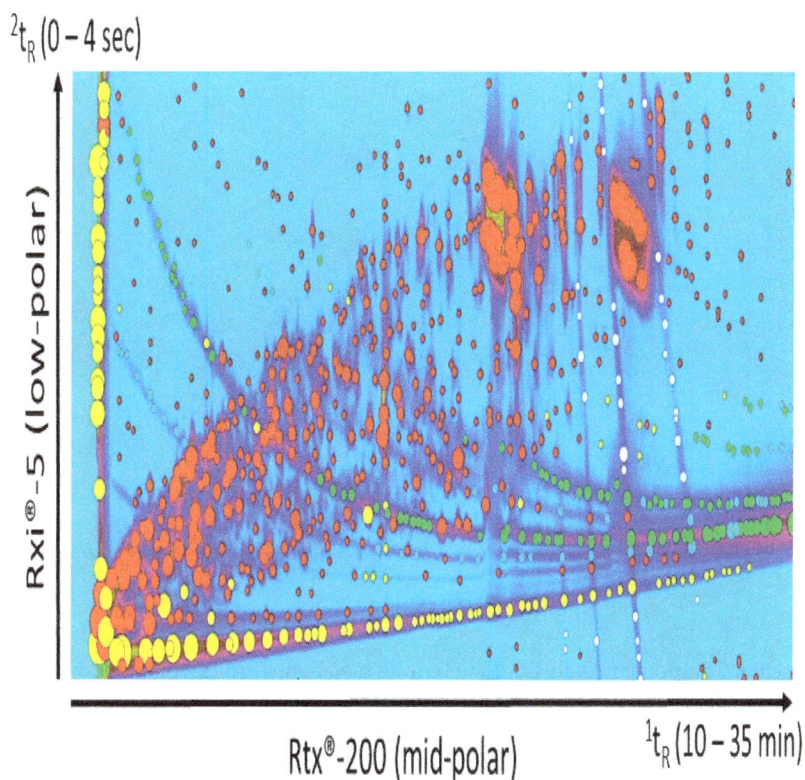

Figure 2. Computer Language for Chemical Identification (CLIC expression) was used to quickly select and remove unwanted peaks from the chromatogram. The "trifluoro" keyword was used to select the peaks corresponding to the derivatization reagent MSTFA (**yellow** markers). Simple scripts were used to recognize features of the other unwanted peaks in the mass spectra, based on two specific fragments with the highest intensities: mass-to-charge ratio (m/z) 59 and m/z 137 (**green** markers), m/z 137 and m/z 215 (**blue** markers), m/z 294 and m/z 309 (**white** markers). The **red** markers correspond to peaks that remained after the cleaning process.

3.3. Principal Component Analysis of Metabolite Profiles of Termites Cultured on Different Carbon Sources

Samples of termites fed with wood, microcrystalline cellulose (Avicel) or xylan were analyzed and compared. Samples of five termite guts were collected in triplicate for each diet. Comparison of the multiple chromatograms was carried out towards the creation of a composite image and resulted in the highlight of 314 compounds compared in all the samples. These compounds were mainly amino acids, amines, sugars, sugar derivatives, and organic acids. It included metabolites involved in major pathways like tricarboxylic acid (TCA) cycle, purine metabolism and fermentation. Table 1 lists detected metabolites and compounds, based on their chemical structures. Retention times, peak volume and percent response data are available in supplementary information (Tables S1–S3). Only the compounds with a library match score identification >700 (or identified with standards), and not found in blank procedures were reported in the table. The principal component analysis (PCA) demonstrated a trend towards the separation of the samples related to the carbon source used by the termites (Figure 3). The first principal component (PC-1, 79%) covered a great part of the variation, separating the three different diets. The second principal component (PC-2, 14%) demonstrated a trend to separate the xylan diet from the wood and Avicel diets. This secondary separation could possibly reflect the influence of cellulose as a carbon source in the diet, as Avicel is a pure cellulose powder and cellulose is the main constituent of the wood (45%). The multivariate analysis demonstrated that the metabolite profile in the termite was dependent of the diet.

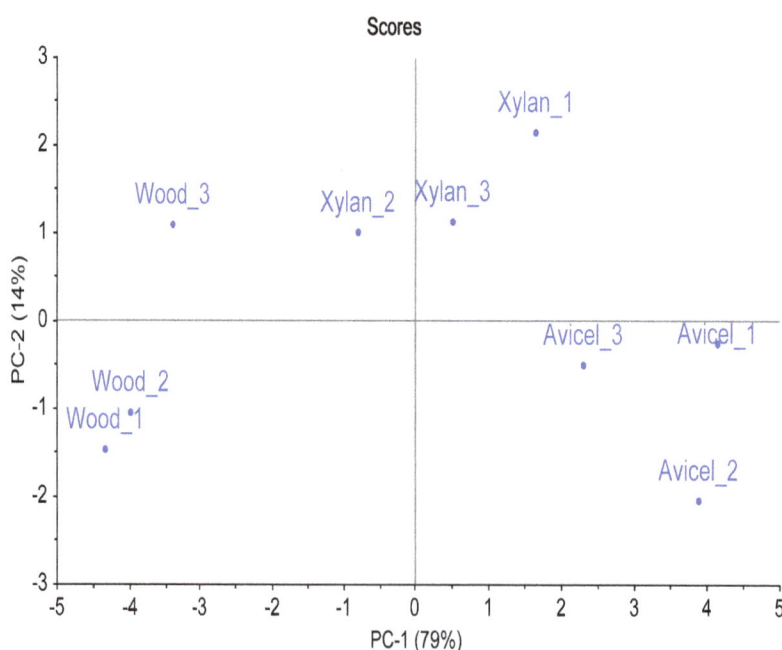

Figure 3. Principal Component Analysis of the termite metabolite profiles obtained with different diets. Three-hundred and fourteen compounds detected in the samples were used as variables (using Percent Response). The termite samples collected from the same diet sat together.

Table 1. Metabolites identified in the termite guts. The identification is based on a National Institute of Standards and Technology (NIST) library match factor > 700. Sugars were identified using specific standards.

Amino Acids	Amines	Organic Acids	Sugars	Others
Alanine	2-Piperidone	2-Ketogluconic acid	Arabitinol	1,2,4-butanetriol
Asparagine	Cadaverine	2-Ketoglutaric acid	Erythrose	Phosphate
Aspartic acid	Nicotinamide	2-Ketovaline	Fructose	Phosphite
Beta-Alanine	N-Methylvaleramide	Benzoic acid	Galactose	
Glutamic acid	Putrescine	Caproic acid	Glucose	
Glutamine		Citric acid	Inositol Isomer	
Glycine		Fumaric acid	Inositol myo-	
Isoleucine		Glyceric acid	Meso-Erythritol	
Leucine		Hydroxyphenylacetic acid	Rhamnose	
N-acetyl-lysine		Lactic acid	Xylose	
5-oxoproline		Malic acid		
Ornithine		Oxalic acid		
Phenylalanine		Phenylacetic acid		
Serine		Phenylpropanoic acid		
Threonine		Pipecolinic acid		
Tyrosine		Succinic acid		
Valine		Uric acid		

3.4. Identification of Diet-Specific Compounds

The analysis of variance was used to check for significant differences in compound levels between the three different diets. The one-way ANOVA compared the variance, explained by differences between sample means, to the unexplained variance within the samples. Among the 314 compounds compared in all samples, 41 compounds presented a significant F-value ($p < 0.05$), ranging from 5.1 to 59.2. The highest F-values were obtained for the compounds phenylpropanoic acid (F = 59.2) and benzoic acid (F = 45.3). The F-test indicates if there is a difference between the groups but does not say who is different from whom. Pair-wise comparisons were achieved with the Tukey's test, to identify which group was different from another. Only the compounds with good library identification (MF > 700) were reported.

3.4.1. Wood Diet

Main characteristics of the wood diet *versus* the artificial diets were a lower level of uric acid (F-value = 25.3), and higher levels of phenylpropanoic acid (F = 59.2), benzoic acid (F = 45.3), hexopyranoside (F = 33.0) and a pentose isomer (F = 10.6) (Figure 4). Uric acid is a product of the metabolic breakdown of purine nucleotides. It is the major waste product of nucleic acid and protein metabolism in termites [24]. In wood-feeding termites, uric acid is stored in the uricocytes of the fat body. It was demonstrated that in *R. flavipes*, uric acid may accumulate to considerable concentrations, and that the gut microbiota plays an important role in helping termites recycle uric acid [25]. The reason why termites keep accumulating uric acid during their entire life, in addition to the lack of symbionts able to help them remobilize uric acid was discussed by Brune and Ohkuma [26]. They made the hypothesis that uric acid is accumulated in bodies to be reingested by nestmates. The transfer of food or other fluids among members of a community through anus-to-mouth feeding is called proctodeal trophallaxis. This behavior is highly developed among termites, and it was observed that the frequency increased with the nitrogen deficiency of the diet [27]. In this study, we observed a higher accumulation of uric acid with low nitrogen diets (cellulose and xylan). This supports the hypothesis that termites accumulate acid uric to be reingested and recycled by nestmates to complement their diet.

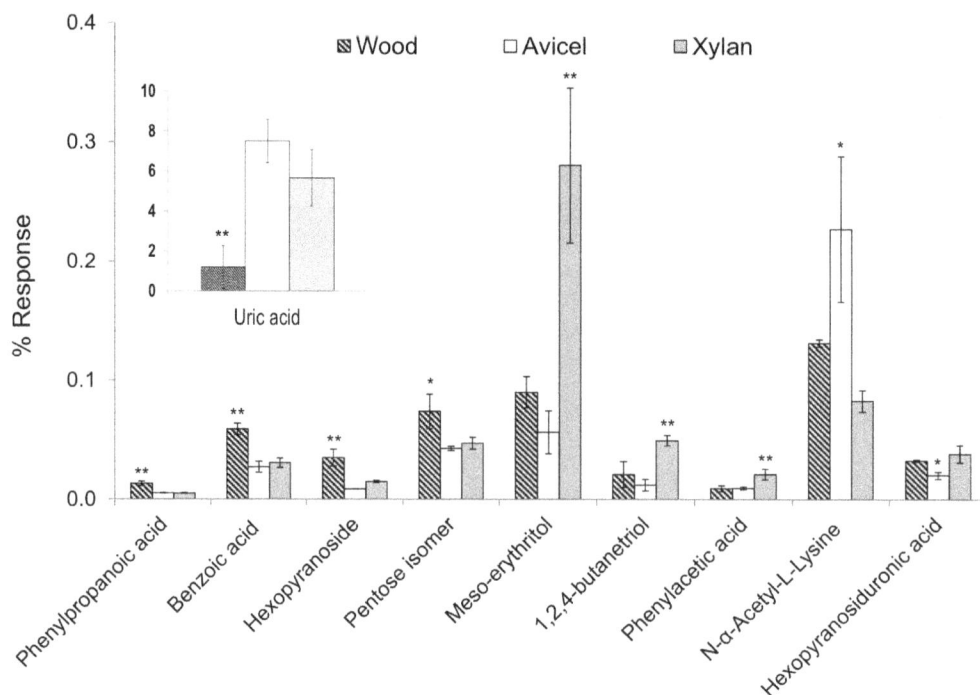

Figure 4. One way-ANOVA and the pair-wise Tukey's test were used to identify significant differences of compound levels between the different diets. For each compound reported, the diet marked with an asterisk presented a significantly different level than the two others, with $p < 0.05$ (*) or $p < 0.01$ (**).

Phenyl propanoic acid (PPA) and benzoic acid are lignin-derived aromatic molecules. The detection of these compounds supports the hypothesis that the polymeric backbone of lignin is depolymerized during passage through the termite gut. In a study of Brune *et al.* [28], it was demonstrated that *R. flavipes* mineralized ring-labeled benzoic or cinnamic acid only if oxygen was present. In the absence of oxygen, benzoate compounds were not attacked and cinnamate was reduced to phenylpropionate. In addition, a large number of anaerobic ring-modifying microorganisms were present. The detection of benzoic acid and phenylpropanoic acid showed that aromatic compounds are not metabolized efficiently in *R. flavipes*. In a study of Brune *et al.* [6], it was demonstrated that, in the termite gut, the oxygen diffuses from the gut wall to the lumen, and the oxygen is consumed before reaching the central part of the gut. The lumen can be therefore considered as an anoxic environment. As aromatic compounds are mostly degraded in aerobic conditions, our results showing intact lignin-phenolic metabolites support the hypothesis of a significant anoxic environment in the termite gut, limiting further lignin oxidation. Higher levels of a hexopyranoside and a pentose isomer in the wood diet is likely to be related to the poplar wood composition, a natural wood that contains other components than glucose and xylose, especially from the hemicellulose content.

3.4.2. Xylan Diet

The main characteristics of the xylan diet *versus* others were higher levels of meso-erythritol (F = 27.8), 1,2,4-butanetriol (F = 21.3), and phenyl acetic acid (PAA) (F = 14.6) (Figure 4). Meso-erythritol and 1,2,4-butanetriol are 4-carbon sugar alcohols (or polyols). These compounds are likely to be derived from xylose, the major constituent of xylan, as they were emphasized with the xylan diet. An increase of polyol production from xylose metabolism was observed with *Aspergillus niger*, under oxygen-limited conditions [29]. *Aspergillus niger* is a filamentous fungus, well-known and studied for its cellulolytic activity and ability to break down lignocellulose from plants. However, the reported polyols included erythritol but not 1,2,4-butanetriol. Butanetriol can be formed from xylose by biosynthetic pathway, combining different bacterial activities [30,31]. This could reflect synergic or multiple enzymatic levels in termites. The detection of these polyols is very interesting as they are likely to be involved in carbohydrate reserves, osmoregulation or storing reducing power strategies. Phenylacetic acid (PAA) was emphasized with the xylan diet. PAA is involved in phenylalanine metabolism pathway. Phenethylamine is produced from phenylalanine biosynthesis, and it can be metabolized into phenylacetic acid, which can be used by many bacteria [32]. The higher production of PAA on xylan diet could be related to a modification of the microflora. The higher production of PAA on the xylan diet could also be related to phenolic contamination present in the beechwood xylan substrate used in this study [33].

3.4.3. Cellulose Diet

The characteristics of the Avicel diet compared to the other diets were a higher level of N-acetyl-lysine (F = 12.7) and a lower level of a derivative of glucopyranosiduronic acid (F = 12.5) (Figure 4). N-acetyl-lysine is an acetyl derivative of the amino acid lysin. The acetylation of lysine residues in protein is involved in an important mechanism of regulation of gene expression. Glucuronic acid is a compound characteristic of xylan composition, also present in hemicelluloses in wood, which could explain why a lower level was detected only with the cellulose diet.

3.5. Influence of Pentoses Versus Hexoses on Metabolite Profile

From pair-wise comparisons, several compounds with a significant F-value were different between only the Avicel and the xylan diets. These results were used to evaluate the influence of pentoses compared to hexoses as a carbon source. No significant differences were observed when compared to the wood diet, which included both pentoses and hexoses from cellulose and hemicellulose content. Higher levels of glyceric acid (F value = 8.2), 5-oxoproline (F = 5.0), and fructofuranose-P (F = 17.4) were found with the Avicel diet, whereas higher levels of arabitinol (F = 7.2) and erythrose (F = 7.1) were found with the xylan diet (Figure 5). For each compound, the wood diet presented an intermediate level.

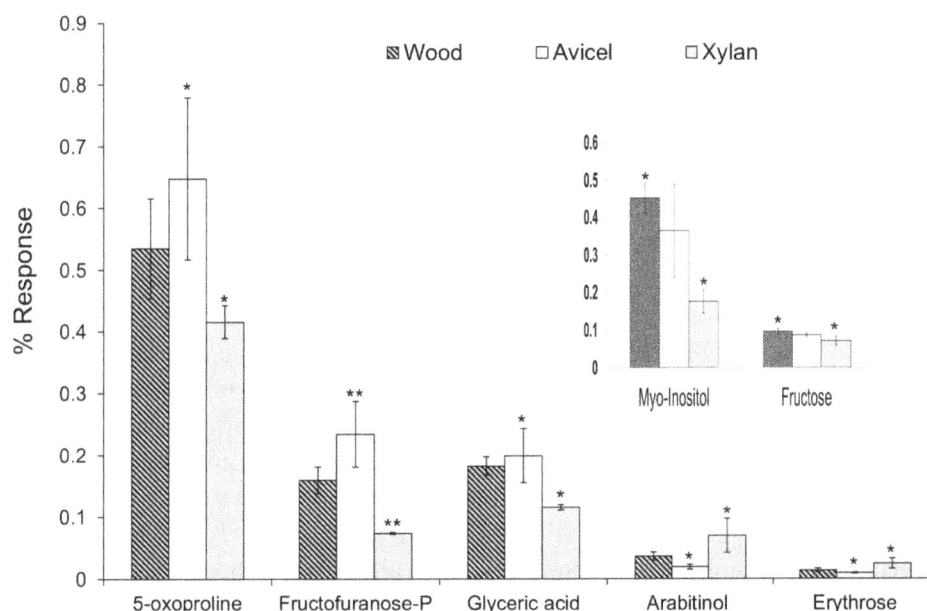

Figure 5. One way-ANOVA was used to identify significant differences of compound levels between the different diets. Pair-wise comparisons, using the Tukey's test, helped to identify a significant difference only between two diets, specifically, with $p < 0.05$ (*) or $p < 0.01$ (**). The diet not marked with an asterisk is not significantly different from the two others.

Glyceric acid and fructofuranose-P are likely to be related to the glycolysis pathway. This could reflect the production of ethanol through glucose fermentation because a higher level was detected when Avicel, a polymer of glucose only, was given as carbon source. Pyroglutamic acid (5-oxoproline) is an amino acid derivative. It is the cyclic lactam of glutamate and an indication of active glutathione metabolism. Proposed functions of free oxoproline in living cells include a role as analogue or reservoir of glutamate and possible osmoprotection [34].

Arabitinol can be produced by reduction of arabinose. The xylan substrate is a polysaccharide of mainly xylose (>90%) but also contains arabinose. No significant difference was obtained with the wood diet that contains hemicelluloses. Erythrose can be produced from dephosphorylation of erythrose-4-P, an intermediate of the pentose phosphate pathway. In the xylan diet, pentoses are the main carbon source.

Some differences of compound levels were significant only between the wood and xylan diets. No significant differences were obtained when compared with the Avicel diet, whereas an intermediate level was observed. Higher levels of myo-inositol (F = 13.9) and fructose (F = 6.6) were detected with the wood diet (Figure 5). Myo-inositol is a 6-carbon polyol, with each carbon hydroxylated. It is involved in the phospholipid metabolism. Phospholipids are complex molecules that are essential for membrane integrity and intracellular signaling. The detection of myo-inositol in the samples is interesting in the field of biofuels. The membranes of yeast and other microorganisms are a target for ethanol damage [35]. In the yeast *Saccharomyces cerevisiae*, it was demonstrated that the phospholipid composition plays an important role in its ability to tolerate ethanol [36]. In bioreactor environments, a supplementation of inositol in the growth media increased the concentration of phosphatidylinositol (PI) in the cellular membrane, increased the membrane tolerance to ethanol and also increased the concentrations of ethanol produced [37,38]. Inositol can also be used as a way to store carbon source. A number of bacteria can utilize inositol as a carbon source, such as *Bacillus subtilis* [39], a bacterium already isolated from *R. flavipes* [40]. The significant decrease of myo-inositol level observed with xylan diet could reflect the utilization of the inositol as a carbon source, to compensate the lack of 6-carbon molecules. Fructose can be produced from dephosphorylation of fructose-6-P. Fructose-6-P is an intermediate of the glycolysis, but it is also an intermediate of the pentose phosphate pathway, in its open-ring form. Wood and Avicel diets both contain glucose. Wood contains both hexoses and pentoses, which could explain why a higher level of fructose was detected.

4. Conclusions

We have investigated the effect of different diets on *Reticulitermes flavipes* metabolism. The aim was to study the adaptation of the termite-symbiont digestion system depending on the carbon source. The main challenges for the analysis of these complex samples were the low fluid volume available from the termite gut, the large level range of compounds present in the samples and the background noise level related to the sample preparation and derivatization. The GC×GC-TOFMS method using mid-polar/low-polar column combination, comparative data processing and CLIC expressions allowed detection of more than 300 compounds that were compared between the different samples. Multivariate analysis demonstrated a trend towards the separation of the samples related to the carbon source used by the termites. Essential and non-essential amino acids, amines, sugars, organics acids and other compounds were identified. Univariate analysis was used to compare the relative levels of the metabolites. The main variations included polyols and uric acid production, potentially related to redox imbalance, carbon storage and nitrogen recycling. This report demonstrated that in-depth study of metabolites could complement other approaches [11,41] and contribute to a better understanding of how the termite gut ecosystem degrades lignocellulose.

Acknowledgments: This work was supported by a CRA contract (Concerted Research Action; University of Liège agreement No. ARC 08-13/02). The GC columns were kindly provided by Restek (Restek Corp., Bellefonte, PA, USA) and Supelco Sigma-Aldrich (Sigma-Aldrich, St. Louis, MO, USA).

Author Contributions: The manuscript was written through contributions of all authors, and all authors have given approval for the final version. J.B. coordinated the culture and collection of termite samples. J.B., C.T., C.M., C.M., P.T., J.D., F.F, E.H., D.P., M.V., E.D.P. and J.-F.F participated in design and coordination of the study and drafted manuscript.

Conflicts of Interest: The authors declare no conflict of interest.

References

1. Ohkuma, M. Termite symbiotic systems: Efficient bio-recycling of lignocellulose. *Appl. Microbiol. Biotechnol.* **2003**, *61*, 1–9. [CrossRef] [PubMed]
2. Chandel, A.K.; da Silva, S.S. Sustainable degradation of lignocellulosic biomass—Techniques, applications and commercialization. *InTech* **2013**. [CrossRef]
3. Breznak, J.A.; Brune, A. Role of microorganisms in the digestion of lignocellulose by termites. *Ann. Rev. Entomol.* **1994**, *39*, 453–487. [CrossRef]
4. Kudo, T. Termite-microbe symbiotic system and its efficient degradation of lignocellulose. *Biosci. Biotechnol. Biochem.* **2009**, *73*, 2561–2567. [CrossRef] [PubMed]
5. Hongoh, Y. Toward the functional analysis of uncultivable, symbiotic microorganisms in the termite gut. *Cell. Mol. Life Sci.* **2011**, *68*, 1311–1325. [CrossRef] [PubMed]
6. Brune, A.; Emerson, D.; Breznak, J. The termite gut microflora as an oxygen sink: Microelectrode determination of oxygen and pH gradients in guts of lower and higher termites. *Appl. Environ. Microbiol.* **1995**, *61*, 2681–2687. [PubMed]
7. Sethi, A.; Slack, J.M.; Kovaleva, E.S.; Buchman, G.W.; Scharf, M.E. Lignin-associated metagene expression in a lignocellulose-digesting termite. *Insect Biochem. Mol. Biol.* **2013**, *43*, 91–101. [CrossRef] [PubMed]
8. He, S.; Ivanova, N.; Kirton, E.; Allgaier, M.; Bergin, C.; Scheffrahn, R.H.; Kyrpides, N.C.; Warnecke, F.; Tringe, S.G.; Hugenholtz, P. Comparative metagenomic and metatranscriptomic analysis of hindgut paunch microbiota in wood- and dung-feeding higher termites. *PLoS ONE* **2013**, *8*, e61126. [CrossRef] [PubMed]
9. Bauwens, J.; Millet, C.; Tarayre, C.; Brasseur, C.; Destain, J.; Vandenbol, M.; Thonart, P.; Portetelle, D.; De Pauw, E.; Haubruge, E.; *et al.* Symbiont diversity in *Reticulitermes santonensis* (Isoptera: Rhinotermitidae): Investigation strategy through Proteomics. *Environ. Entomol.* **2013**, *42*, 882–887. [CrossRef] [PubMed]
10. Brune, A. Symbiotic digestion of lignocellulose in termite guts. *Nat. Rev. Microbiol.* **2014**, *12*, 168–180. [CrossRef] [PubMed]
11. Xie, S.; Syrenne, R.; Sun, S.; Yuan, J.S. Exploration of Natural Biomass Utilization Systems (NBUS) for advanced biofuel—From systems biology to synthetic design. *Curr. Opin. Biotechnol.* **2014**, *27*, 195–203. [CrossRef] [PubMed]
12. Tokuda, G.; Tsuboi, Y.; Kihara, K.; Saitou, S.; Moriya, S.; Lo, N.; Kikuchi, J. Metabolomic profiling of ^{13}C-labelled cellulose digestion in a lower termite: Insights into gut symbiont function. *Proc. R. Soc. B.* **2014**, *281*. [CrossRef] [PubMed]
13. Ralston-Hooper, K.; Jannasch, A.; Adamec, J.; Sepulveda, M. The use of two-dimensional gas chromatography-time-of-flight mass spectrometry (GC×GC-TOF-MS) for metabolomic analysis of polar metabolites. *Methods Mol. Biol.* **2011**, *708*, 205–211. [PubMed]
14. Kamleh, M.A.; Dow, J.A.; Watson, D.G. Applications of mass spectrometry in metabolomic studies of animal model and invertebrate systems. *Brief. Funct. Genom. Proteom.* **2009**, *8*. [CrossRef] [PubMed]
15. Patterson, D.G., Jr.; Welch, S.M.; Turner, W.E.; Sjödin, A.; Focant, J.-F. Cryogenic zone compression for the measurement of dioxins in human serum by isotope dilution at the attogram level using modulated gas chromatography coupled to high resolution magnetic sector mass spectrometry. *J. Chromatogr. A* **2011**, *1218*, 3274–3281. [CrossRef] [PubMed]
16. Brune, A. Termite guts: The world's smallest bioreactors. *Trends Biotechnol.* **1998**, *16*, 16–21. [CrossRef]
17. Evans, T.A.; Forschler, B.T.; Grace, J.K. Biology of invasive termites: A worldwide review. *Ann. Rev. Entomol.* **2013**, *58*, 455–474. [CrossRef] [PubMed]
18. Lewis, V.R.; Sutherland, M.; Haverty, M.I. Pest Notes: Subterranean and Other Termites. UC ANR Publication 7415 (2014). Available online: http://www.ipm.ucdavis.edu/PDF/PESTNOTES/pntermites.pdf (accessed on 9 March 2016).

19. Schauer, C.; Thompson, C.L.; Brune, A. The bacterial community in the gut of the cockroach *Shelfordella lateralis* reflects the close evolutionary relatedness of cockroaches and termites. *Appl. Environ. Microbiol.* **2012**, *78*. [CrossRef] [PubMed]

20. Shin, M.H.; Lee, D.Y.; Skogerson, K.; Wohlgemuth, G.; Choi, I.G.; Fiehn, O.; Fiehn, O.; Kim, K.H. Global metabolic profiling of plant cell wall polysaccharide degradation by *Saccharophagus degradans*. *Biotechnol. Bioeng.* **2010**, *105*, 477–488. [CrossRef] [PubMed]

21. Wachsmuth, C.J.; Vogl, F.C.; Oefner, P.J.; Dettmer, K. Gas Chromatographic Techniques in Metabolomics. In *Chromatographic Methods in Metabolomics*; Hyotylainen, T., Wiedmer, S., Eds.; RCS Publishing: Cambridge, UK, 2013; pp. 87–105.

22. Magel, E. Physiology of cambial growth, storage of reserves and heartwood formation. In *Trends in European Forest Tree Physiology Research*; Huttunen, S., Heikkila, H., Bucher, J., Sundberg, B., Jarvis, P., Matyssek, R., Eds.; Springer: Dordrecht, The Netherlands, 2001; pp. 19–32.

23. Austin, J.; Szalanski, A.; Scheffrahn, R.; Messenger, M.; Dronnet, S.; Bagnères, A.G. Genetic evidence for the synonymy of two Reticulitermes species: Reticulitermes flavipes and Reticulitermes santonensis. *Ann. Entomol. Soc. Am.* **2005**, *98*, 395–401. [CrossRef]

24. Breznak, J.A. Ecology of prokaryotic microbes in the guts of wood- and litter-feeding termites. In *Termites: Evolution, Sociality, Symbiosis, Ecology*; Abe, T., Bignell, D.E., Higashi, M., Eds.; Kluwer Academic Publishers: Dordrecht, The Netherlands, 2000; pp. 209–231.

25. Potrikus, C.J.; Breznak, J.A. Gut bacteria recycle uric acid nitrogen in termites: A strategy for nutrient conservation. *Proc. Natl. Acad. Sci. USA* **1981**, *78*, 4601–4605. [CrossRef] [PubMed]

26. Brune, A.; Ohkuma, M. Role of the Termite gut Microbiota in Symbiotic digestion. In *Biology of Termites: A Modern Synthesis*; Bignell, D.E., Roisin, Y., Lo, N., Eds.; Springer: Dordrecht, The Netherlands, 2011; pp. 439–475.

27. Machida, M.; Kitade, O.; Miura, T.; Matsumoto, T. Nitrogen recycling through proctodeal trophallaxis in the Japenese damp-wood termite Hodotermopsis japonica (Isoptera). *Insectes Soc.* **2001**, *48*, 52–56. [CrossRef]

28. Brune, A.; Miambi, E.; Breznak, J.A. Roles of oxygen and the intestinal microflora in the metabolism of lignin-derived phenylpropanoids and other monoaromatic compounds by termites. *Appl. Environ. Microbiol.* **1995**, *61*, 2688–2695. [PubMed]

29. Meijer, S.; Panagiotou, G.; Olsson, L.; Nielsen, J. Physiological characterization of xylose metabolism in Aspergillus niger under oxygen-limited conditions. *Biotechnol. Bioeng.* **2007**, *98*, 462–475. [CrossRef] [PubMed]

30. Abdel-Ghany, S.E.; Day, I.; Heuberger, A.L.; Broeckling, C.D.; Reddy, A.S.N. Metabolic engineering of Arabidopsis for butanetriol production using bacterial genes. *Metab. Eng.* **2013**, *20*, 109–120. [CrossRef] [PubMed]

31. Niu, W.; Molefe, M.N.; Frost, J.W. Microbial synthesis of the energetic material precursor 1,2,4-butanetriol. *J. Am. Chem. Soc.* **2003**, *125*, 12998–12999. [CrossRef] [PubMed]

32. Ramos, J.L.; Filloux, A. *Pseudomonas-A Model System in Biology*; Springer: New York, NY, USA, 2007.

33. Buslov, D.K.; Kaputski, F.N.; Sushko, N.I.; Torgashev, V.I.; Solov'eva, L.V.; Tsarenkov, V.M.; Zubets, O.V.; Larchenko, L.V. Infrared spectroscopic analysis of the structure of xylans. *J. Appl. Spectrosc.* **2009**, *76*, 801–805. [CrossRef]

34. Kumar, A.; Bachhawat, A.K. Pyroglutamic acid: Throwing light on a lightly studied metabolite. *Curr. Sci.* **2012**, *102*, 288–297.

35. Ingram, L.O.; Buttke, T.M. Effects of alcohols on micro-organisms. *Adv. Microb. Physiol.* **1984**, *25*, 253–300. [PubMed]

36. Chi, Z.; Kohlwein, S.D.; Paltauf, F. Role of phosphatidylinositol (PI) in ethanol production and ethanol tolerance by a high ethanol producing yeast. *J. Ind. Microbiol. Biotechnol.* **1999**, *22*, 58–63. [CrossRef]

37. Furukawa, K.; Kitano, H.; Mizoguchi, H.; Hara, S. Effect of cellular inositol content on ethanol tolerance of Saccharomyces cerevisiae in sake brewing. *J. Biosci. Bioeng.* **2004**, *98*, 107–113. [CrossRef]

38. Krause, E.L.; Villa-García, M.J.; Henry, S.A.; Walker, L.P. Determining the effects of inositol supplementation and the *opi1* mutation on ethanol tolerance of *Saccharomyces cerevisiae*. *Ind. Biotechnol.* **2007**, *3*, 260–268. [CrossRef] [PubMed]

39. Yoshida, K.; Yamaguchi, M.; Morinaga, T.; Kinehara, M.; Ikeuchi, M.; Ashida, H.; Fujita, Y. Myo-inositol catabolism in Bacillus subtilis. *J. Biol. Chem.* **2008**, *283*, 10415–10424. [CrossRef] [PubMed]

40. Tarayre, C.; Brognaux, A.; Brasseur, C.; Bauwens, J.; Millet, C.; Mattéotti, C.; Destain, J.; Vandenbol, M.; Portetelle, D.; De Pauw, E.; *et al.* Isolation and cultivation of a xylanolytic *Bacillus Subtilis* extracted from the gut of the termite *Reticulitermes santonensis*. *Appl. Biochem. Biotechnol.* **2013**, *171*, 225–245. [CrossRef] [PubMed]

41. Scharf, M.E. Omic research in termites: An overview and a roadmap. *Front. Genet.* **2015**, *6*. [CrossRef] [PubMed]

Permissions

List of Contributors

Katelynn A. Perrault, Pierre-Hugues Stefanuto, Lena Dubois and Jean-François Focant
Organic and Biological Analytical Chemistry Group, Chemistry Department, University of Liège, Allée du 6 Août 11 (Bât B6c), Quartier Agora, Sart-Tilman, Liège 4000, Belgium

Dries Cnuts and Veerle Rots
Traceolab/Prehistory, University of Liège, Quai Roosevelt 1B (Bât. A4), Liège 4000, Belgium

Lucia Valverde-Som, Alegría Carrasco-Pancorbo, Cristina Ruiz-Samblás, Natalia Navas and Luis Cuadros-Rodríguez
Department of Analytical Chemistry, Faculty of Science, University of Granada, C/Fuentenueva s/n, E-18071 Granada, Spain

Saleta Sierra, Soraya Santana and Javier S. Burgos
Neuron Bio, P.T.S. Granada, C/Avicena 4, E-18016 Granada, Spain

Somchai Rice
Department of Agricultural and Biosystems Engineering, Iowa State University, Ames, IA 50011, USA
Interdepartmental Toxicology Graduate Program, Iowa State University, Ames, IA 50011, USA
Midwest Grape and Wine Industry Institute, Iowa State University, Ames, IA 50011, USA

Murlidhar Dharmadhikari
Midwest Grape and Wine Industry Institute, Iowa State University, Ames, IA 50011, USA
Department of Food Science and Human Nutrition, Iowa State University, Ames, IA 50011, USA

Nanticha Lutt
Genetics and Plant Biology Program, University of California at Berkeley, Berkeley, CA 94704, USA

Jacek A. Koziel
Department of Agricultural and Biosystems Engineering, Iowa State University, Ames, IA 50011, USA
Interdepartmental Toxicology Graduate Program, Iowa State University, Ames, IA 50011, USA
Department of Food Science and Human Nutrition, Iowa State University, Ames, IA 50011, USA

Anne Fennell
Plant Science Department, South Dakota State University and BioSNTR, Brookings, SD 57007, USA

Jerson Veiga
Chemical Engineering and Biotechnology Research Center, Instituto Superior de Engenharia de Lisboa, IPL, 1959-007 Lisboa, Portugal

José Coelho
Chemical Engineering and Biotechnology Research Center, Instituto Superior de Engenharia de Lisboa, IPL, 1959-007 Lisboa, Portugal
Centro de Química Estrutural, Instituto Superior Técnico, Universidade de Lisboa, 1049-001 Lisboa, Portugal

Beatriz Nobre and António Palavra
Centro de Química Estrutural, Instituto Superior Técnico, Universidade de Lisboa, 1049-001 Lisboa, Portugal

Amin Karmali
Chemical Engineering and Biotechnology Research Center, Instituto Superior de Engenharia de Lisboa, IPL, 1959-007 Lisboa, Portugal
CITAB-Centre for the Research and Technology of Agro-Environmental and Biological Sciences, University of Trás-os-Montes and Alto Douro, 5000-801 Vila Real, Portugal

Marisa Nicolai
CBiOS, Research Center for Biosciences & Health Technologies, ULHT, 1749-024 Lisboa, Portugal

Catarina Pinto Reis
iMED, ULisboa, Research Institute for Medicines, Faculty of Pharmacy, ULisboa, 1749-003 Lisboa, Portugal
IBEB, Biophysics and Biomedical Engineering, Faculty of Sciences, ULisboa, Campo Grande, 1749-016 Lisboa, Portugal

Carlos Sanz, Araceli Sánchez-Ortiz and Ana G. Pérez
Department of Biochemistry and Molecular Biology of Plant Products, Instituto de la Grasa, CSIC, 41013-Seville, Spain

Angjelina Belaj
IFAPA, Centro Alameda del Obispo, 14004-Cordoba, Spain

Juan Pablo Fernández-Trujillo and Mohamed Zarid
Department of Agricultural and Food Engineering, Regional Campus of International Excellence "Campus Mare Nostrum" (CMN), Technical University of Cartagena (UPCT), Paseo Alfonso XIII, 48, ETSIA, E-30203 Cartagena, Murcia, Spain

María Carmen Bueso
Department of Applied Mathematics and Statistics, CMN, UPCT, Doctor Fleming s/n, ETSII, E-30202 Cartagena, Murcia, Spain

José Gustavo Ronderos-Lara and Hugo Saldarriaga-Noreña
Centro de Investigaciones Químicas, Instituto de Investigación en Ciencias Básicas y Aplicadas, Universidad Autónoma del Estado de Morelos, Av. Universidad 1001, 62209 Cuernavaca, Mexico

Mario Alfonso Murillo-Tovar
Cátedras, Consejo Nacional de Ciencia y Tecnología, Av. Insurgentes Sur 1582, Colonia Crédito Constructor, Del. Benito Juárez, 03940 Ciudad de México, Mexico

Josefina Vergara-Sánchez
Laboratorio de Análisis y Sustentabilidad Ambiental, Escuela de Estudios Superiores de Xalostoc, Universidad Autónoma del Estado de Morelos, 62715 Ayala, Morelos, Mexico

Eleni Michopoulou and Georgios Theodoridis
Laboratory of Analytical Chemistry, Department of Chemistry, Aristotle University of Thessaloniki, Thessaloniki 54124, Greece

Adamantios Krokos
Laboratory of Analytical Chemistry, Department of Chemistry, Aristotle University of Thessaloniki, Thessaloniki 54124, Greece
Laboratory of Forensic Medicine & Toxicology, Department of Medicine, Aristotle University of Thessaloniki, Thessaloniki 54124, Greece

Elisavet Tsakelidou, Nikolaos Raikos and Helen Gika
Laboratory of Forensic Medicine & Toxicology, Department of Medicine, Aristotle University of Thessaloniki, Thessaloniki 54124, Greece

María José Aliaño-González, Marta Ferreiro-González, Gerardo F. Barbero, Miguel Palma and Carmelo G. Barroso
Department of Analytical Chemistry, Faculty of Sciences, ceiA3, IVAGRO, University of Cadiz, 11510 Puerto Real, Cadiz, Spain

Kevin E. Eckert and Katelynn A. Perrault
Laboratory of Forensic and Bioanalytical Chemistry, Forensic Sciences Unit, Division of Natural Sciences and Mathematics, Chaminade University of Honolulu, 3140 Waialae Ave., Honolulu, HI 96816, USA

David O. Carter
Laboratory of Forensic Taphonomy, Forensic Sciences Unit, Division of Natural Sciences and Mathematics, Chaminade University of Honolulu, 3140 Waialae Avenue, Honolulu, HI 96816, USA

Katie D. Nizio and Shari L. Forbes
Centre for Forensic Science, University of Technology Sydney, Broadway, NSW 2007, Australia

Naysla Paulo Reinert, Camila M. S. Vieira and Eduardo Carasek
Departamento de Química, Universidade Federal de Santa Catarina, Florianópolis 88040-900, SC, Brazil

Cristian Berto da Silveira
Departamento de Engenharia de Pesca e Ciências Biológicas, Universidade do Estado de Santa Catarina, Laguna, Santa Catarina 88790-000, Brazil

Dilma Budziak
Departamento de Ciências Naturais e Sociais, Universidade Federal de Santa Catarina, Curitibanos 89520-000, SC, Brazil

Robin J. Abel and James J. Harynuk
Department of Chemistry, University of Alberta, Edmonton, AB T6G 2G2, Canada

Grzegorz Zadora
Institute of Forensic Research, Westerplatte 9, 31-033 Krakow, Poland
Department of Analytical Chemistry, Institute of Chemistry, The University of Silesia, Szkolna 9, 40-006 Katowice, Poland

P. Mark L. Sandercock
Royal Canadian Mounted Police, National Forensic Laboratory Services-Edmonton, 15707-118th Avenue, Edmonton, AB T5V 1B7, Canada

María del Carmen Alcudia-León, Mónica Sánchez-Parra, Rafael Lucena and Soledad Cárdenas
Departamento de Química Analítica, Instituto de Química Fina y Nanoquímica IUIQFN, Universidad de Córdoba, Campus de Rabanales, Edificio Marie Curie (anexo), E-14071 Córdoba, Spain

Lingshuang Cai
Department of Agricultural and Biosystems Engineering, Iowa State University, Ames, IA 50011, USA
DuPont Crop Protection, Stine-Haskell Research Center, Newark, DE 19711, USA

Somchai Rice
Department of Agricultural and Biosystems Engineering, Iowa State University, Ames, IA 50011, USA
Interdepartmental Toxicology Graduate Program, Iowa State University, Ames, IA 50011, USA

Jacek A. Koziel
Department of Agricultural and Biosystems Engineering, Iowa State University, Ames, IA 50011, USA
Interdepartmental Toxicology Graduate Program, Iowa State University, Ames, IA 50011, USA
Department of Food Science and Human Nutrition, Iowa State University, Ames, IA 50011, USA

Murlidhar Dharmadhikari
Department of Food Science and Human Nutrition, Iowa State University, Ames, IA 50011, USA

Pierre-Hugues Stefanuto, Katelynn A. Perrault and Jean-François Focant
Organic and Biological Analytical Chemistry Group, Chemistry Department, University of Liège, Allée du 6 Août 11 (Bât B6c), Quartier Agora, 4000 Liège (Sart-Tilman), Belgium

Silke Grabherr
Forensic Anthropology and Imaging Unit, University Center of Legal Medicine, Chemin de la Vulliette 4, CH-1000 Lausanne 25, Switzerland

Vincent Varlet
Forensic Toxicology and Chemistry Unit, University Center of Legal Medicine, Chemin de la Vulliette 4, CH-1000 Lausanne 25, Switzerland

Jean-François Focant
Organic and Biological Analytical Chemistry, Department of Chemistry, University of Liège, Sart-Tilman, Liège B-4000, Belgium

Catherine Brasseur
Organic and Biological Analytical Chemistry, Department of Chemistry, University of Liège, Sart-Tilman, Liège B-4000, Belgium
Mass Spectrometry Laboratory, Department of Chemistry, University of Liège, Sart-Tilman, Liège B-4000, Belgium

Edwin De Pauw
Mass Spectrometry Laboratory, Department of Chemistry, University of Liège, Sart-Tilman, Liège B-4000, Belgium

Julien Bauwens, Frédéric Francis and Eric Haubruge
Department of Functional and Evolutionary Entomology, Gembloux Agro-Bio Tech, University of Liège, Gembloux B-5030, Belgium

Cédric Tarayre, Philippe Thonart and Jacqueline Destain
Department of Bio-Industries, Gembloux Agro-Bio Tech, University of Liège, Gembloux B-5030, Belgium

Catherine Millet, Christel Mattéotti, Daniel Portetelle and Micheline Vandenbol
Department of Microbiology and Genomics, Gembloux Agro-Bio Tech, University of Liège, Gembloux B-5030, Belgium

Index

www.ingramcontent.com/pod-product-compliance
Lightning Source LLC
Chambersburg PA
CBHW080624200326
41458CB00013B/4495